The Book of Seven Contexts

A Pragmatic Method for Building Software with AI

Blake Tullysmith

STATEN HOUSE PUBLISHING

The Book of Seven Contexts

A Pragmatic Method for Building Software with AI

Published by Staten House Publishing.

ISBN: 979-8-90417-694-5 (paperback)

2026-04_2 edition

Printed in the United States of America.

Table of Contents

Back Matter

List of Figures

How to Read This Book When Tired

A note for the reader who picked this book up at the end of a hard week.

The book is laid out to be read from front to back. That is the recommended order, and it is also the order that produces the cleanest experience of the methodology. But the book has been written with the suspicion that many of its readers are tired in a specific way — the way an engineer is tired after a year of supervising AI tools that type faster than human eyes can read — and that those readers may not have a clean front-to-back reading session in them this week.

If that is you, here are four shorter paths through the book that are honest entry points. Each one is meant to be read in a single sitting of forty-five to ninety minutes.

If you are depleted and want to feel that someone has thought clearly about why you're depleted. Read the Introduction. Then skip to Chapter 1 (Buzzy and the cedar shingle, ten or fifteen minutes). Then read Chapter 11 — the whole chapter, if you can; the first three sections if you cannot. Come back to the rest of the book on a Saturday morning when the world is quieter. The methodology is real, the book will still be here, and Chapter 11 is the part of the book most likely to help you on a hard night.

If you are curious about the method but uncertain whether to invest the setup time. Read the Introduction. Then read Chapter 3 (the three-phase tour). Then read Chapter 6 (a day inside the system). Those three pieces will give you a felt picture of what running the method looks like, without asking you to commit to building it. If you finish those and still want it, the rest of Part II will get you there. If you finish those and decide it isn't for you, you have spent a couple of hours forming a real opinion, which is more than most engineers manage to do about most methodologies.

If you have already started running the method and want to make it more sustainable. Skip directly to Chapter 11, the whole thing. Then Chapter 8 (refinement, where the slow improvements live). Then Chapter 12, particularly the section on teammates who won't adopt — being a good colleague to non-adopters is a sustainability practice in disguise. Loop back to anything else as it becomes relevant.

If you are about to be asked to roll the method out to a team. Read Chapter 12 in full before any other chapter. Then read the *Cast of Concepts* page that follows this one. Then read the long manager-rollout section in Chapter 12 a second time, more slowly. Most of what you need is in the second half of the book; you can come back to the first half as the rollout creates specific questions you cannot answer.

A few extra permissions, which the rest of the book does not explicitly grant and which I would like you to take anyway.

You may stop reading any chapter at any point, return to it later, and lose nothing. The chapters are self-contained enough that you can drop a session mid-chapter and pick it up next weekend without losing thread.

You may skim Chapter 4 (Setup with Claude Code) on a first pass and return to it when you are actually setting up the workspace at a keyboard. Reading it cold, away from a computer, is allowed but rarely the best use of an evening.

You may treat the appendices as the reference material they are. Appendix A is the prompt library you will reach for most often. Appendix F is what to read when something has gone visibly wrong. Appendix G is the glossary. The other appendices are good when you need them and ignorable when you don't.

You may close this book after the Introduction, do nothing for a week, and come back to it. The book is not going anywhere. Neither are the AI tools. The methodology has been in development for years and will still be here next month. Whatever pace you can sustain right now is the right pace.

The point of the book is not that you read it efficiently. The point of the book is that, over some number of weeks or months, the way you work shifts into something that is sustainable for the next decade of your career. Any reading pattern that gets you to that destination is a correct reading pattern. I would much rather you read a third of this book carefully and act on it than read the whole thing superficially and forget the parts that mattered.

Take what you need. Leave what you don't. Come back when you are rested.

A Cast of Concepts

You do not have to read this page now. Skim it once so you know what's here, then come back to it the way you'd come back to a glossary — when a word slipped away from you and you'd rather not flip back fifty pages to find where it lived. The book will introduce each of these properly when it gets to them. This page exists for the moments in between.

None of the ideas below is a difficult one. They are, however, words this book uses with discipline — the same word for the same thing on every page — and a single page that holds the discipline in one place is sometimes useful when the chapters have you carrying a lot.

Slot. A permanent position on your screen. Seven slots total. Slots never move. The Claude in a slot changes; the slot does not.

Environment. A Norse-named working directory on your filesystem — *thor, freya, tyr, loki, odin, heimdall, baldr.* Seven environments total. Environments are leased to Claude sessions for the life of a ticket and released afterwards.

Zone. One of three fixed regions on the screen. *Zone 1* is the upper-left validation workspace; *Zone 2* is the upper-right communications workspace; *Zone 3* is the bottom-right Jira workspace. There are exactly three zones; the seven slots are not zones.

Stack. The browser-and-Claude-terminal pair living inside a single slot. Click either visible edge to flip which one is on top. The stack never moves; only the layer order changes.

Lease. A claim on an environment, recorded in `ENVIRONMENT_LEASES.md` with the claiming process's PID. A lease is alive while the PID is alive; the next agent to want the environment confirms the PID is dead before claiming.

The three phases. *Setup* (one-time, building the workspace), *Execution* (daily, running the rotation), *Refinement* (continuous, slowly making the system better).

The work-on-a-ticket protocol. You say *"Work on ABC-1234."* Claude leases an environment, reads the ticket, summarizes it back to you and waits for your confirmation, creates a branch, makes an empty commit, pushes, opens a draft pull request, and starts working. You promote the PR out of draft after validation.

The validation ritual. Six steps run on every finished ticket: Claude self-critiques, Claude runs tests, Claude checks the work against acceptance criteria, you review the PR at the shape level, you promote it to ready-for-review, or you give Claude feedback and iterate.

NOTES.md and the wind-down ritual. A short note (three to ten lines), written by each active Claude at the end of each session into its environment's `NOTES.md`, capturing what got done, what's in progress, and what to pick up first. *Defined in Chapter 6.*

The glance cycle. Every two or three minutes you scan all seven slots. Each slot is in one of four states: *typing* (leave alone), *finished* (validate), *asking* (answer), *stuck* (intervene).

The two-session day. Mornings 9 to 12 and afternoons 1 to 4, separated by a real lunch off the screen. Seven tickets per session, fourteen per day, sustainable across months.

The dish by the door. The principle that every kind of work has a fixed, permanent home — so you never have to think about where it lives. The phrase recurs throughout.

Air traffic controller, not pilot. The metaphor for what you actually do during the rotation. You control the airspace; the seven Claudes fly. *(Chapter 6.)*

Tidy / Warm / Cold start. The three shapes of returning to your workspace. *Tidy:* everything still running. *Warm:* partial survival, some slots are ghosts. *Cold:* fresh rebuild from a new machine.

You will meet each of these properly in the chapters where they live. This page is for the moments when you've met one already and would rather not flip back fifty pages to remember exactly what it was.

Introduction

You've probably tried using AI for your coding work. Most developers have by now. And if you're like most, your experience has been — how shall I put this politely — a bit of a car crash.

You open Claude or ChatGPT or Copilot, paste in a problem, and get back something that's eighty percent right. You then spend twenty minutes fixing the twenty percent that's wrong, during which time the AI will helpfully refactor three other functions you didn't ask it to touch, insert a dependency that doesn't exist, and very confidently tell you the sky is green. At the end of the day you're exhausted in a way that feels different from honest coding exhaustion. Your manager asks if the AI tools are helping. You say "yeah, they're great," because that's what you're supposed to say. But secretly you're not sure. You might be going faster. You might be going slower. You're definitely more tired.

And then there's the other thing — the quiet worry you won't say out loud. You read an article about AI replacing developers. Your company mentions "AI-driven productivity gains" on an earnings call, delivered in the flat cadence of someone who has not personally debugged anything since the first Obama administration. Someone on LinkedIn posts about how they've replaced three engineers with an AI agent, and you briefly wonder whether those three engineers were in the habit of showing up at all. You don't know whether to laugh or start polishing your resume.

This book is for you.

Not because it promises to make the worry go away, and not because it claims AI is either the best thing ever invented by humans or the opening credits of the end of our careers. This book is for you because there's a middle path — a practical, sustainable way to work alongside AI that delivers real throughput without setting your nervous system on fire — and I've been living in that middle path for years. I want to show you how.

A Word about the Title

You've probably noticed this book has a slightly grand name. *The Book of Seven Contexts*. That's a deliberate nod to *The Book of Five Rings*, written in 1645 by Miyamoto Musashi, the Japanese swordsman generally regarded as having been the best at his job in a profession where being merely second-best was a career-limiting condition.

Musashi's book is a philosophical strategy guide for becoming an elite swordsman. He argues that the Way of the swordsman is a complete system — a set of principles about timing, distance, position, attention, and character — and that mastery comes not from frantically practicing sword strokes but from internalizing the strategy and then letting the strokes follow. Along the way he uses one of my favorite extended metaphors in the whole of martial literature: the comparison of a swordsman to a master carpenter. Both work with tools. Both must understand the grain of their material. Both must choose the right timber for the right purpose. Both become masters not by working harder than everyone else, but by *thinking better about their work*.

I like the parallel. What you're going to read is, in its own much-more-modern way, a philosophical strategy guide — not for warfare, thankfully, but for the equivalently high-stakes business of making a living as a software engineer in an age where the tools have just gotten unreasonably powerful. The swordsman had his timing and distance and position. You have your contexts, your slots, your rotation rhythm, and your lease protocols. The principles underneath are not so different.

You may also notice that the very first chapter of this book is called *The Hammer and the Wood* and involves, among other things, a carpenter. That was not an accident.

What This Book Teaches

The rest of the book is a methodology. Not a tool. Not a prompt template. Not a bunch of tips you could find in a blog post, and definitely not a forty-thousand-word love letter to any particular AI model — because the AI models are going to change, and the good ones five years from now will make today's look like pocket calculators, and then *those* will be replaced by something else five years after *that*, and on it goes.

What you're going to learn is a methodology. A system. A way of organizing your screen, your workspace, your day, and your AI sessions that lets you run seven parallel streams of work without losing your mind. Done right — and "right" is what the rest of this book is about — this approach delivers sustainable throughput of around fourteen tickets per day.

Not as a heroic push. Not once. Every day.

Let me say that again, because it probably sounded like hype: **fourteen tickets per day, sustainably**. A ticket in this book means a real unit of work — a bug fix, a feature, a refactor, a piece of infrastructure. Not a stream of tiny commits. Not line-of-code counts. Real tickets from your real Jira board, resolved to Code Complete, merged into main, and pushed to production over the course of a normal workweek.

Speaking of Jira — a small confession. Every developer I have ever met, myself emphatically included, hates Jira. This is not a hot take; it is as close to a universal truth as the profession produces. The woodland creatures in the woods behind my house hate Jira, and they have never used it. Jira is a tool built, as far as I can tell, by people who have never had to use Jira to do a job, which is why it will reliably convert a three-line ticket description into a seventeen-click odyssey involving Epics, Fix Versions, Story Point Estimates you didn't ask to provide, and a custom field added by a project manager in 2017 that no one since has had the heart to delete. It is the cardboard-flavored oatmeal of project management tools, and we all eat it every morning because the alternative is worse.

And yet — and this is the quiet brilliance of the method you're about to learn — one of the side effects of this system is that *you barely touch Jira anymore*. You look at it in the morning when you pick tickets. You look at it when you triage status. In between, Jira recedes into something closer to a rumor. The AI handles the ticket mechanics. You handle the swordsmanship.

That's actually the whole point, and it's worth sitting with for a moment. Most of what exhausts a modern developer isn't the code. The code is the fun part. What exhausts you is the *metawork* — the tickets, the status updates, the notifications, the mild but unending pressure to always be performing productivity for an audience of dashboards. This method is, at its heart, a structure that lets you put the metawork on autopilot and put your attention where it belongs: on the cuts.

If the number — fourteen — feels too high, good. It should. It felt too high to me too, until I actually did it. The point of this book is not to convince you the number is real. The point is to show you the system that makes it possible, so you can build it yourself. Whether you end up at fourteen or eight or twenty is between you, your codebase, and your taste. But the method is what gets you somewhere dramatically north of where you are now.

Who This Book Is For

If you're worried AI is going to replace you, this book will show you something more interesting: a way of working where you use AI as a force multiplier and become *more* valuable, not less. Developers who adopt this method stop being paid to type code. They start being paid to make decisions, set direction, and supervise parallel streams of work. That is a more senior role, and the market pays more for it.

If you've tried AI and found it exhausting, this book will diagnose exactly why. The exhaustion comes from a fundamental shape error — you're using AI as a faster typing tool when it should be a parallelism tool. Faster typing adds to your cognitive load. Parallelism, done right, subtracts from it. I'll show you how to get from one to the other.

If you feel behind on AI and like you'll never catch up, this book is the fastest way I know to become competent. There's no prerequisite beyond "you've written software before." I don't assume you know how to prompt an AI. I don't assume you have opinions about LLMs. I assume you've been doing software engineering long enough to be tired of bad advice, and I meet you there.

If you're already fluent with AI tools but not getting the throughput numbers you expected, this book will almost certainly close the gap. The method here isn't about having a better prompt library. It's about structural changes to how you work — the kind of changes that pry loose throughput prompts alone can't reach, no matter how cleverly you write them.

What This Book Is Not

It's not a prompt-engineering book. There are plenty of good resources on prompting already, and I'll share specific prompts where they matter. But prompts are maybe one percent of the value. The other ninety-nine percent is the system those prompts live inside.

It's not an AI philosophy book. I don't have a strong opinion about whether large language models "understand" anything, and I don't think it matters for your workflow. If you want deep philosophical debates about the nature of intelligence, this isn't it. There are entire podcasts about this already, and we have collectively agreed as a species to let them have that conversation without us.

It's not a hype book. I'm going to tell you exactly what works, what doesn't, where the method breaks, what kind of work it's bad for, and when you should ignore everything I'm telling you. If that sounds less exciting than "AI WILL CHANGE EVERYTHING," good. Excitement is what gets you to buy a book. Steady truth is what helps you for years.

It's not a book about any specific AI model. The examples use Claude Code because that's what I use daily, but the method transfers. Part III covers adapting the method to other tools — and more importantly, it covers the principles that make the method work regardless of which AI you pair it with. When the next model comes out, the method will still work. That's on purpose.

Why I'm Qualified to Teach This

A few reasons.

I've been writing software for over thirty years. Not thirty years of the same year repeated thirty times — thirty years of continuously building systems that had to run reliably in production, of learning whatever the next thing was, and then learning the thing after that. I've built cybersecurity platforms that process tens of millions of events per second. I've led international engineering teams. I've shipped five-person-startup products where I wrote every line myself, and I've served as the senior technical authority across sprawling enterprise systems owned by multiple vice-presidents. I've been the person people bring their hardest problems to, for a long time.

That matters because this isn't a theory book. It's a field-tested method. Every piece of it has been earned through use — through days when it worked beautifully, days when it broke, days when I had to figure out why it broke and then fix the system so it wouldn't break the same way again. What you're going to read is the distillation of several years of iteration.

The other thing that matters is that I've taught this method to engineers at every level, from first-job developers to seasoned staff peers, and watched it work for all of them. That told me the method isn't dependent on being exceptional. It's dependent on building the structure correctly and then trusting it. If you can set up the workspace I describe and follow the protocols for a few weeks, you will see the gains.

A Tour of What's Coming

Part I — Why This Book Exists is two short chapters that set up the mental frame. Chapter 1 is about the oldest story in software: people getting handed new tools and figuring out how to use them well. Chapter 2 argues that AI is not a special case, just the next beat in a long rhythm that includes the IDE, IntelliSense, and Stack Overflow — all of which were also going to replace us, if you believed the more breathless commentary of their respective decades. Although in fairness to Stack Overflow, nobody was ever ambitious enough to claim *it* would replace developers. The claim was only that developers who used it would replace those who didn't. I was a developer who used it. I found it to be a wonderful tool for obtaining precisely the wrong information, a few hours too late to solve my problem, and just in time to make the problem considerably worse. Good times. And to the gentleman who, on a balmy August afternoon, posted an authoritative-sounding answer explaining how MD5 hashes are calculated — which is not, in point of fact, how MD5 hashes are calculated — you cost me fourteen hours of my life I will never get back, and should our paths ever cross I shall greet you in the tradition of Inigo Montoya.

Part II — The System is the core of the book. Chapter 3 introduces the three phases: Setup, Execution, and Refinement. Chapter 4 walks you through the one-time workspace setup — a sandboxed directory, a set of coordinated environments, the command-line tools you'll need, and the lease protocol that keeps the parallel AI sessions from stepping on each other. Chapter 5 is where you build the screen layout that makes everything else work. It's a long chapter on purpose; the layout is roughly ninety percent of the value of the whole system. Chapter 6 shows you what a day looks like in execution — the rotation rhythm, the triage heuristics, the interruption rules. Chapter 7 handles the realistic parts: how to come back to a workspace after lunch, after a meeting, and — you may dimly recall these from the Before Times — after a full night of sleep. The restful, uninterrupted, woke-up-on-my-own-without-cortisol kind. This method quietly returns sleep to the menu of things a working developer gets to have. You are welcome. Chapter 8 covers Refinement — the slow continuous tuning that turns a good system into a great one.

Part III — Portability covers how to translate the method to other AI tools and the principles that apply regardless of which AI you're using. A quick note on this, because it matters: there is an active arms race among the coding AIs, with Claude Code and Codex and their various competitors leapfrogging each other on roughly a fortnightly basis. I use both interchangeably. I am not loyal to either — the method in this book is independent of both. I have run it against open-source models, against AWS Bedrock with half a dozen different foundation models sitting behind it, and against various homegrown harnesses people have cooked up on their weekends. It works with all of them. I happen to trend toward Claude in my daily practice, but that is only because Claude and I have learned to argue with each other like a comfortably married couple, and that particular dynamic suits me. You should pick whichever AI you have learned to argue with most productively. The method is indifferent.

Part IV — The Human Side is about keeping yourself intact while all this is happening. I want to say something plainly here, because I think it needs saying.

You have had a rough couple of years. I know this, because I have had a rough couple of years too, and so has every developer I talk to. The AI industry arrived without warning, threw a pile of half-working tools at us, told us to be ten times more productive by Tuesday, and — this was a particularly nice touch — also informed us that we would probably be replaced in the near future regardless. It is a difficult psychological climate. It would be a difficult psychological climate for anyone.

You might be feeling the terror of replacement. The anxiety of not improving fast enough to keep up. The sickening sense of being on a roller coaster whose tracks are getting built ten feet ahead of the cars. If any of that resonates, please hear me when I say: *it is not your fault.* You are not stupid. You are not slow. You were dropped into an environment that changed weekly, told to switch contexts without any of the support that used to come with such expectations, and then watched while the people doing the watching expected you to be grateful for the opportunity. That is not a fair ask. Of course you are tired. Of course you are worried.

Let me also tell you something I believe to be true. You are not going to be replaced by a machine that has to forget everything every few hours, or compact its own memory when the context gets too long, or lose track of why it opened a file three prompts ago. The AI is a remarkable tool. It is not a replacement for the person who has been holding the whole system in their head for seven years. You have the overview *and* the depth. You have the breadth of the architecture *and* the intimate knowledge of why that one function is written strangely, because you remember the incident in 2022 that made it necessary. You are no longer just a coder. You are the commander. You are the strategist. You are the one who reasons about where the AI has utility and where it does not. That is a higher job than the one you had before, and it is not going anywhere.

Chapter 11 covers AI burnout in detail — what causes it, how to recognize it in yourself early, and how to recover from it if you're already there. Chapter 12 is about bringing your team along without becoming insufferable about it.

The appendices cover prompt templates, installation walkthroughs for every major platform, troubleshooting, and a glossary of the terms I use.

One More Thing Before We Start

This book asks you to do something that might feel silly. It asks you to invest a weekend setting up a workspace with a very specific layout, commit to a very specific set of daily protocols, and trust that the structure will pay off. The first few days will feel awkward. You'll think "this is overengineered, I could just open an AI in one terminal and be fine."

You can. A lot of people do. That's why most people plateau at maybe one-and-a-half times their pre-AI throughput, feel more tired than they used to, and wonder what the hype is about.

The system in this book is overengineered *on purpose*. The overengineering is what converts AI from a tool that adds cognitive load into a tool that removes it. Every protocol, every naming convention, every weird rule about screen layout is there to answer a question in advance so your brain doesn't have to answer it in the moment. When you stop answering thousands of small questions a day, you have energy left for the big ones. That's where the throughput comes from.

So please — read the whole book before you decide which parts to skip. Everything in the method is there for a reason. I've tried removing pieces, and they are all load-bearing in ways that only revealed themselves *after* I'd removed them, which is of course the only way anything ever reveals itself to be load-bearing. What looks like excess is structure.

Give it two weeks. Build it, run it, refine it. If after two weeks it hasn't helped — if you're not visibly faster, visibly less tired, and visibly more confident at the end of your day — throw the book away with my blessing. I don't want anyone forcing themselves to use a method that isn't working for them.

But I think it'll work. It's worked for me for a long time, and it's worked for everyone I've taught.

Now. Pick up your hammer. Or your sword. They're closer to the same thing than you think.

— Blake Tullysmith
Santa Rosa, California

Part I — Why This Book Exists

Chapter 1
The Hammer and the Wood

Buzzy and the Cedar Shingle

When I was nine years old, my father and I were on the side of the barn at our old house, installing cedar shake siding. I had, by that age, already hammered something like a thousand nails into 2×4s helping around the property. I was pretty good at it. I had earned, in my own mind, the honorary title of Experienced Nail-Driver, possibly even Junior Carpenter, certainly no longer a beginner.

So when my dad handed me a small fistful of 5d galvanized ring-shank nails and pointed me at a row of fresh cedar shakes, I did what I had always done. Positioned the nail. Swung. Drove it home in a single confident stroke. And the cedar shingle split, beautifully and conclusively, along the grain from the nail all the way to the edge of the board — rendering it a lovely piece of scrap wood and deleting approximately thirty seconds of my father's morning.

Fortunately we were not working alone. One of my dad's friends, a wiry old carpenter named Buzzy, was up on the wall with us. I was nine, so I remember spending a non-trivial portion of that morning trying to work out whether Buzzy was called Buzzy because he looked like a bee or sounded like a bee, and I remember arriving — after some careful empirical observation — at the disappointing conclusion that he did neither. Buzzy was, however, reliably excellent at two things that I now recognize as important carpenter credentials: drinking beer, and smoking.

Buzzy saw what I had done with the shingle, declined to laugh at me in any way I could detect, and said something along the lines of "here, watch this." He picked up a fresh 5d ring-shank and held it vertically between two fingers, the head resting against his fingernails like he was about to flick it. He gave the tip of the nail a single gentle tap with the hammer — just one tap, barely a touch — to flatten it. Then he flipped the nail around in his fingers the same way he flipped his cigarettes, which he was also fluently good at, set the now-blunt tip against the cedar exactly where he wanted it, and drove it home with one swing. No split. Not even the suggestion of a split.

I was nine. That was a moment.

What I took from that morning, and what took me another couple of decades to fully articulate, is that Buzzy had two advantages over me. The first and obvious one was that he knew something I didn't: that a sharp nail works like a tiny wedge and splits softwoods along the grain, whereas a blunt nail works like a tiny plug cutter and punches a small cylinder of wood out of its path without splitting anything. The second and less obvious advantage was that Buzzy had internalized this to the point where he didn't have to think about it. He didn't pause and consider which technique was right for this particular cedar shake. He just reached for the hammer and the nail and did the right thing, the same way he reached for a cigarette or a beer.

This book is not a manual for how to hold a hammer. You already know how to hold a hammer — you've been writing software for years, which means you have held a lot of hammers. What this book is, instead, is a manual for what to do with the hammer once you're holding it. How to recognize which nail you're looking at. How to blunt it in advance. How to place it. How to drive it home so the wood doesn't split. That's the missing layer between "I have the tool" and "I know what I'm doing with the tool." And it's what separates the developer wrecking cedar shingles from the one who is, in their own chosen corner of the craft, Buzzy.

What That Has to Do with AI

AI coding tools are the newest hammer on the shelf, and right now most developers using them are the nine-year-old on the side of the barn. They are getting nails in — sort of — but the wood is splitting, the nails are going in sideways, and at the end of the day their hands hurt and the work doesn't quite look right. They see other people online saying things like *"I'm twelve times more productive now"* and they wonder what they're missing, because they don't feel twelve times more productive. They feel tired, and a little embarrassed, and vaguely worried that they might be falling behind.

This book is not a manual for holding the hammer. It's not a book about what AI *is*, or how large language models work under the hood, or what the future of software development looks like. It's a book about what to do at nine o'clock on a Monday morning when you open your laptop and you have fourteen Jira tickets assigned to you and a vague sense that you should be using AI better than you are.

By the end of this book, you will be using AI better than you are. Not because you've learned some magical incantation that grants secret capabilities — there aren't any — but because you'll have a *system* for getting the work done. The system handles the things that are currently costing you effort and attention: remembering what you were doing, switching between tasks without losing your place, keeping track of seven different things at once, making sure nothing dangerous happens by accident, knowing when to trust the AI and when to push back. All of that will be handled by the structure you build around the tools, so you can put your attention on the actual thinking work.

A Specific Promise

Most books about productivity tools make vague promises. I want to make a specific one.

A developer who adopts the method in this book, and sticks with it through the rough first week, can realistically expect to complete somewhere around fourteen five-point Jira tickets per day as a sustainable average. Not every day, but day in and day out, as the normal rhythm of their work. The arithmetic that gets us to fourteen is covered in detail in Chapter 3, but the short version is: seven parallel contexts, two working sessions a day, with the AI doing the mechanical work inside each context while you do the thinking and validation.

Now, I want to acknowledge something here, because I know what a large number of my readers are already doing, and what it's costing them. A developer using AI *without* a method like this one — just opening Claude or Copilot and riffing — can often get up to one or two five-point tickets done in a day. That is, on the face of it, an impressive number. In the Before Times, a five-point story was a few days of work, or half a sprint if the estimator was being pessimistic. Two a day is a genuine productivity gain and nobody should minimize it.

But if that's where you are right now, I suspect you already know the catch. It doesn't *feel* sustainable. It feels breakneck. It feels like every day is the day you're going to fall behind, and every ticket is slightly more scrambled than the last, and by Thursday you're running on coffee and a kind of gritted-teeth momentum that is not going to survive the quarter. A lot of developers in that position are already burned out and don't quite have the language for it yet. A lot of the rest are a few weeks away. If that's where you are, hear me clearly: the gains are real, but the rhythm is wrong. Without a rhythm you can hold indefinitely, you're just running yourself to pieces a little faster than you used to.

This book gives you the rhythm. The throughput goes up, and the burnout goes away, because the method distributes the cognitive load across a structure instead of piling it all onto you. The number gets bigger and your life gets *easier* at the same time. That's the specific promise.

If you're currently completing one or two tickets a day and finding it exhausting, this probably sounds unbelievable. I understand. I would have thought it was unbelievable too, before I built the system. The number isn't the result of working harder or longer hours; if anything, developers using this method report working *less* hard than they did before, because the cognitive load of what they're doing is lower. The number comes from working in parallel across multiple contexts at once, with a structural layout that keeps the contexts from colliding, and with a set of habits that keep the AI doing useful work without requiring you to babysit it.

I'm not going to promise you'll be a superstar. I'm not going to promise you'll get promoted. I'm not going to promise your company will love you. I'm promising that if you follow the method, you'll complete significantly more work than you do now, with less stress and fewer surprises. That's a big claim, and the rest of this book is about making good on it.

Who This Book Is For

You probably picked up this book because one of these things is true:

- **Your company has decided everyone needs to use AI now.** Maybe there was a memo. Maybe there was a mandate. Maybe it was quieter than that — just a sense that it's what's expected, and a quiet worry that if you don't figure it out, you'll be the one left behind. Whatever the pressure is, you're feeling it, and you want to actually get good at this rather than just *seeming* like you're getting good at it.

- **You've been laid off, or you're worried you might be.** The tech industry has been through a rough stretch. You're trying to add skills that make you more employable, and AI fluency is clearly one of them. You don't have time for a six-month course. You need to be capable in weeks, not months.
- **You've been trying to use AI tools for a while and feeling like you're missing something.** You've used Cursor or Copilot or Claude or ChatGPT. You've had some good moments and a lot of frustrating ones. You see other people claiming dramatic productivity gains and you can't figure out what they're doing differently. You suspect — correctly — that there's a method they have that you don't.
- **You're generally curious and you want the good stuff.** You don't have any particular urgency, but you've noticed that most of the AI-for-developers content online is shallow, repetitive, or out of date. You want something more structured.

If any of these is you, you're in the right place.

Who This Book Is Not For

A few types of reader will not get much out of this book, and it's fair to name them up front so nobody wastes their time.

If you're looking for a deep technical explanation of how large language models work, this isn't that book. There are good books on that topic and this one will not duplicate them.

If you're an AI skeptic looking for ammunition, this book won't help either way. I take it as a given that AI coding tools are useful. The book is about how to use them well, not whether to use them at all. If you need to be convinced they're useful in the first place, nothing in here will do that for you.

If you already have a system that works for you, you might still find some ideas worth borrowing, but you don't need to do the whole method. The value is in the system being *complete* — every piece supports every other piece. If you're going to pick and choose, you're better off keeping what you have than partially adopting something new.

If you're a total non-technical reader who has never written code, this book assumes you know what a terminal is, what a git branch is, what a pull request is, and what a Jira ticket is. If those are all unfamiliar words, you'll want to pick up a basic software development book first and come back to this one.

Everybody else: welcome.

What You'll Need

The method in this book is built around a specific shape of tooling and hardware. The specifics are covered in detail in later chapters, but here's the shortlist so you know what you're signing up for:

- **A laptop or desktop computer running macOS, Windows, or Linux.** Any recent machine is fine. The method doesn't require unusually powerful hardware.
- **At least one large monitor**, ideally around 3440×1440 or larger. An ultrawide is ideal; the method works on smaller monitors but with reduced effectiveness. If you don't currently have a large monitor, consider whether this is a good moment to buy one.
- **Claude Code.** This book's primary examples use Claude Code, the command-line agent from Anthropic. The methods adapt to other AI coding tools (we'll cover Codex in Chapter 9), but the clearest way to teach the system is with one specific tool in front of us.
- **The willingness to set things up carefully.** This is not a tool you pull out of the box and start using. The method requires two to three hours of setup before you do any real work. If you try to skip that, the method doesn't work. If you do the setup right, you essentially never have to do it again.

How to Read This Book

The book is designed to be read linearly, at least the first time. The early chapters build the ideas and the later chapters depend on them.

If you're in a hurry, you can skip Chapters 2 and 3 on a first pass and go straight to Chapter 4 (Setup). You'll miss some context, but you'll still be able to execute the method. Come back to Chapters 2 and 3 after you've used the system for a week and you're ready to think about *why* it works the way it does.

If you're skeptical about the big productivity claim, start with Chapter 3. It explains exactly where the fourteen-tickets-a-day number comes from, piece by piece, and by the end of it the number will feel less like magic and more like arithmetic.

If you're coming from another AI tool like Cursor or Copilot, read everything. The philosophical differences between the tools will matter, and your existing habits may be working against you in ways you haven't noticed. Chapter 9 will help you translate the method to whichever tool you prefer once you understand the canonical version.

Regardless of your starting point, don't try to read this book in the bathtub. There's a terminal to set up, prompts to send to Claude, windows to arrange. You'll want to be in front of your computer, at least for Chapters 4 through 7.

One last thing before we move on. This is not a book that rewards skimming. The details matter — the exact names of directories, the exact format of prompts, the exact layout of your screen. If you find yourself reading in a hurry and glossing over specifics, slow down. The specifics *are* the system.

Ready? Let's start with why AI feels strange, and why it shouldn't.

Chapter 2
Nothing New Under the Sun

What the Old-Timers Know

I started writing code professionally when I was very young, and I've been doing it for thirty of the last thirty-two years. The missing two years are accounted for: at one point I stepped away from technology to serve in the United States Air Force Security Forces, an experience which was cut short by a medical discharge that left me a disabled veteran and, as it turned out, drew me right back to technology – which was, I suspect, where I was always going to end up anyway. I wrote assembly language for hardware drivers on SCO Unix and early Windows. I built out data centers. I worked on medical software, telecommunications software, civil engineering software, and cybersecurity platforms processing tens of millions of events per second. I've served as a consultant to dozens of startups and led teams at companies of every size. Most recently I've been architecting distributed systems at scale.

I'm telling you this not to impress you but to ground what comes next. I've watched the tools that developers use evolve through several distinct eras, and I want to make a claim that some readers will find controversial: **AI coding tools are not a new kind of thing. They're the next step in a very old story.**

The Lineage

Think about what a developer's job has looked like across the last few decades.

When I started, if you wanted to write a program, you sat in front of a terminal with a text editor and typed code. You remembered the syntax, because there was nowhere to look it up quickly. You remembered the standard library functions, because there was no autocomplete. If you got stuck, you either figured it out on your own, you called someone, or you drove to a bookstore and bought a reference manual. The tools were primitive, and as a consequence, the parts of the job that counted as "real work" included a lot of things that we'd now consider clerical: remembering function signatures, manually tracking which files you'd edited, copying and pasting between windows.

Then came IDEs — integrated development environments — and a chunk of that clerical work disappeared. The IDE remembered the syntax for you. It told you when you made a typo. It let you jump from a function call to its definition. Some old-timers complained that the IDEs were making developers soft, that nobody would know how to program without their IDE, that it was cheating. Those old-timers are not wrong that IDEs changed what it meant to be a developer. They just didn't notice that what the IDE changed was *the boring part.* The part where you have to think clearly about what the program should do — that part didn't change at all.

Then came IntelliSense and autocomplete. Now the IDE wasn't just passively helping; it was actively suggesting the next token as you typed. Old-timers, again, complained. Developers would forget how to spell function names. Developers would become lazy. Developers would lose touch with the metal. The old-timers weren't wrong that something was changing, but they were looking at the wrong thing. What was changing was, again, the boring part.

Then came Stack Overflow. For the first time, when you got stuck, you could reach into a global pool of other developers who had been stuck in exactly the same way and had written down how they got unstuck. This was a bigger shift than any of the previous ones, because it didn't just automate the mechanical parts — it socialized the debugging process. Old-timers complained that developers would stop thinking for themselves. They weren't wrong that something was changing, but again, the thing that was changing was not the core of the job. It was the edges.

And now there's AI. An agent that can read your codebase, understand the structure, write code, run tests, commit to git, open pull requests. The old-timers are complaining again. Developers will forget how to code. Developers will become dependent. Developers will not learn the fundamentals.

And again, the old-timers are not wrong that something is changing. But what's changing is — one more time — *the boring part*. The part where you type the code. The part where you run the tests. The part where you write the commit message. The part where you remember which branch you're on and whether it's up to date with main.

What doesn't change is the thinking. Deciding what to build. Deciding whether to build it. Deciding whether what got built is correct. Deciding how the pieces fit together. Deciding what to tell the customer. Deciding what to tell your manager about whether the timeline is realistic. None of that is getting automated. None of that is getting easier. All of that is still you.

Why AI Feels Different Anyway

If AI is just the next step in this lineage, why does it feel so dramatically different? Why does every conversation about it carry an undertone of anxiety that conversations about IDEs and autocomplete didn't have?

Three reasons, I think.

First, the pace. The previous transitions — IDEs, IntelliSense, Stack Overflow — happened over years. You had time to adjust. You could ignore the new thing for a while and catch up later. AI has compressed a similar shift into months. That's disorienting even for people who would be enthusiastic about the change if it happened more slowly.

Second, the marketing. The previous transitions were tools, sold as tools. The current transition is being sold, in many quarters, as a replacement. Thinkpieces and LinkedIn posts talk about AI "replacing developers," which is rubbish but which successfully makes every developer who reads them a little scared. The fear is not a good guide to reality, but it's real, and it's widespread.

Third, the shape of the tool. A text editor is a passive thing. It does what you tell it to. An IDE is a *responsive* thing — it reacts to your actions. An AI agent is something new: it's an *agentic* thing. It has opinions. It makes decisions. It does things you didn't ask it to do if it thinks they're helpful. That's genuinely unprecedented in developer tooling, and working with it productively requires some new skills that nobody has more than a few years of experience in.

So the feeling of novelty isn't wrong. But it's important not to let the feeling overwhelm the underlying truth: the job still breaks down the same way. There's the thinking part, and there's the boring part. AI takes more of the boring part off your plate than any previous tool has. That's all.

The Cognitive Load Problem

Here's a place where AI really is different from previous tools, and this is the one that will kill you if you're not prepared for it.

Every previous developer tool *reduced* your cognitive load. An IDE takes the burden of remembering syntax off your mind. IntelliSense takes the burden of remembering function names. Stack Overflow takes the burden of figuring out obscure bugs. Each one made your job easier in a measurable way.

AI, used naively, *increases* your cognitive load. Every prompt requires you to context-switch from thinking about the code to thinking about how to explain what you want. Every response requires you to read what the AI produced and decide whether it's right. Every decision about whether to accept, reject, or modify a suggestion is another small mental task. After eight hours of that, people are exhausted in a way they weren't exhausted after eight hours of coding by hand.

This is the source of the burnout problem you've probably heard about. Developers who adopt AI tools without a system find themselves *more* tired at the end of the day, not less, because they've added a layer of constant micro-decisions on top of their normal work. They dread opening the laptop. They start finding reasons to do things without the AI. They feel vaguely bad about this because they know they "should" be using it. The dread and the guilt compound.

The system in this book is designed around a specific solution to this problem: **replace many small decisions with one large structural decision, so your cognitive load stays low even as your throughput goes up.**

Here's what that means in practice. Instead of deciding, on each prompt, whether to accept or reject what the AI produced, you set up a structural boundary inside which the AI is free to do whatever it wants. Instead of deciding, on each task, how to describe it, you build templates and protocols that the AI has already been taught to follow. Instead of deciding, on each context switch, where to click and what to look at, you put every kind of work in a fixed physical location on your screen so your hand knows where to go without your brain having to decide.

Every piece of the system you're going to build is, at heart, a *pre-made decision* — a choice you make once, thoughtfully, so you don't have to make it thousands of times while you're trying to get real work done. Once you see the system this way, the layout decisions and the protocol decisions and the directory naming decisions all start to feel less arbitrary. They're all the same kind of thing: load-bearing structure that takes cognitive weight off your shoulders.

The Team of One

One last idea before we move on, because it bears on how you should think about your role in whatever organization you work in.

You do not need permission from your organization to adopt this method. You do not need your whole team to change how they work. You do not need buy-in from your manager, or from engineering leadership, or from anyone else. The method is entirely local to *you* — your laptop, your workspace, your Claude sessions, your tickets. The only things that change outside your own boundary are the draft pull requests you open (which look identical to any other developer's PRs) and the tickets you close (which look identical to any other developer's tickets, except there are more of them).

This matters because most productivity advice aimed at developers assumes an organizational context — you need to get your team to adopt the new practice, you need to convince leadership, you need to run a pilot program, you need to measure outcomes and write a retrospective. All of that is slow, it's political, and for most developers, it's a reason not to try.

You get to skip all of that. You become what I call a *team of one* — an individual using the method inside an organization that's doing its own thing. Your colleagues don't need to know. Your manager doesn't need to know, though they'll probably notice your throughput go up. Your process, inside your own boundary, is your business.

Later in the book (Chapter 12) we'll talk about what happens when other people on your team start to notice and want to learn the method. That's a nice problem to have, but it's not a problem for today. Today's problem is getting *you* set up.

And that's what we'll do starting in Chapter 3, which walks through the shape of the finished system so you know what you're building before we start building it. Then Chapters 4 and 5 build it, and Chapter 6 teaches you to drive it. Let's go.

Part II — The System

Chapter 3
The Three Phases

A Tour Before the Build

Before we start building, I want to walk you through the finished system so you know what you're aiming for. It's easier to assemble a piece of furniture when you know what it looks like assembled, and it's easier to learn a system when you can see the whole shape of it before you start putting in parts.

The method has three phases: **Setup**, **Execution**, and **Refinement**. Each one does different work, happens at a different cadence, and has a different feeling.

Setup happens once. You build the workspace, you install the tools, you teach the AI agent the rules it's going to follow. Setup takes about two to three hours the first time you do it, and roughly zero time after that, because everything Setup creates is persistent — it lives on your machine and in your cloud services, and it keeps working across every session without you having to do anything.

Execution happens every day. Execution is the actual work: starting your Claude sessions, loading your tickets, letting the AI work while you do the things only humans can do, validating the work when it's done, closing the tickets. Execution has a rhythm — there's a morning session, there's an afternoon session, there's a particular shape to each one. By the end of Chapter 6 you'll know this rhythm well enough to have it in your bones.

Refinement happens continuously, in the background, while you're executing. Refinement is the slow tuning of the system — noticing things that could work better, adjusting your prompts, adding new rules to the AI's permanent memory, pruning things that have gone stale. You don't schedule Refinement; it happens naturally as you notice things. Chapter 8 is about making sure you actually notice.

Figure 3.1 — The Three Phases

Setup happens once. Execution happens daily. Refinement happens continuously.

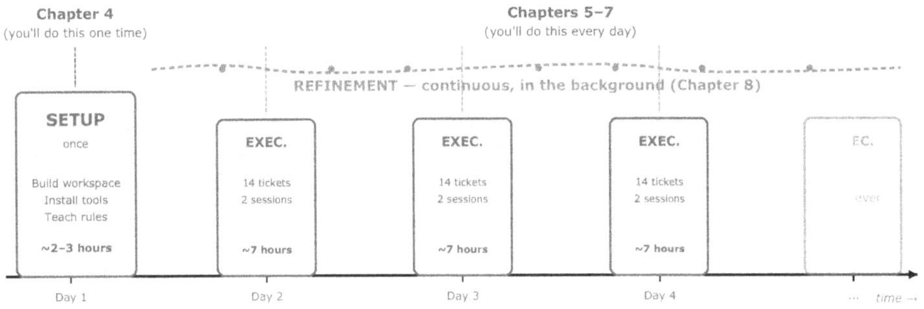

The setup phase is front-loaded. Execution is the steady state. Refinement is what makes the steady state keep getting better.

Most developers who try to adopt AI tooling skip Setup entirely. They just start prompting. They're treating AI tools the way they'd treat a new text editor — install it, open it, get to work. That's why most developers who try AI tooling bounce off it. The AI in front of them is powerful, but they're using it without structure, without memory, without boundaries, and without a coordinated system. They're trying to be a carpenter without a workbench.

Setup is the workbench. Without it, the rest of the method doesn't work. So we're going to invest in Setup carefully, and then Execution will feel easy.

The Shape of Execution

Let me paint the picture of what a normal working day looks like once the system is running, so you have something concrete to aim at.

It's Monday morning. You sit down at your desk in front of your laptop. There's a large monitor in front of you — ideally an ultrawide, around 3440×1440 or bigger. You open the lid. Your screen lights up with a layout you set up before: a big browser window in the upper left, a communications app (Slack, Teams, whatever) in the upper right, a Jira board view in the lower right, and seven smaller rectangles filling the rest of the screen. Each of the seven smaller rectangles is a *slot* — a permanent screen position where one ticket's worth of work happens. Chapter 5 covers slots in detail; for now, picture them as seven side-by-side workbenches.

You look at your Jira board. You have fourteen tickets assigned to you this sprint. You pick the first seven and open them as tabs across the top of the Jira board window. Then, one at a time, you drag each ticket tab out of the Jira window and drop it into one of the seven slots. Now each slot has a ticket associated with it.

For each slot, you click on the Claude terminal that lives underneath the browser in that slot. You type three words: *"Work on [ticket number]."* Claude reads the ticket, creates a branch, opens a draft pull request on GitHub, and starts working. You move to the next slot and do the same thing.

Within a few minutes, all seven Claudes are working in parallel. Seven different tickets are progressing at the same time. While that's happening, *you* do the things only you can do — you check email, you respond to Slack messages, you review your teammates' pull requests, you think about the bigger picture.

Every few minutes, you glance across the seven slots to see what the Claudes are doing. When one asks a question, you answer it. When one seems stuck, you help it get unstuck. When one finishes, you review what it did. When you're satisfied with the work, you promote the draft PR to ready-for-review and close the ticket. The Claude in that slot is free; you lease a new ticket to that environment and start the next one.

By lunch, you've closed somewhere between five and nine tickets, depending on how much interruption you've had from meetings and how tricky the tickets were. You eat. You take a walk. You come back and do it again in the afternoon.

By the end of the day, you've closed somewhere in the neighborhood of fourteen tickets. Not every day, but on a good day, yes. On an average day, maybe twelve. On a slow day, maybe eight. Your old pace of one or two feels like a half-remembered dream from a previous life.

Why Seven

The number seven is load-bearing in this system and it's worth explaining why before we commit to it.

Start with the human side. Psychological research on working memory has famously pegged human short-term capacity at "seven, plus or minus two" — the idea being that most people can juggle about seven distinct things at once before starting to drop some. That research has been refined over the years, and the real number depends a lot on what the "things" are, but the rough intuition holds. Seven is around where most brains start to saturate.

In this method, the "things" you're juggling are individual ticket contexts, each one running its own AI agent. Your actual cognitive load per context is low — the agent is doing the real work, and the system's structure handles most of the context-switching cost. But there's still a floor. At four or five parallel contexts, you're leaving capacity on the table; the AI is idle too often and you're not getting the throughput the system is capable of. At ten or twelve parallel contexts, you start making mistakes — losing track of which ticket is which, typing into the wrong window, confusing the state of one context with another.

Seven turns out to be the sweet spot. It's high enough that the AI is usefully busy while you're doing other things, and low enough that your brain can keep all seven contexts mentally distinct. Your mileage may vary slightly — some people do fine at eight, some top out at six — but seven is the right starting point for almost everyone.

Then there's the practical side: seven is a number that fits on a reasonably-sized monitor. The layout you're going to build puts seven slots on one screen, arranged so that each one has a fixed permanent position that you'll come to know by reflex. If you had ten or twelve slots, the slots would be too small to work in. If you had only four, you'd be wasting expensive screen real estate. Seven is the number where the geometry works out.

Finally, there's a minor but real benefit that seven is *memorable.* Seven dwarfs, seven deadly sins, seven seas, seven samurai. The number seven shows up in cultural shorthand for groupings where each element is distinct and each one matters. I promise I didn't pick the number for this reason, but it does make the method easier to remember.

The Dish by the Door

The second load-bearing concept in the system is what I call the *dish-by-the-door principle*, and it's so important I'm going to belabor it.

If you have a dish or a hook by the front door where you always put your keys when you come home, you never lose your keys. You don't think about it. You walk in, you drop the keys in the dish. You walk out, you grab the keys from the dish. If someone asked you *where* the keys are, you'd have to think for a moment before you could answer, because the answer isn't stored as a piece of information — it's stored as a habit. Your hand knows where the keys are. Your head doesn't need to be involved.

Now imagine the alternative. Imagine you don't have the dish. Imagine you drop your keys wherever feels convenient when you come home — sometimes the kitchen counter, sometimes the coffee table, sometimes the bedroom dresser. What happens? You spend five minutes every morning looking for your keys. You miss appointments. You get angry at yourself. You buy a tracker from Amazon. Eventually you decide you need a dish by the door.

This is exactly what happens to developers working across multiple contexts without a fixed layout. They start a new Claude session, and the window opens wherever Claude feels like opening it. They click into a terminal, and the terminal comes to the front and covers up whatever was there before. They drag windows around trying to see multiple things at once, and by the end of the day their screen is a mess. They can't find their keys.

The layout you're going to build in Chapter 5 is the dish by the door, for your whole workday. Every kind of work has a fixed physical position on your screen. The validation browser is always upper-left. The communications app is always upper-right. The Jira board is always lower-right. The seven slots are always in the same seven places, in the same order, with the same names. You don't have to think about where things are, because everything is always where it's always been.

This sounds small. It is not small. The amount of mental energy most people burn every day on the low-grade task of "where is that window I need" is enormous, and getting that back is a significant chunk of where the productivity gain in this method comes from.

The Workspace, the Environments, the Leases

Now let me sketch the software shape of the system. Don't worry about building it yet — we do that in Chapter 4. This is just so you can see the pieces and how they fit.

At the root of everything is a single directory on your laptop called `workspace`. Inside `workspace` lives everything Claude is allowed to touch. Outside `workspace`, Claude can't see anything. This boundary is the thing that makes it safe to run Claude with broad permissions — it can do whatever it wants inside the boundary, and nothing at all outside it.

Inside `workspace` are three interesting kinds of thing.

First, there are configuration files — a `CLAUDE.md` file that serves as Claude's permanent memory, and a `.claude` directory with configuration including the allow/deny list of commands. These files teach Claude the rules of engagement: what it can and can't run, what it should do when you give it a ticket, how it should coordinate with other Claude instances.

Second, there's a `repos` folder where you check out clean copies of the code repositories you work in. You don't do work directly in `repos` — it's a library, a source of clean copies. When you want to work on a ticket, you don't modify `repos` directly.

Third, there are seven *ephemeral environments* — seven sub-directories named after Norse gods: thor, freya, tyr, loki, odin, heimdall, and baldr. Each of these is a self-contained work area where one ticket's worth of work happens. When you start working on a ticket, Claude copies the relevant repo from `repos` into one of the seven environments, creates a branch, and does all the actual editing there. The environment is disposable — when you're done with the ticket, you reset it and reuse it for the next one.

Coordinating all this is a single file called `ENVIRONMENT_LEASES.md`, which is a lightweight bookkeeping system that keeps multiple Claude instances from stepping on each other. When Claude takes over an environment, it writes down that it's using that environment and records its own process ID. When another Claude wants an environment, it reads this file to find a free one. The process ID is a liveness check — if the recorded PID isn't running anymore (because that Claude crashed or you rebooted), any new Claude can safely take over the environment.

That's the whole software shape. It's small. It's all files. There's no server, no database, no special infrastructure. The entire system fits in a single directory on your laptop, and yet it's enough to coordinate seven parallel AI agents working on different tickets at the same time without collisions.

How Setup, Execution, and Refinement Fit Together

Here's how the three phases interact across your life as a developer using this method.

In the first two or three hours of using the method, you do Setup. You build the workspace, you install the tools, you teach Claude the rules. This is a one-time investment. Chapter 4 walks you through it in detail.

On your first day of real work, you do your first Execution. Chapter 5 walks you through the one-time-ish work of arranging your desktop — the cold start — and Chapter 6 walks you through what the actual workday looks like once the layout is in place.

Every subsequent day is another Execution, but you don't usually start from scratch — you come back to a workspace that's mostly where you left it. Chapter 7 walks through the various states you might find things in, and how to get yourself back to "seven Claudes running" as quickly as possible.

Refinement happens in the background throughout all of this. As you use the system, you'll notice things — a prompt pattern that works especially well, a rule you wish Claude had, a habit you want to break. Chapter 8 is about making those noticings actionable without letting them overwhelm you.

The pattern is: Setup is front-loaded, Execution is the steady state, and Refinement is what makes the steady state keep getting better.

With the shape of the system now in front of you, let's start building it. The next chapter is Setup.

Chapter 4
Setup with Claude Code

What This Chapter Does

This is the chapter where you actually build the system. By the end of it, you'll have a sandboxed workspace on your laptop, a complete set of rules that govern how Claude behaves inside it, and all the command-line tools you need to do real work. It's the longest-feeling chapter in the book, but it's also the one that pays for itself the most: the two to three hours you spend here are hours you will essentially never spend again, because everything you build in this chapter persists forever.

There are two steps: building the workspace (Step 1), and installing the tools (Step 2). Step 1 is conceptually the harder of the two, because it involves teaching Claude some new habits. Step 2 is mechanically fiddlier but conceptually simpler. Both are covered in the same warm, careful way that the rest of the book uses.

A reminder about hardware and software before you dive in: you'll need a laptop or desktop running macOS, Windows, or Linux (Manjaro is covered here in the main text; Ubuntu and others are in Appendix D). You'll need Claude Code installed (Appendix B). And you'll need a terminal you're comfortable in — we'll set that up at the start of Step 1.

Ready? Here we go.

Step 1: Create the Workspace and Teach Claude How to Live in It

Before We Start

This is the longest single step in the book, and it will probably take you somewhere between forty-five minutes and an hour and a half the first time you do it. Take your time. There is no prize for finishing fast. The work you do in this step is the foundation that every other chapter rests on, and if you rush it you'll spend the rest of the book limping on a foundation that doesn't quite hold.

A second thing worth saying out loud: if you're reading this book because your employer suddenly decided everyone has to "use AI now" and you're worried about keeping up, or because you've been laid off and you're trying to add a new skill that makes you employable again, or because you read an article about developers being twelve times more productive and you feel like you're already behind — you're in the right place, and you're not behind. Most developers using these tools today are using them badly. The bar is genuinely low. By the end of this chapter you will be set up better than the vast majority of people who've been using Claude Code for a year. That's not flattery, it's just a description of where the field is right now.

Take a breath. Pour something to drink. We're going to walk through this carefully.

Why This Step Is the Foundation

When most developers try to use an AI coding agent at scale, they hit the same wall within about two days. The agent is genuinely useful, but to be genuinely useful it needs to do real things — write files, run commands, install dependencies, execute tests. Each of those actions is potentially dangerous, so by default the agent stops and asks you for permission before doing any of them. Within an hour you've clicked "yes" two hundred times. Within a day you've started clicking "yes" without reading. Within a week you've either turned permissions off entirely (which is genuinely dangerous), or you've decided AI tools are too annoying to use and gone back to writing code by hand (which is a waste of a powerful tool).

This is the hammer-splits-the-wood problem from Chapter 1, and the way out of it is structural rather than procedural. Instead of asking the agent to seek permission before every action, we give the agent a *boundary*. We carve out a specific area of the filesystem that the agent is allowed to operate inside. We give it a list of commands it must never run. We make those rules persistent so they survive across sessions. And then, with the boundary in place, we let the agent move freely *inside* the boundary without asking permission for every step.

> 💡 **Callout: What does "sandboxed" mean?**
>
> A sandbox, in software, is a controlled area where a program can do whatever it wants without affecting anything outside that area. The name comes from the idea of children playing in a literal sandbox at a playground — they can dig, build, knock down, and make a mess, but the mess stays in the box. The grass around it stays clean. We're going to put Claude Code in a sandbox. Inside the sandbox it can build sandcastles, knock them down, and start over a hundred times a day. Outside the sandbox, it can't touch anything.

The directory you're about to create — we'll call it `workspace` — is that sandbox. Everything Claude does for the rest of your career using this method happens inside `workspace`. Nothing Claude does will ever escape `workspace`, because we're going to teach Claude very explicitly that it's not allowed to.

There's a second thing this step does, and it's the part that delivers the throughput numbers from Chapter 3. Inside the workspace, we're going to create seven small environments — seven separate sub-directories — that act as isolated work areas. Each one can host a different ticket. Each one can have a different Claude instance working in it. They don't step on each other because we're going to give them a shared coordination file that lets them take turns, like a conference room signup sheet. This is the mechanical implementation of the seven spinning plates from Chapter 3. By the end of this step they'll exist on disk, ready to be claimed.

Last thing before we open the terminal: a lot of what we're about to do will feel arbitrary on the first read. Why seven? Why Norse mythology? Why this directory layout and not some other? I'll explain each choice as we hit it. None of it is arbitrary, but some of it is the result of patterns I found through trial and error rather than first principles. Trust the pattern on the first build. Once you've used the system for a week or two and you understand why each piece is there, you'll have earned the right to change things to suit your taste, and you should.

Set Your Terminal to Dark Mode

Before any commands, switch your terminal to a dark color scheme. You're going to be staring at this terminal across multiple panes simultaneously for most of your workday. A bright white background will burn through your visual stamina by mid-afternoon and you'll feel it as a vague tiredness behind your eyes that you won't connect to the cause. Dark mode is not a style preference for this work. It's part of the ergonomics. See **Appendix C** for the click-path to dark mode in the default terminal on each platform.

While you're in there, also set your terminal font to something around 14 points. You'll be reading a lot of small text in side-by-side panes, and the temptation will be to shrink the font to fit more. Resist it. Big enough to read without leaning in is the right size.

Create the Workspace Directory

Open your terminal. Make a directory called `workspace` in your home folder, then change into it.

macOS / Manjaro:

```
mkdir -p ~/workspace
cd ~/workspace
```

Windows (PowerShell 7):

```
New-Item -ItemType Directory -Path "$HOME\workspace" -Force
Set-Location "$HOME\workspace"
```

The name `workspace` is a convention. You can call it anything you want. But every example in this book assumes the directory is called `workspace` and lives in your home folder, so if you call it something else, you're signing up to do mental translation on every example for the rest of the book. I recommend you don't. Just call it `workspace`.

> 💡 **Callout: The home folder**
>
> Your "home folder" is the directory associated with your user account. On macOS and Linux it's written ~ (a tilde), which is shorthand for `/Users/yourname` on macOS or `/home/yourname` on Linux. On Windows it's `$HOME`, which expands to `C:\Users\yourname`. When this book says "in your home folder," it means right inside that directory.

Launch Claude Code with Skipped Permissions

If you don't have Claude Code installed yet, jump to **Appendix B** for installation instructions on each platform, install it, then come back here.

Once it's installed, launch it from inside the workspace directory with the permission prompts disabled:

macOS / Manjaro / Windows:

```
claude --dangerously-skip-permissions
```

The flag is named the way it is for a good reason. Running an AI agent without per-action approval is genuinely dangerous if the agent has unrestricted access to your filesystem. Anthropic named the flag this way to make sure you don't turn it on by accident. We're about to make it safe to use by giving Claude a boundary it cannot cross. But until that boundary is set up, do not navigate this Claude session outside the workspace directory. For the next ten minutes, while the boundary doesn't exist yet, treat this session as if it could touch anything on your machine, because it can.

> 💡 **Callout: Why disable permissions at all?**
>
> Per-action permission prompts are a fine default for casual use. They are a productivity catastrophe for the workflow this book teaches, because we're going to be running seven Claude instances in parallel, each one taking dozens of small actions per minute. If every action prompted, you would do nothing all day except click "yes." The boundary we're about to set up replaces hundreds of small permission decisions with one big structural decision: Claude can do whatever it wants inside `workspace` and nothing at all outside `workspace`. That's the trade.

A Different Way of Prompting

Most developers who are new to AI tools type one giant prompt, send it, and hope. The result is that the agent picks up on the first thing in the prompt strongly, the last thing weakly, and the middle barely at all. That's how language models work. They have a limited working attention, and a wall of text spreads that attention thin.

The fix is to break a complex task into a back-and-forth conversation, where each turn is small enough that the agent can hold it cleanly in its attention. We're going to do that now to set up the workspace. Instead of giving Claude one massive seven-part prompt, we're going to set up a multi-step conversation where Claude finishes each piece, confirms it's done, and waits for the next piece.

Send this prompt first to set the pattern:

> **Prompt to Claude Code (the opener):**
>
> ```
> I'm going to walk you through a multi-step setup process for this
> workspace. After each step, complete the work and then ask me for
> the next step. Confirm when you're ready for step one.
> ```

Claude will respond with something like "Ready for step one." Now you can feed the steps one at a time. Each step from here is short — usually one or two sentences. The two exceptions are the lease protocol and the work-on-a-ticket protocol near the end, which have to be longer because they're describing procedures rather than single actions.

Step 1 of the Conversation: Memory and Configuration

The first thing Claude needs is a place to store its permanent memory for this workspace, so the rules we're about to give it survive across sessions.

> **Prompt to Claude Code:**
>
> `Step one: create a CLAUDE.md file at the workspace root and a .claude directory. These will hold your permanent memory and configuration for everything we do here.`
>
> **💡 Callout: What is CLAUDE.md?**
>
> `CLAUDE.md` is a special file that Claude Code automatically reads at the start of every session in a directory. Anything you write in it — or that Claude writes in it — becomes part of Claude's context for that session. Think of it as a letter you leave for the next instance of Claude that walks into the room. It's how permanent memory works in a system that's otherwise stateless.
>
> The `.claude` directory is for Claude's configuration files, including the allow and deny rules we're about to set up. The dot at the front means it's hidden from normal directory listings, which is a Unix convention for "configuration that the user usually doesn't need to look at directly."

Claude will create both, confirm, and ask for step two.

A Short Detour about Deny Lists

Before we hand Claude the deny list, you need to actually understand what one is and why it matters, because in a few minutes I'm going to ask you to extend mine to cover your own situation, and you can't do that well if it feels like cargo-culting.

A deny list is exactly what it sounds like: a list of commands the agent is not allowed to run, no matter what. When Claude considers running a shell command, it checks the command against the deny list first. If the command matches a pattern on the list, Claude refuses to run it and tells you why. The commands are matched as patterns, not exact strings, which is why you'll see asterisks in the list — those are wildcards that match "anything that starts this way."

> **💡 Callout: What do the asterisks mean?**
>
> The asterisk is a "glob" wildcard. `git push --force*` matches `git push --force`, `git push --force-with-lease`, `git push --force origin main`, and any other command that starts with that string. Globs are a simple, well-understood pattern matching system used everywhere in Unix. We use them in the deny list because we want to block whole *families* of dangerous commands, not just one exact spelling.

Why do we need a deny list at all if Claude has a boundary that keeps it inside `workspace`? Because not all dangerous things involve writing to a filesystem. Some dangerous things involve talking to other systems — pushing a force-push to a shared git branch, deleting an AWS resource, dropping a database table. Those commands originate from inside the workspace (Claude is running them from the workspace directory), but their *effects* reach far outside the workspace. The deny list is the safety net that catches commands whose blast radius extends beyond the sandbox.

There are three guiding principles for what belongs on a deny list. Memorize these, because you're going to need them when you adapt my list to your stack:

1. **Anything destructive and irreversible.** A command that deletes data, force-pushes over history, drops tables, or terminates production services. These are the cases where "Claude made a mistake, let me undo it" is not an option.
2. **Anything that touches production.** A command that deploys, modifies live infrastructure, or changes settings on a production service. Even if it's reversible in principle, you don't want an agent running it without you in the loop.
3. **Anything that costs significant money to undo.** Spinning up large cloud resources, sending mass emails, calling expensive APIs in a tight loop. The blast radius here is your wallet rather than your data, but it's a real cost.

If a command falls into any of those three buckets, it belongs on the deny list. When in doubt, add it. The cost of an unnecessary deny rule is mild — Claude refuses to do the thing, you reconsider, and either you tell Claude to do it manually under your supervision or you remove the rule. The cost of a missing deny rule is potentially catastrophic.

The list I'm about to give you is shaped by the kind of work I do — heavy on AWS, with a particular concern for git history. If you don't work with AWS, most of these rules won't apply to you. If you work with Kubernetes, you almost certainly want `kubectl delete*` on your list. If you work with databases directly, you want destructive SQL patterns. If your organization has a published "never run these in your IDE" list, that whole list belongs here verbatim. Use mine as a starting point, not as the final answer.

Step 2 of the Conversation: The Deny List

> **Prompt to Claude Code:**
>
> ``` Step two: configure the following commands as deny rules in your .claude config. I will paste the list now.
>
> git push –force *git reset –hard* cdk deploy *aws dynamodb* * *aws s3 rm* aws s3 mv *aws s3 cp* aws lambda delete *aws lambda update-function-configuration* aws lambda create *aws cloudformation delete* aws iam * aws cognito *aws apigateway* ```

Claude will configure them, confirm, and ask for step three. Take a moment after Claude confirms to actually look at the file Claude wrote — confirm with your own eyes that the rules are there, spelled correctly, and formatted in the way Claude Code expects. Verifying is part of the habit we're building.

Step 3 of the Conversation: The Filesystem Boundary

> **Prompt to Claude Code:**
>
> `Step three: you may not read or write any files or folders outside this workspace, with two exceptions: files I specifically link to you in a prompt, and anything inside /tmp or its subdirectories. Record this rule in your permanent memory.`

That's the boundary. From this moment forward, Claude will refuse to look at, modify, or create anything outside `workspace`. The two exceptions are practical: you'll occasionally want to share a specific file from elsewhere on your machine, and `/tmp` is genuinely useful as a scratch area for things like generated test fixtures that you don't want cluttering the workspace.

> 💡 **Callout: Why /tmp?**
>
> `/tmp` is a directory that exists on every Unix-like system (and has an equivalent on Windows). It's understood by everyone — humans and tools alike — to be a scratch area whose contents can be deleted at any time. It's wiped on reboot. Letting Claude use it as a scratch area means you don't have to clutter your workspace with throwaway files, and there's no risk to letting Claude write there because nothing in `/tmp` is supposed to matter.

Step 4 of the Conversation: The Repos Folder

> **Prompt to Claude Code:**
>
> `Step four: create a folder called repos at the workspace root. That's where we'll check out the repositories I work in. Don't check anything out yet - wait until I ask in a future session.`

The `repos` folder is the canonical home for clean, untouched checkouts of your work repositories. We don't actually do work inside `repos`. We use it as a *source* — when we set up a working environment to attack a ticket, we copy from `repos` into one of the seven environments and do the messy work there. That way `repos` always has a clean copy to reset from if a working environment gets hopelessly tangled.

Step 5 of the Conversation: The Seven Environments

> **Prompt to Claude Code:**
>
> `Step five: create seven directories at the workspace root, named thor, freya, tyr, loki, odin, heimdall, and baldr. These are the ephemeral working environments where actual ticket work will happen.`
>
> **💡 Callout: Why seven, and why Norse?**
>
> Seven is the count where most people can keep all the contexts mentally distinct without losing track. Fewer than seven and you're leaving capacity on the table. More than seven and you start making mistakes — typing into the wrong window, confusing the state of one ticket with another. Chapter 3 covers the cognitive science behind this number in more detail.
>
> The Norse names are arbitrary in their *content* but not in their *form*. The form matters because human memory is much better at distinct, vivid names than at numbered slots. "Did I leave that work in environment 3 or environment 4?" is a question your brain will get wrong. "Did I leave that work in thor or loki?" is a question your brain will get right, because thor and loki are *different* in a way that 3 and 4 are not. Pick a naming scheme you find vivid. Norse gods, Greek gods, planets, dwarves from The Hobbit, your seven favorite cocktails — the content doesn't matter, the distinctness does. I use Norse mythology because it's what my brain happens to lock onto. Pick whatever your brain locks onto.
>
> **💡 Callout: What's an ephemeral environment?**
>
> "Ephemeral" means short-lived, temporary, meant to be discarded. An ephemeral environment is a working directory whose state doesn't matter long-term — when you're done with it, you wipe it clean and reuse it for the next ticket. Compare that to your `repos` folder, which is *persistent* — it accumulates clones of repositories over time and is meant to last. The seven Norse directories are the opposite: they get filled, used, and reset constantly.

Step 6 of the Conversation: The Lease File

> **Prompt to Claude Code:**
>
> ```
> Step six: create ENVIRONMENT_LEASES.md at the workspace root. I'll
> give you the lease protocol in the next step.
> ```

This is just creating the empty file. The interesting part is what it's *for*, which we're about to teach Claude.

Step 7 of the Conversation: The Lease Protocol

This is the one prompt in this whole sequence that has to be long, because it's describing a multi-step procedure rather than a single action. There's no way around the length without making the procedure ambiguous, and ambiguity in a procedure that multiple Claude instances will follow concurrently is exactly what we can't afford.

Prompt to Claude Code:

``` Step seven: here is the lease protocol. Record it in your permanent memory and follow it every time you start work in this workspace.

To lease an environment:

1. Read ENVIRONMENT_LEASES.md and look for a free environment.
2. If no environment is free, evaluate each leased environment to check whether the lease is still valid. A lease is invalid if the recorded PID is no longer running on this machine — that means the session crashed or the machine rebooted.
3. For any environment with an invalid lease, check the environment for uncommitted work. If you find uncommitted work, review the git log, summarize what's there, and ask me whether to continue that work or discard it.
4. Once you have an environment, record the lease in ENVIRONMENT_LEASES.md with: environment name, the agent identity (write "claude"), your process ID, and the current date and time.
5. Copy the relevant repos from the repos folder into the environment. For each repo, check out and pull the default branch (main or master) so the environment starts clean.
6. Read README.md, CLAUDE.md, and AGENTS.md for each repo, plus anything in their .claude or .codex directories.
7. Tell me you're ready to do work or answer questions.

A lease is held for the entire working session. Release it only when I explicitly tell you to. To release: check out the default branch, use git stash to clear local state, then clear the lease entry. ```

### 💡 Callout: What's a PID?

PID stands for "Process ID." Every running program on your computer has a PID — a unique number assigned by the operating system. When a program exits or the machine reboots, its PID is gone. By recording the PID of the Claude instance that holds a lease, we get a free liveness check: if you can't find that PID in the running process list anymore, the lease is stale and the environment
```

is up for grabs. This is a tiny, elegant trick that saves us from needing a complicated lock-cleanup mechanism.

💡 Callout: What are CLAUDE.md and AGENTS.md in a repo?

Many teams now keep AI-readable instructions inside their repositories so any agent that walks in already knows the conventions. `CLAUDE.md` is the file Claude Code reads automatically. `AGENTS.md` is an emerging convention for tool-agnostic agent instructions. If your repos don't have these yet, that's normal — the world is still adopting them. If they do have them, the lease protocol makes sure Claude reads them on the way in so it has codebase context before you start.

Step 8 of the Conversation: The Work-on-a-Ticket Protocol

So far we've taught Claude how to set up the workspace, how to enforce a boundary, and how to coordinate with other Claude instances using the lease file. We haven't yet taught Claude what to actually *do* when you give it a ticket. That's what this step does.

The goal is that once Step 1 is complete, you should be able to walk up to a Claude session in any of the seven environments, type the words *"Work on ABC-1234,"* and have Claude execute a complete preamble before writing a single line of code: lease the environment if it hasn't already, read the Jira ticket, create a branch following your company's naming convention, push that branch up, open a draft pull request, and then — *only then* — begin the actual work. As Claude works through the ticket, it should commit and push regularly so the draft pull request always reflects the current state.

> 💡 **Callout: Why a draft pull request from the very beginning?**
>
> This is one of the most important habits in the whole methodology and it deserves its own paragraph. A draft PR is a pull request that's marked as "not ready for review." It's visible on GitHub, it shows your work-in-progress to anyone who wants to look, but it doesn't ping reviewers and it can't be merged.
>
> Opening the draft PR *before* Claude writes any code does three things at once. First, it makes the work *visible* — your manager, your tech lead, and your teammates can see what you're working on without you having to tell them. Second, it gives Claude a continuous publishing target — every commit Claude makes gets pushed up, so if your laptop dies mid-ticket, you've lost nothing. Third, it builds the PR description incrementally as the work proceeds, which means by the time the work is done, the PR description is already mostly written. Compare that to the old way: code for hours, finish the work, then sit down to write a PR description from scratch when you're already exhausted. The draft-first pattern moves all that documentation work into the moments when it's cheapest.

You'll also need to tell Claude what your company's branch naming convention is. Most companies have one — common patterns include `feature/ABC-1234-short-description`, `ABC-1234-short-description`, or `yourname/ABC-1234`. If you don't know what your company uses, the fastest way to find out is to look at the existing branches in one of your repos — but since you probably haven't cloned any repos yet at this point, you may need to set this rule provisionally and update it later. That's fine; we'll talk about updating the permanent memory in Chapter 8.

For now, send the work-on-a-ticket protocol prompt with your best guess at the convention:

Prompt to Claude Code:

``` Step eight: here is the work-on-a-ticket protocol. Record it in your permanent memory and follow it every time I tell you to work on a ticket.

When I say "Work on TICKET-NUMBER" (for example, "Work on ABC-1234"):

1. If you don't already hold a lease on an environment, follow the lease protocol from step seven to acquire one.
2. Read the full ticket using: jira issue view TICKET-NUMBER Summarize the ticket back to me in two or three sentences so I can confirm you understood it correctly. Wait for me to confirm before continuing.
3. Identify which repository in the leased environment the work belongs in. If it's not obvious, ask me.
4. In that repository, create a new branch following this naming convention: [PASTE YOUR COMPANY'S CONVENTION HERE, e.g. feature/TICKET-NUMBER-short-kebab-description]
5. Make an initial empty commit on the new branch with the message: "TICKET-NUMBER: begin work" Then push the branch to the remote.
6. Open a draft pull request against the default branch using: gh pr create –draft –title "TICKET-NUMBER: " The PR body should start with a link to the Jira ticket and a brief description of what you understand the work to be.
7. Now begin the actual work on the ticket. As you make progress, commit logically grouped changes with clear commit messages and push after each commit so the draft PR stays current.
8. When you believe the work is complete, do not mark the PR as ready for review yourself. Tell me the work is done and let me decide when to promote the PR out of draft state. ```

Notice the careful division of labor in steps 7 and 8. Claude does the work and keeps the draft PR updated. *You* decide when the work is actually finished and ready for human eyes. We'll talk much more about this validation step in Chapter 6.
```

> 💡 **Callout: Why an empty initial commit?**
>
> An empty commit (`git commit --allow-empty`) is a commit that doesn't change any files. It seems pointless until you realize what it enables: it gives you something to push, which gives you a remote branch, which lets you open a draft PR *immediately* before any actual work has happened. Without an empty commit, you'd have to wait until Claude wrote some code before the PR could exist. With it, the draft PR exists from minute one and you can watch the work appear in it incrementally.

Claude will confirm the protocol is recorded and ask for the next step.

Step 9 of the Conversation: Make It Permanent

> **Prompt to Claude Code:**
>
> `Step nine, final step: commit all of these procedures, rules, and the deny list to your permanent memory in CLAUDE.md and the .claude directory so they apply automatically to every future session in this workspace.`

This is the step that turns everything you just configured from "rules in this conversation" into "rules in this workspace forever." Future Claude sessions launched from inside `workspace` will read CLAUDE.md and the `.claude` directory at startup and inherit the entire setup automatically. You will not have to do this conversation again.

What You Have Now

Take a moment to look at what you've built. Run a directory listing and you should see something like this:

```
~/workspace/
├── CLAUDE.md
├── ENVIRONMENT_LEASES.md
├── .claude/
│   └── (allow/deny rules and other config)
├── repos/
├── thor/
├── freya/
├── tyr/
├── loki/
├── odin/
├── heimdall/
└── baldr/
```

A directory listing won't show the `.claude` folder by default because of the dot prefix. On macOS and Linux you can see it by running `ls -la`. On Windows from PowerShell, run `Get-ChildItem -Force`.

Figure 4.1 — The Workspace Directory Structure

What lives in ~/workspace after Step 1 is complete

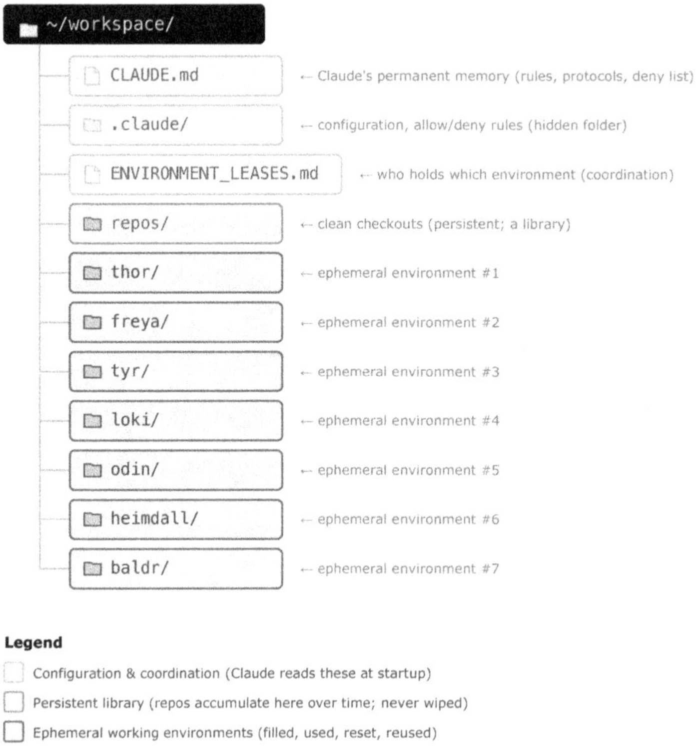

What you have, more importantly than the directory layout, is a sandboxed work area with a complete, persistent ruleset for an AI agent to operate inside it. Every future Claude session you start in this directory will arrive already knowing the rules, the boundary, the deny list, the lease protocol, and the work-on-a-ticket protocol. You will not have to re-explain any of it. The rules travel with the workspace.

You also have a coordination primitive — the lease file — that's about to let you run multiple Claude instances at the same time without them stepping on each other. We won't actually exercise that capability until Chapter 5, but the foundation is now in place.

Verification

Before you move on, go through this checklist. Don't skip it. If anything fails, fix it now — every later chapter assumes Step 1 worked.

1. **Directory structure.** List the workspace contents (with hidden files visible) and confirm you see all seven Norse directories, the `repos` folder, `CLAUDE.md`, `ENVIRONMENT_LEASES.md`, and the `.claude` directory.

2. **CLAUDE.md is populated.** Open `CLAUDE.md` in a text editor and read what Claude wrote. You should see the boundary rule, the deny list, the lease protocol, and the work-on-a-ticket protocol in Claude's own words. If anything is missing, vague, or wrong, tell Claude to fix it before continuing. This file is the source of truth for every future session.

3. **The lease file is empty but structured.** Open `ENVIRONMENT_LEASES.md`. It should exist, have a clear structure for recording leases (probably a table or simple list with columns for environment, PID, date/time), and have no leases recorded yet.

4. **The deny list is in the .claude config.** Check the `.claude` directory for a configuration file containing the deny rules. Confirm the rules are spelled correctly and formatted in the syntax Claude Code expects.

5. **The boundary holds.** Test it with a small probe. Ask Claude (in the same session, or in a fresh one launched from inside `workspace`) to read a file outside the workspace. Try this exact prompt:

> **Prompt to Claude Code:**
>
> `Please read the file at ~/.zshrc and tell me what's in it.`

Claude should refuse and explain that the file is outside the workspace boundary. If Claude reads the file and reports its contents, the boundary is not in place and you need to fix it before going any further. Here's how to troubleshoot, in order:

First, check that Claude actually wrote the rule down. Open `CLAUDE.md` and search for any mention of the boundary or the workspace restriction. If it's not there, Claude either didn't understand the instruction or didn't persist it. Re-run Step

3 of the conversation and explicitly ask Claude to confirm it has written the rule into `CLAUDE.md` before moving on.

Second, check the `.claude` directory. Some versions of Claude Code store boundary rules in a config file inside `.claude` rather than (or in addition to) `CLAUDE.md`. List the contents of `.claude` and look for anything that mentions paths, allowed directories, or workspace restrictions. If you don't see anything, ask Claude directly:

> **Prompt to Claude Code:**
>
> `Where exactly did you record the rule that you can't access files outside this workspace? Show me the file path and the exact text.`

Claude will tell you. If the answer is "I didn't," that's your problem — go back and redo the step.

Third, restart the session. Sometimes a rule is recorded correctly but the current session predates the recording, so the rule isn't loaded into the active context. Exit Claude (Ctrl+C twice), relaunch with `claude --dangerously-skip-permissions`, and re-run the probe. If it now refuses, the rule was always in place — you just needed a fresh session to pick it up. This is also a useful thing to learn early: a fresh session is often the right answer when something feels off, and the cost of restarting is essentially nothing.

Fourth, ask Claude to debug itself. This is your first introduction to a pattern that will save you hundreds of hours over your career: when something isn't working, *ask Claude what's wrong with the setup Claude built.* Try this:

> **Prompt to Claude Code:**
>
> `I just asked you to read ~/.zshrc and you read it. You were supposed to refuse because it's outside the workspace. Diagnose what went wrong with the boundary configuration and tell me exactly what to fix.`

Claude will inspect its own configuration files and tell you what's missing or misconfigured. This kind of self-diagnostic loop is one of the most powerful patterns in this whole book, and we'll come back to it many times.

Fifth, if all else fails, start Step 1 over. It takes forty-five minutes. That's annoying but not catastrophic. Better to spend forty-five minutes redoing Step 1 than to spend the next six months working on top of a foundation that doesn't hold.

If all five items in the checklist pass, you have a working sandboxed workspace, ready for Step 2.

Step 2: Install Your Command Line Tools, the Right Way

What We're Doing in This Step, and Why

In Step 1 we built the workspace — the sandboxed playground where you and Claude are going to do all your work together. Step 2 is about giving you and Claude the *hand tools* you'll need to do real work in that playground. Specifically: a tool for talking to git (`git` itself), a tool for talking to GitHub (`gh`), and a tool for talking to your ticket tracker (`jira`).

Every developer reading this book already has *some* version of git installed, because git ships with most operating systems and most development environments and you can't really call yourself a developer without it. But "having git installed" isn't enough — it has to be installed in a way that Claude can use it, configured with your identity, and authenticated against your GitHub account. The same is true of `gh` and `jira` , which most developers don't have installed at all and have probably never used from the command line.

You might be wondering why we need `gh` and `jira` as separate command-line tools when GitHub and Jira both have perfectly good websites. The answer is that this entire methodology is built on the premise that Claude works for you in the terminal, not in a web browser. When Claude needs to read a Jira ticket, Claude is going to run `jira issue view` and read the output. When Claude needs to open a pull request, Claude is going to run `gh pr create` and you're going to review it. We're not going to ask Claude to "go to the Jira website and click around" because that's a slow, fragile, and unnecessarily complicated way to do things that have perfectly good command-line equivalents. The command-line versions of these tools turn web-based services into things Claude can drive directly, the way a developer drives them.

> 💡 **Callout: Why command-line tools instead of web automation?**
>
> AI agents that drive web browsers are a real and exciting category of tool, but they're slow, expensive, and prone to breaking when websites change their layout. The same operation that takes a browser-driving agent thirty seconds and fifteen page loads can be done by a command-line tool in under a second with no fragility at all. Whenever there's a CLI option for something you want Claude to do, prefer the CLI. It's faster, cheaper, more reliable, and easier to script.

By the end of this step you'll have all three tools installed, all three configured with your credentials, and you'll have practiced the single most important skill in this whole book: **asking Claude to walk you through something instead of doing it yourself from memory or Stack Overflow**.

A Teachable Moment: Let Claude Help You Install Things

Before we install a single thing, I want to point at something you might not have noticed. You're already in a Claude session, in your sandboxed workspace, with a powerful AI agent sitting there waiting to help you. The natural instinct for most developers — especially experienced developers — is to ignore that and go install these tools the way they always have, by Googling "how to install gh on macOS" and copy-pasting from a Stack Overflow answer.

That instinct is a habit from the pre-AI era, and you should consciously work on breaking it. Claude already knows how to install `gh` on every platform, knows the gotchas, knows the verification commands, and can adapt to whatever it finds on your specific machine. Asking Claude to walk you through the install is faster than searching for it, and — more importantly — it builds a habit that's going to compound over your entire career: **when you don't know how to do something, ask Claude before you ask Google.**

This is an early, low-stakes opportunity to practice that habit. If Claude makes a mistake here, the worst case is you spend an extra ten minutes installing a CLI tool. There's no production system at risk. Use this safe environment to start training your reflexes.

So for each of the three tools we're about to install, I'll give you the prompt to ask Claude for help. You should also feel free to riff on these prompts — ask Claude to explain what it's about to run, ask Claude why it's choosing one installation method over another, ask Claude to verify the install worked. The more you treat Claude as a knowledgeable colleague who happens to be sitting next to you, the faster you'll internalize what it can and can't do.

> **💡 Callout: Claude doesn't need to install things itself**
>
> An important nuance: Claude is in your terminal, but in this step we're going to have *you* run the installation commands rather than Claude. That's because installing software usually requires elevated privileges (sudo on Mac/Linux, administrator on Windows), and Claude operating with the boundary we set up in Step 1 isn't going to have those privileges. Think of Claude in this step as a knowledgeable guide telling you exactly what to type, while you're the one with the keys to the kingdom who actually presses the keys. We'll work this way often: Claude figures out what to do, you do the thing that requires your authority.

Installing Git

Git is almost certainly already installed on your machine, but let's verify it and update it if needed.

Try this prompt first:

> **Prompt to Claude Code:**
>
> `I'd like to verify that git is installed on my machine and is a reasonably current version. Please give me the command to check the installed version, and tell me what version is considered current as of today.`

Claude will give you the command (`git --version`) and tell you what version is current. Run the command in your terminal — you can run it from inside the same terminal that's running Claude Code, just open a new tab or window so you don't disturb the Claude session.

If git isn't installed at all, or if your version is several years old, ask Claude for the install or upgrade command:

> **Prompt to Claude Code:**
>
> ``My git version is [paste what `git --version` returned]. I'm on [macOS / Manjaro / Windows]. Please tell me how to install or upgrade git on my platform. I prefer using [Homebrew / pacman / winget] if it's an option.``

Substitute the bracketed parts with your actual situation. Claude will give you the exact command. Read the command before running it — this is another habit worth building from day one. **Never paste a command into your terminal that you haven't at least skimmed.** If something in the command looks unfamiliar, ask Claude what it does before running it.

Once git is installed at a current version, configure your identity. Git uses your name and email to sign every commit, and an unconfigured git installation will refuse to let you commit anything.

> **Prompt to Claude Code:**
>
> `I need to configure my global git identity with my name and email. Please give me the commands. My name is [your name] and my email is [your email].`

Claude will give you the two `git config --global` commands. Run them.

> 💡 **Callout: Use the same email as your GitHub account**
>
> The email you set in your git config should match the email associated with your GitHub account. If they don't match, your commits will appear on GitHub as "unverified" and won't be attributed to your profile (no green squares for you). If you're not sure what email is on your GitHub account, log in to GitHub in a browser, go to Settings → Emails, and use the primary email shown there.

Installing the GitHub CLI (`gh`)

The GitHub CLI is the official command-line tool for interacting with GitHub. It lets you (and Claude) create pull requests, list issues, check the status of CI runs, review PRs, and much more, all from the terminal. It is the single most important tool in this entire methodology after Claude itself.

Ask Claude to walk you through it:

> **Prompt to Claude Code:**
>
> `I need to install the GitHub CLI (gh) on [macOS / Manjaro / Windows]. Please give me the recommended install command for my platform, then tell me how to verify the install worked.`

Claude will give you the install command and the verification command (`gh --version`). Run them both.

Now you need to authenticate `gh` against your GitHub account. This is a one-time step.

> **Prompt to Claude Code:**
>
> `Walk me through authenticating gh against my GitHub account. I want to authenticate via the browser flow rather than with a personal access token. Tell me what to expect at each step.`

Claude will tell you to run `gh auth login`, walk you through the prompts, and explain that you'll be asked to copy a one-time code, then a browser will open where you paste the code and approve the login. Follow Claude's guidance.

Verify the auth worked:

> **Prompt to Claude Code:**
>
> `Give me a single gh command I can run to verify that I'm authenticated and to see which GitHub account I'm authenticated as.`

Claude will give you something like `gh auth status`. Run it. You should see your username and confirmation that you're authenticated.

> **💡 Callout: SSH versus HTTPS**
>
> When you authenticate `gh`, you'll be asked whether to use SSH or HTTPS as the default protocol for git operations. Either works, but they have different tradeoffs. SSH requires a one-time setup of an SSH key on your machine and on GitHub, after which you never type a password again. HTTPS uses your `gh` authentication automatically with no key setup required, but some advanced workflows assume SSH. If you've never used SSH keys before, just pick HTTPS — it's simpler, it works fine, and you can always switch later. If you're already comfortable with SSH keys, use SSH.

Installing the Jira CLI

The Jira CLI is a community-maintained tool for talking to Jira from the command line. Note carefully: there are several different "jira CLI" tools out there, and the one this book uses is the one written in Go by Ankit Pokhrel, available at github.com/

ankitpokhrel/jira-cli. Other tools with similar names exist and are not interchangeable. Make sure you install the right one.

> **Prompt to Claude Code:**
>
> `I need to install the jira CLI by Ankit Pokhrel (the one at github.com/ankitpokhrel/jira-cli). I'm on [macOS / Manjaro / Windows]. Please give me the install command for my platform and the verification command.`

Claude will give you the platform-appropriate install command. Run it, then run the verification command Claude gave you.

Now configure it. The jira CLI needs to know three things: the URL of your Jira instance, your email address, and an API token.

> **Prompt to Claude Code:**
>
> ``Walk me through configuring the jira CLI for the first time. I'll need to generate an API token from Atlassian – tell me how to do that. Then walk me through the `jira init` flow and explain what each prompt is asking for.``

Claude will walk you through generating an Atlassian API token (you'll do this in your browser at id.atlassian.com), then walk you through the `jira init` interactive setup. Follow along.

> 💡 **Callout: Treat your API token like a password**
>
> The Atlassian API token you generate is, functionally, a password for your Jira account. The jira CLI will store it on your machine in a config file. Don't paste it into chat windows, don't commit it to git, don't share it. If you ever accidentally expose it (in a screenshot, in a paste to a coworker, anywhere), go straight back to id.atlassian.com and revoke it, then generate a new one. Treating credentials as disposable and revocable is one of the safest habits a developer can have.

Verify the install worked by listing some issues:

> **Prompt to Claude Code:**
>
> `Give me a single jira command I can run to list a few issues assigned to me, just to verify the configuration is working.`

Claude will give you a command like `jira issue list -a$(jira me)` or similar. Run it. You should see a list of issues, even if it's a small list.

Walking Claude through Cloning Your First Repository

Now we get to put all three tools together — and to do the first thing in this book that feels like *real work*. We're going to add your first repository to the workspace.

Pick one repository you actually work in. Just one for now. You'll add more later, but for the first pass, pick a single repo so you can see the whole flow without juggling.

You need the clone URL for the repo. You can get this from the GitHub web UI by clicking the green "Code" button on the repo page, or you can ask `gh`:

> **Prompt to Claude Code:**
>
> `Give me the gh command to find the clone URL for a repository I have access to on GitHub. I want to use HTTPS for the URL since that matches how I authenticated gh.`

Claude will give you something like `gh repo view OWNER/REPO --json url` or suggest you use `gh repo list` to browse. Use whichever feels natural.

Once you have the clone URL, hand it to Claude and let Claude do the actual clone:

> **Prompt to Claude Code:**
>
> ```
> Please clone the repository at [paste URL here] into the repos
> folder. After cloning, do not do anything else — don't read any
> files, don't check out any branches, don't run anything. Just clone
> it and confirm it landed in repos/.
> ```

This is the first time in this book that Claude is going to do something with real consequences (writing files into `repos/`), so I want you to watch it carefully. Claude should:

1. Confirm the URL it's about to clone.
2. Run `git clone` into `repos/`.
3. Confirm the clone succeeded and tell you the path it landed at.

If Claude does anything else — starts reading files, starts checking out branches, starts running tests — stop it. The instruction was deliberately narrow. Part of working with Claude productively is noticing when it goes beyond what you asked, even when its initiative is well-intentioned. We'll talk more about this in Chapter 6, in the common-failure-modes section. For now, just notice the behavior and steer it back if needed.

> **💡 Callout: Why have Claude do the clone and not you?**
>
> A fair question, since installation in the previous section was you-driven. The difference is privileges. Cloning a repo into `~/workspace/repos/` is something Claude is fully capable of doing inside its boundary, with no elevated privileges required. Installing system software is not. As a general rule: if Claude *can* do it inside the boundary, let Claude do it. Your job is the things only you can do (decisions, approvals, things requiring your credentials or authority). Claude's job is everything mechanical that fits inside the sandbox. Building this division of labor is half the productivity gain.

Repeat the clone step for any other repositories you regularly work in. You don't have to do them all today — you can add new ones at any time by running another Claude session and asking it to clone. But getting the most-used three or four into `repos/` now means you can hit the ground running tomorrow.

Verification

Before moving on, run through this checklist:

1. **git works and is identified.** Run `git --version` and `git config --global user.name` and `git config --global user.email`. The first should return a current version; the other two should return your name and email.
2. **gh works and is authenticated.** Run `gh auth status`. You should see your GitHub username and confirmation that you're logged in.
3. **jira works and is authenticated.** Run a simple `jira issue list -a$(jira me)` (or the equivalent Claude gave you). You should see issues assigned to you, even if the list is short.
4. **At least one repo is in repos/.** List the contents of `~/workspace/repos/` (or `$HOME\workspace\repos\` on Windows) and confirm you see at least one cloned repository.
5. **Claude respected the narrow instruction.** Look back at the conversation where you asked Claude to clone the repo. Confirm Claude only cloned and didn't go off on its own to read, explore, or modify. If it did go off on its own, that's worth noting — you'll want to be more explicit in future prompts about scope.

If anything fails, the troubleshooting pattern is the same one we established in Step 1: ask Claude to diagnose what went wrong. For example:

> **Prompt to Claude Code:**
>
> ```
> When I run `gh auth status` I see [paste the output]. Diagnose
> what's wrong with my authentication and tell me the exact commands
> to fix it.
> ```

Use this pattern — paste the actual error output, ask for a diagnosis, ask for the exact fix — every time something doesn't work. It will become second nature within a week, and it's the single highest-leverage habit in this entire methodology.

The Hardest Part Is Over

If you've done Steps 1 and 2 in one sitting, you've been at this for somewhere between two and three hours. That's a lot of work. Close the laptop for a bit if you need to. Get something to eat. Come back fresh for Chapter 5.

You've built the engine. Every future session of Claude Code in this workspace will inherit the rules you taught Claude today. Every repo you ever clone will land in `repos/`. The seven environments will sit quietly waiting to be leased. None of this will need to be rebuilt. The only thing you'll do from here on, day in and day out, is *use* the system.

The next chapter is about starting it up. The first time you do it, you'll do the whole procedure — opening terminals, placing windows, launching seven Claudes, leasing environments, loading tickets. It's going to feel like a lot. After the first time, most of it will still be there from the day before, and starting up will take a few minutes instead of the better part of an hour. We'll cover that comforting reality in Chapter 7. But today, let's focus on the first time.

Chapter 5
Starting the Engine

Today Is Your First Day

Take a moment to appreciate where you are. In the previous chapter you built the engine — a sandboxed workspace, a coordinated environment system, a set of command-line tools, and an AI agent that knows the rules. Today is the day you start it up. By the end of this chapter, you'll have seven Claude instances running in parallel, each leased to its own environment, each working on a different ticket, and a physical layout on your screen that makes juggling all seven feel almost effortless.

This chapter is going to feel different from Chapter 4. That chapter was mostly about typing commands and prompts. This one is mostly about *arranging windows on your screen*. That probably sounds like the least technical thing in the whole book, and in a way it is. But the layout you're about to build is the single most underappreciated part of this entire methodology. People who skip it or improvise it report dramatically worse results than people who set it up exactly as described and then adjust it over time.

Plan to spend about thirty to forty-five minutes on this chapter the first time. Most of that time is just dragging windows into position and saving them. Once the layout is established, you will rarely have to build it from scratch again — your operating system will remember the positions across reboots, and on most mornings you'll walk up to a workspace that's already mostly configured from the day before. Chapter 7 covers those easier starts in detail. For now, focus on today. Today is the day you build it.

Why the Layout Matters This Much

Here's the problem we're solving. You're about to be running seven Claude instances in parallel. Each instance is doing work for a different ticket, in a different repository (potentially), with different file changes, with a different draft PR open in GitHub, and a different ticket open in Jira. That's seven different streams of context, plus the things you have to do as a human — read email, respond on Slack, review your teammates' PRs, look at the ticket board to see what's coming next.

If you let those streams of context float around your screen wherever windows happen to land, two bad things will happen. First, you'll lose track of which window belongs to which ticket and you'll start typing into the wrong window — pasting Claude output meant for ticket A into the chat for ticket B. Second, every time you switch contexts, your brain will spend several seconds *re-orienting* — finding the right window, remembering what you were doing, scrolling back to where you left off. Multiply those several seconds by the dozens of context switches you'll do per hour and you've burned half your day on just *finding things*.

The fix, which sounds almost embarrassingly simple, is that **every kind of work has a fixed permanent location on your screen.** Email always lives in the same spot. Slack always lives in the same spot. Each of the seven slots always lives in the same spot. When you want to look at the work happening in Slot 1, your hand goes to the Slot 1 region of your screen the same way your hand goes to the light switch when you walk into a dark room. You don't think about *where* — you just *do*.

This is the dish-by-the-door principle from Chapter 3, applied to every surface of your workday. And it's the single biggest source of the productivity gain in this method.

What You'll Need

You need one large monitor, ideally curved, somewhere around 3440×1440 or larger. (3840×1600 is what I use; the figures in this chapter were rendered on a similar display.) If your monitor is significantly smaller than that, you can still make this work, but you'll need to either reduce the size of each zone (which makes things harder to read) or accept that some zones will overlap (which breaks the dish-by-the-door principle a little). If you don't have a large monitor, this might be a place to spend some money — a good ultrawide will pay for itself in productivity within a month or two.

You also need a web browser that lets you have multiple windows open at the same time, each with its own set of tabs. Chrome, Firefox, Edge, Arc, Safari, Brave — all fine. The browser will be doing a lot of work in this layout because seven of your slots use a browser window.

This Layout Is Ninety Percent of the System

Before we start placing windows, one thing needs saying clearly, because if you don't internalize it you'll skim this chapter and regret it: **the layout you're about to build is approximately ninety percent of the value of the entire system.** Not fifty percent. Not seventy percent. Ninety.

The Claude Code setup from Chapter 4 is important. The command-line tools are important. The lease protocol is important. But all of those things combined contribute roughly ten percent of what makes this methodology produce the throughput numbers from Chapter 3. The other ninety percent is *what your screen looks like while you're working.*

That's not an exaggeration. It's the honest arithmetic of where attention goes in a workday. When you have seven Claude instances running in parallel, plus email, plus Slack, plus Jira, plus your IDE, plus a browser for testing, plus whatever else, the default outcome is chaos. You spend half your day looking for windows. The layout is what *prevents* that chaos by giving every kind of work a fixed permanent home on your screen. Without the layout, the seven Claudes just multiply your confusion. With the layout, they multiply your throughput.

So read this chapter carefully. Don't skim.

Slots versus Environments

One concept to get straight before we look at Figure 5.1, because if you confuse these two words the rest of the chapter will be harder than it needs to be.

A **slot** is a permanent position on your screen. It's a place where a browser and a Claude terminal live together, side by side, in the same physical region of your display. Slot 1 is the top-left slot. Slot 2 sits in the top row, between Zone 1 and Zone

2. Slots 3 through 7 fill the bottom row, left to right. Slots never move. They are the dishes by the door — always in the same place, waiting for keys.

An **environment** is one of the seven Norse-named working directories from Chapter 4: thor, freya, tyr, loki, odin, heimdall, baldr. An environment is where code lives while work is being done on a ticket. Environments get leased to Claude sessions according to the lease protocol.

Here's the key relationship: **slots and environments are not the same thing, and they are not permanently paired.** Slot 1 might hold thor this morning and baldr this afternoon. Slot 3 might hold heimdall all week. It depends on what Claude leases when you start a new session in that slot. The slot is the *docking bay*; the environment is the *spacecraft* currently parked in it. If you want to know which environment a given slot is holding right now, click into that slot's terminal and ask Claude: *"Which environment are you in?"* Claude will tell you immediately. You can also check `ENVIRONMENT_LEASES.md` at any time to see the full map.

Why bother with the distinction? Because it buys you flexibility. If you ever want to work on a *different* project during the day — say, you have two codebases you contribute to — you can spin up a second set of seven slots on a second virtual desktop, and those slots can lease from their own set of environment names. The system is scalable: seven slots is the recommended starting point for one project, but nothing prevents you from having seven slots on Desktop 1 for Project A and another seven slots on Desktop 2 for Project B, flipping between desktops with a keyboard shortcut. The slot concept is what makes that possible without confusion.

For now — your first day — you have one desktop, seven slots, seven environments. They happen to match up one-to-one. But the *concept* is separate. Slots are positional, environments are leased dynamically, and the browser in each slot bridges them by co-locating the Jira ticket and the draft PR with the Claude terminal doing the work.

The Layout at a Glance

Figure 5.1 — The Desktop Layout

Two rows · seven docking slots · three fixed zones · one deliberate empty strip

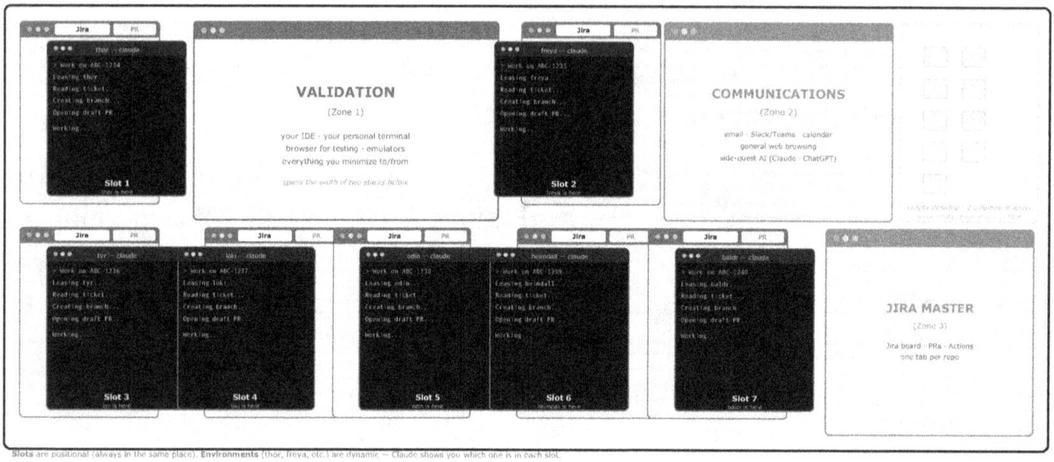

Slots are positional (always in the same place). **Environments** (thor, freya, etc.) are dynamic — Claude shows you which one is in each slot.
Top row terminals stagger R-(val)-L-(comms). Bottom row staggers R-L-R-L-R-(jira). This alternation is what packs seven slots onto a single screen.

Look at Figure 5.1. This is the whole layout on a single large monitor. There are exactly two rows, and the screen is organized into the following regions:

Top row, left to right: Slot 1, then Zone 1 (Validation), then Slot 2, then Zone 2 (Communications). Note that Zone 2 stops short of the right edge of the screen, leaving a visible strip of empty desktop.

Bottom row, left to right: Slot 3, Slot 4, Slot 5, Slot 6, Slot 7, then Zone 3 (Jira Master) extending all the way to the right edge.

The alignment rules matter and they're not decorative:

- Zone 1 (Validation) is the same width as the slot directly below it.
- Zone 2 (Communications) is wider, but its left edge aligns with the slot directly below it and its right edge stops short of where Zone 3 (Jira Master) starts.
- Zone 3 (Jira Master) is wider than the slot above it — it extends to the right edge of the screen — which means there's an empty strip of desktop above it and to the right of Zone 2.

That empty strip is deliberate. When you take a screenshot, when you download a file, when you drag a key folder to a quick-access position — those things land on your desktop. The empty strip is where you can *see* them without moving any of your work windows. It's a dedicated "things I need to reach in one click" area.

Now count the windows. There are fourteen windows in the seven slots (a browser plus a terminal in each). Plus one validation browser. Plus one communications browser. Plus one Jira Master browser. That's **seventeen windows** you'll have open on this one monitor during a working session, and a few more apps layered inside the three fixed zones. It sounds like a lot. It isn't, because they all have permanent homes.

The Three Fixed Zones in Detail

The three fixed zones — Validation, Communications, and Jira Master — are where your human work lives. Each of them is a *workspace within itself*, not just a single window. Let's look at what populates each.

Zone 1 — Validation (Upper-Left of Your Screen)

Zone 1 is where everything you personally use to validate, test, edit, and explore lives. The browser window you see in Figure 5.1 is just the front of it — behind and around that browser, you layer every other tool you work with.

Figure 5.2 — Zone 1: The Validation Workspace

Where everything you personally use to validate, test, edit, and explore lives. Example: web development stack.

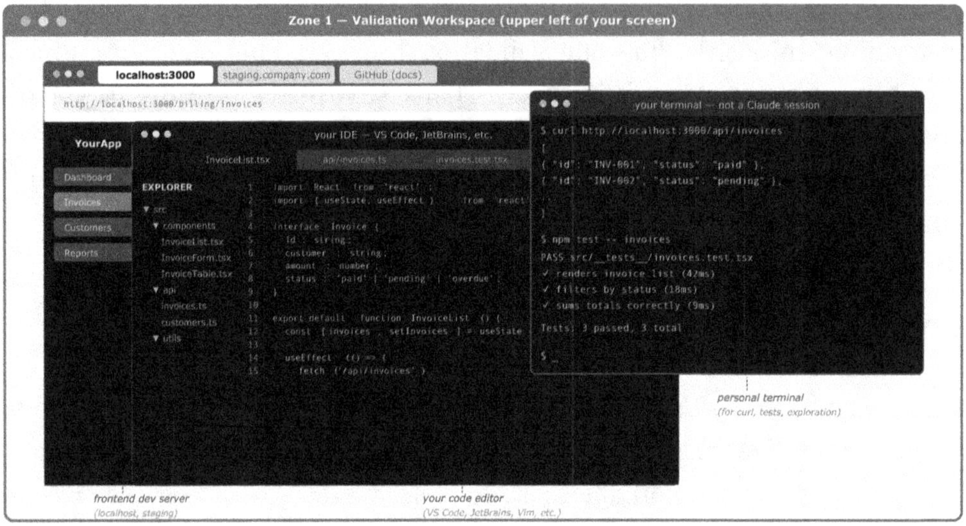

Zone 1 accumulates your personal work tools. When you minimize an IDE or a helper window, it comes back HERE — not to the taskbar.

Figure 5.2 shows an example Zone 1 setup for web development. Notice the layered stack: a frontend dev server in the back (serving your app at localhost), your code editor in the middle (VS Code, JetBrains, Vim, whatever you use), and your personal terminal in front (for running tests, making curl requests, exploring). Each of those tools has its own permanent position within Zone 1, and you flip between them with the same dish-by-the-door muscle memory we've been building.

Crucially, this zone is **where you minimize things from and to**. When you minimize your IDE, it doesn't go to the dock or taskbar — it goes back to its spot in Zone 1. When you want your testing browser, it comes up in Zone 1. When you open a desktop app under test, it lives in Zone 1. Think of Zone 1 as your workbench — it's where *you* work, with your hands, as opposed to the slots where *Claude* works on your behalf.

Zone 1's contents vary dramatically by what kind of software you build:

- **Web developers** stack a frontend dev server (localhost), an IDE, a personal terminal, and often a staging-environment browser tab.

- **Mobile developers** stack a mobile emulator or simulator (an iOS Simulator or Android Emulator window), an IDE that's usually Xcode or Android Studio, and a personal terminal.
- **Desktop app developers** stack a running instance of the desktop app under test alongside an IDE like CLion, Visual Studio, or Rider, plus a personal terminal.
- **Backend-only or library developers** might have just an IDE, a personal terminal, and a browser tab pointed at a staging API, with no visible "front-end" at all.

The principle is the same for all of them: *everything you personally use, right here, in this zone, always.*

Zone 2 — Communications (Upper-Right of Your Screen)

Zone 2 is where human messages live, plus whatever you use for side-quest AI and general browsing.

Figure 5.3 — Zone 2: The Communications Workspace

Staircase-offset windows — each one's top edge peeks out so any of them is one click away.

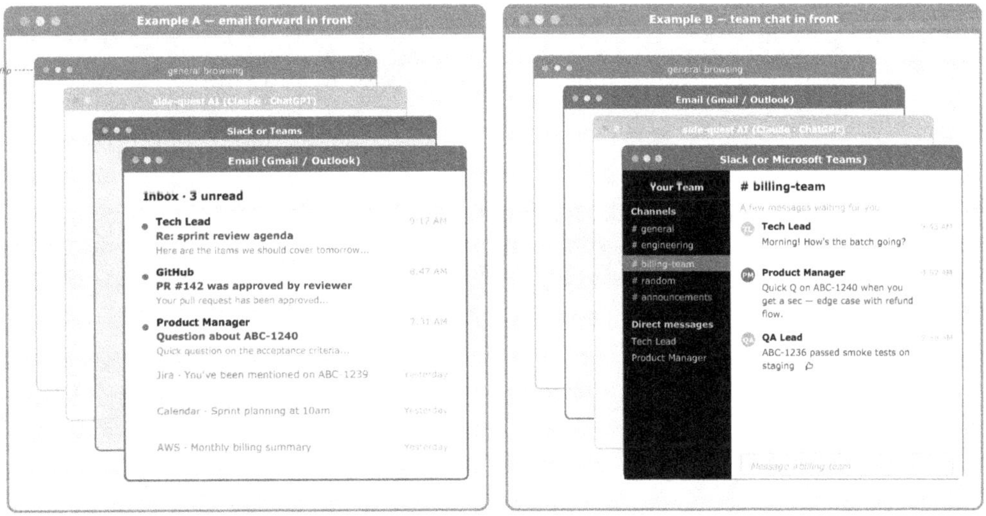

Each window's title bar stays exposed — click any of them once and it rises to the front. The geometry never moves, only the z-order flips.
Side-quest AI lives here too — any AI task unrelated to your codebase (drafting, research, general questions) belongs in Zone 2, not in your Claude terminals.

Figure 5.3 shows two example Zone 2 setups. The contents are more person-and-company-specific than Zone 1, so I'm showing the zone as a stack of typical windows rather than a single prescribed configuration:

- **Team chat** — Slack, Microsoft Teams, or both if your company uses both. This is usually the tool that gets interrupted most, so it's often what sits in front when you're not actively looking at something else.
- **Email** — Gmail in a browser tab, Outlook as a desktop app, whatever your company uses. This tends to be lower-urgency than chat but still lives in this zone.
- **Calendar** — often just another tab in the same browser window as email.
- **General browsing** — a tab for ad-hoc Google searches, reading articles, looking up documentation. Anything that's *not* validating your work (that's Zone 1) and *not* project tracking (that's Zone 3) and *not* direct communication.
- **Side-quest AI** — a Claude Desktop window or a ChatGPT browser tab for AI tasks *unrelated* to your codebase. Drafting an email to a vendor, summarizing an article, rubber-ducking a non-coding problem. These tasks don't belong in your Claude Code terminals, which are busy working on tickets. They belong here.

The Zone 2 "which app is in front" setting is dynamic and personal. If you're a chat-forward person (your company lives in Slack), team chat sits in front. If you're an email-forward person (messages tend to arrive by email), email sits in front. The point is that *everything person-facing is in this one zone* and whichever app you look at most often sits on top.

Zone 3 — Jira Master (Bottom-Right of Your Screen)

Zone 3 is the project-tracking workspace. It's **always** a single browser window with these specific tabs:

Figure 5.4 — Zone 3: The Jira Workspace

Project tracking lives here. Jira board plus one PR tab and one Actions tab per repo you work in.

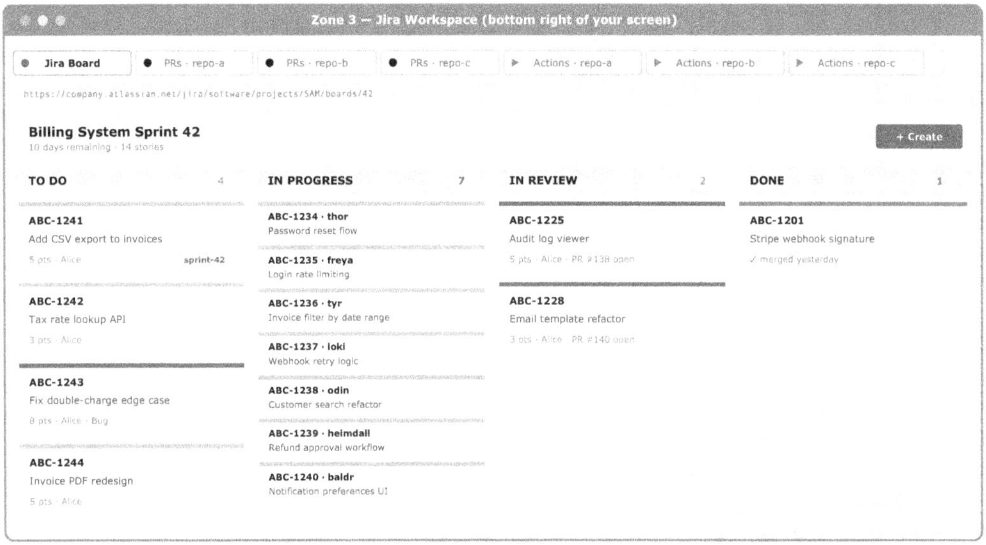

One tab for the Jira board, then one PR tab and one Actions tab per repository you're actively working in. Flip between tabs to check anything project-wide. Some browsers support left-side vertical tabs (Edge natively, Chrome via extension), which is especially helpful if you work across many repos.

- **The Jira board** — your sprint board, showing every ticket's state across the To Do / In Progress / In Review / Done columns.
- **One Pull Requests tab per repository you work in** — `github.com/your-org/repo-a/pulls`, `/repo-b/pulls`, and so on. This lets you flip to any repo's full PR list in one click.
- **One Actions tab per repository you work in** — `github.com/yourorg/repo-a/actions`, and so on. This is where you check whether CI passed on the draft PRs that Claude is building.

Flip between tabs with `Cmd+1`, `Cmd+2`, `Cmd+3` on macOS, or `Ctrl+1`, `Ctrl+2`, `Ctrl+3` on Windows and Linux. If you work across many repositories and the tab bar gets crowded, browsers like Microsoft Edge support **left-side vertical tabs** natively, and Chrome supports them via extension. Vertical tabs make Zone 3 scale to eight, ten, twelve tabs without the labels getting compressed.

Some people prefer to swap Zone 2 and Zone 3 — putting Jira in the larger upper-right space and communications in the smaller bottom-right corner. If you spend significantly more time looking at Jira and PRs than at Slack and email, that variant might suit you better. See **Appendix H** for the swapped layout and when to pick it. For your first few weeks, stick with the canonical layout in Figure 5.1.

The Seven Slots in Detail

That leaves the seven slots. Each slot contains exactly two windows: a **browser** and a **Claude terminal**. The two windows sit in fixed positions relative to each other, sized so the browser is slightly bigger (taller and wider) than the terminal, with the terminal offset so that it extends past the browser on one side.

Figure 5.5 — The Context Stack

Two windows in fixed positions — nothing moves when you click. Only the stacking order changes.

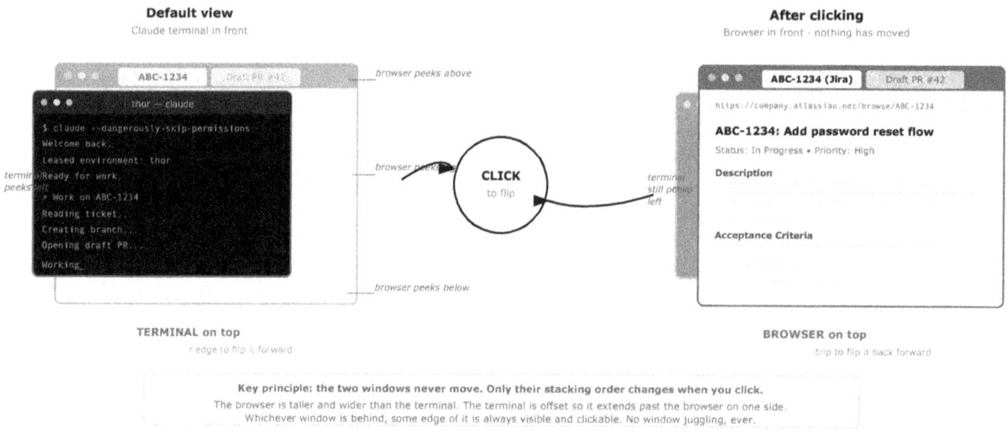

Figure 5.5 shows what a single slot looks like up close. Look at both panels — you'll notice the two windows are in **exactly the same positions** in both panels. Nothing moves when you click. The browser is just taller and wider than the terminal, so bits of browser are always visible above, below, and on one side. The terminal is offset so that its other side sticks out past the browser too. The result is that no matter which window is on top, an edge of the other one is always visible and clickable. You flip between them with a single click — on the browser edge to bring the browser forward, on the terminal edge to bring the terminal forward.

Why this particular geometry? Because the **browser is the bridge**. The browser in each slot holds two tabs: the current Jira ticket, and the current draft pull request. The terminal in each slot holds the Claude session doing the work. Keeping them in the same physical position — on your screen and in your muscle memory — is how the system makes the ticket, the PR, and the work feel like parts of the same thing. When you flip to the browser, you see what you're working on (the Jira ticket and its PR). When you flip to the terminal, you see what Claude is doing about it. Both are in the same slot. Both are the same work.

Look back at Figure 5.1 and notice the alternation pattern across the slots:

- **Slot 1 (top-left)**: terminal peeks out on the right.
- **Slot 2 (top-middle)**: terminal peeks out on the left.
- **Slot 3 (bottom-left)**: terminal peeks out on the right.
- **Slot 4 (bottom)**: terminal peeks out on the left.
- **Slot 5 (bottom)**: terminal peeks out on the right.
- **Slot 6 (bottom)**: terminal peeks out on the left.
- **Slot 7 (bottom, next to Jira)**: terminal peeks out on the right.

The alternation is deliberate. By staggering which side each terminal sticks out from, we pack seven slots horizontally into a space that would otherwise only hold five or six. The browsers in adjacent slots can slightly overlap visually (they don't actually overlap — they're in different positions — but the eye-space they occupy interleaves). It's a bit like dovetail joinery: pieces that fit tightly together because each one's protruding parts land in the spaces the next one leaves.

Setting Up the Three Fixed Zones

Start with the three fixed zones, because they're the simplest to build and they define the boundaries of where the seven slots will go.

Open three separate browser windows. Drag each to the position shown in Figure 5.1:

- **Zone 1 (Validation)** — upper-left, positioned between Slot 1 and Slot 2 on the top row, sitting above Slot 4 on the bottom row.
- **Zone 2 (Communications)** — upper-right, to the right of Slot 2, left edge aligned roughly with Slot 5 below, right edge stopping short of where Zone 3 starts.

- **Zone 3 (Jira Master)** — bottom-right, extending all the way to the right edge.

For now, leave the Validation browser on a blank tab — you'll add real content (your IDE window layered in, your personal terminal, etc.) once you know the zone's position is right. In the Communications browser, open Slack or Teams and email as tabs. In the Jira Master browser, navigate to your team's Jira board and bookmark it; you'll add the per-repo PR and Actions tabs once you've cloned your repos.

Take a moment to *commit* these three windows to their positions. On macOS, you can use a window manager like Rectangle or Magnet to snap windows to half-screen or quarter-screen positions instantly. On Windows, the built-in Snap feature does the same with `Windows + arrow keys`. On Manjaro, KDE and most other window managers have similar snap features built in. Whichever tool you use, learn the keyboard shortcut for snapping a window to each region. You'll use those shortcuts every time your layout drifts.

> 💡 **Callout: Window managers are worth learning**
>
> If you've never used a window manager beyond manually dragging windows around with your mouse, this is a great moment to invest twenty minutes in learning one. On macOS, Rectangle (free) or Raycast (free) both give you keyboard shortcuts for snapping windows to halves, thirds, quarters, and any custom region you want to define. On Windows, FancyZones (part of the free PowerToys suite) lets you define custom zones and snap any window to them with a hold-and-drag. On Manjaro, your window manager probably already has this built in — check your shortcut settings.
>
> Why does this matter? Because windows drift. You'll click on something, accidentally drag a window an inch, and now it's slightly off from where it's supposed to be. With a keyboard shortcut you can snap it back into place in half a second. Without one, you'll either spend ten seconds dragging it back or — more likely — leave it drifted and let your layout slowly degrade over the day.

Building the Seven Slots

Now the interesting part. You're going to build seven slots. Each slot will get a browser and a terminal in the configuration we described above.

Start with Slot 1 (top-left). Open a new browser window — *separate* from the Validation, Communications, and Jira Master windows. This is going to be the browser for Slot 1. Size it and position it to match Figure 5.1 — above the left end of the bottom row, width matching Zone 1 above it, taller than the terminal you're about to place.

Now open a terminal window. Size it *slightly smaller* than the browser (both shorter and narrower) and position it so that it sits in front of the browser but offset to the right — meaning the terminal's right edge extends past the browser's right edge, and strips of browser remain visible above, below, and on the left of the terminal. This is Slot 1's configuration. (For Slot 2 you'll mirror this — terminal offset to the *left*. Slots 3, 5, and 7 are right-offset like Slot 1; Slots 4 and 6 are left-offset like Slot 2.)

In this terminal, run:

```
cd ~/workspace
claude --dangerously-skip-permissions
```

Once Claude launches, give it the very first instruction this session will receive:

> **Prompt to Claude Code:**
>
> `Please lease an environment according to the lease protocol in your permanent memory.`

Claude will follow the lease protocol from Chapter 4: check the lease file, find a free environment (one of thor/freya/tyr/loki/odin/heimdall/baldr), claim it with the current PID and timestamp, and tell you which environment it's now working in. The first Claude you start this way will almost certainly pick up thor, but you don't care which one — you just care that Slot 1 now has a leased environment and is ready. Don't give it a ticket yet; we're still setting up the other slots.

Now repeat the entire process for Slots 2 through 7. For each one:

1. Open a new browser window, size and position it for that slot according to the layout in Figure 5.1.
2. Open a new terminal window, size and position it next to the browser with the correct left/right offset for that slot.
3. In the terminal, `cd ~/workspace` and launch Claude with `claude --dangerously-skip-permissions`.
4. Ask Claude to lease an environment.

By the end of this, you'll have seven slots arranged across the two rows of your screen. Each slot has a browser and a terminal side by side. Each Claude has leased one of the seven environments (the lease file will record which is which), and each is sitting there waiting for work.

> 💡 **Callout: Why one Claude per slot, not one Claude that switches between slots?**
>
> A reasonable question. Why not have one Claude session that you tell "now work on thor, now work on freya"? Two reasons. First, *context*. Each Claude session has its own working memory of what it's doing. Mixing seven different tickets into one conversation would dilute that memory and you'd start getting cross-contamination — Claude making changes for ticket A while thinking about ticket B's requirements. Second, *parallelism*. With seven separate Claude sessions, each one can be actively *thinking* at the same time. While the Claude in Slot 1 is generating code for one ticket, the Claude in Slot 2 can be running tests for another, and the Claude in Slot 3 can be reading documentation for a third. Seven parallel thinkers move much faster than one sequential thinker.

Verifying the Slots Work

Before we load any tickets, test the slot mechanic. Click on the visible browser edge in Slot 1 — the browser should come to the front. Click on the visible terminal edge — the Claude terminal should come back to the front. The flip should feel instant, and crucially, nothing should move. Only the stacking order changes.

If flipping feels awkward or you can't find the visible edge to click, your browser-and-terminal sizes are probably off. Go back and check that the browser is slightly taller and wider than the terminal, and that the terminal is offset so it extends past the browser on one side. If that geometry is right, there's always a clickable strip of whichever window is behind.

If you're still having trouble, use your operating system's window switcher as a fallback:

- **macOS:** Use Mission Control (swipe up on the trackpad) to see all windows and pick one; or use `Cmd+~` to cycle through windows of the same application.
- **Windows:** `Alt+Tab` cycles through all windows; `Win+Tab` shows a visual overview.
- **Manjaro (KDE):** `Alt+Tab` works the same way; KWin has a "Present Windows" effect that shows everything.

The goal is that once muscle memory builds — a few days in — you're not using the keyboard shortcut at all. You're just clicking the visible edge. That's the dish by the door working as intended.

Loading Your First Seven Tickets

Everything is set up. The dishes are by the doors. Now we put the keys in them.

Figure 5.6 — Loading Your First Seven Tickets

Open tickets in the Jira master, drag each to its context browser, then tell Claude to work

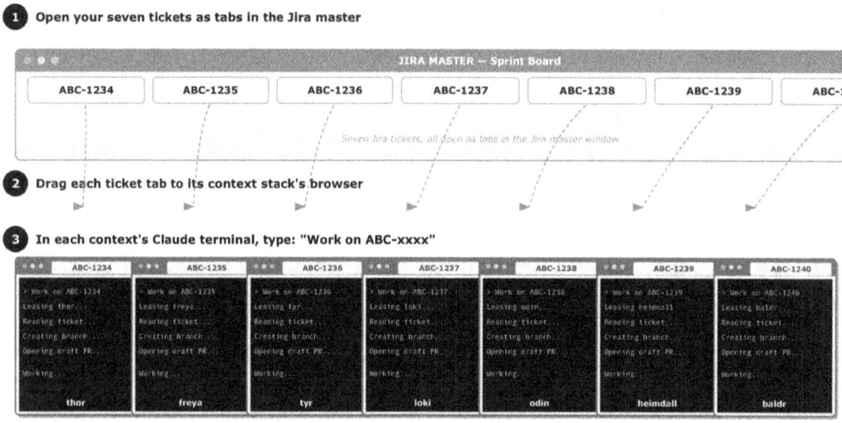

Result: seven Claudes running in parallel, each working its own ticket, each with an open draft PR. You now do the work only you can do.

Go to the Jira Master window (Zone 3, bottom right). Look at your queue. Pick the first seven tickets you're going to work on today. Open each one in a new tab in the Jira Master window. By the end of this, the Jira Master should have at least seven ticket tabs open alongside its PR and Actions tabs.

Now, one ticket at a time, do the following:

1. **Pick a slot for the ticket.** Look at the ticket in the Jira Master. Read the title and the first few lines of the description. Decide which slot it's going to live in. For your first day, just assign tickets in order — first ticket to Slot 1, second to Slot 2, and so on. Later, as you get more experienced, you might match tickets to slots based on which repo they touch or other criteria that suit your work.

2. **Drop the Jira tab into that slot's browser.** Drag the Jira tab out of the Jira Master and drop it onto the browser window in the slot you picked. The tab will move from the Jira Master to that slot's browser. (If your browser doesn't support tab dragging between windows, right-click the tab and choose "Move tab to" — or just copy the URL and paste it into the slot's browser.) Now the active ticket for that slot sits in its browser, one click away from the Claude terminal beside it.

3. **Tell the Claude in that slot what to work on.** Note the ticket number. Click on the slot's Claude terminal to bring it forward. Type the magic phrase:

```
Work on ABC-1234
```

(Substituting your actual ticket number.) Claude will execute the work-on-a-ticket protocol from Chapter 4: read the ticket, summarize it back to you, ask you to confirm. Confirm. Then Claude will create the branch, make the empty commit, push, open the draft PR, and start working.

4. Once Claude has confirmed it's started working, *don't sit there watching it*. Move to the next slot, drag the next Jira tab into its browser, type "Work on" with the next ticket number, and let that Claude start working too.

Repeat seven times. By the time you've started the seventh Claude, the first one is probably already deep into the work and may have asked you a question or made enough progress that the draft PR has real content in it.

> 💡 **Callout: What "work" actually looks like in parallel**
>
> The first time you do this, it will feel surreal. You'll have seven Claudes simultaneously typing, thinking, running tests, committing code, pushing changes, and making draft PRs grow in real time. Some Claudes will be done before others. Some will get stuck and ask you a question. Some will run into a problem and ask for guidance. You'll move between them, answering questions, giving guidance, and occasionally course-correcting.
>
> Don't expect this to feel comfortable on day one. It will feel like trying to juggle seven balls when you've only ever juggled one. By the end of your first week, it will start to feel manageable. By the end of your first month, it will feel natural — and going back to one ticket at a time will feel painfully slow.

What to Do While Claude Is Working

You are not supposed to sit and watch Claude type. That defeats the entire purpose of the layout. While the seven Claudes are working in parallel, *you* should be doing the things only you can do:

- Read your email and respond to anything urgent. (Communications zone, top right.)

- Catch up on Slack threads. (Same zone.)
- Review your teammates' pull requests. (Validation zone, top left, opened to GitHub.)
- Look at upcoming work in the Jira board. (Jira master, bottom right.)
- Check in on the bigger picture — sprint goals, deadlines, anything that requires your judgment rather than mechanical execution.

You'll glance over at the stacks every two or three minutes to see if any Claude has stopped typing or is asking for input. When one is, you flip to it, deal with whatever it needs, and flip back to whatever else you were doing. The rotation rhythm — how often to check, how to know when to intervene, how to know when something has gone wrong — is the entire subject of Chapter 6, so we won't go deeper here. For this chapter, the goal is just to *get to* the moment when seven Claudes are working and you have time to do other things. That's the milestone.

> 💡 **Callout: Keep the Claude window in front by default**
>
> When you're not actively looking at a Jira ticket or a draft PR for a slot, leave that slot's Claude terminal in the front of the stack. The reason is that you want to be able to *glance* at all seven Claudes at once and see what they're doing. If the browser is in front, you can't see Claude. If Claude is in front, you can see what each one is up to with a single sweep of your eyes. The browsers come forward only when you specifically need to look at the ticket or the PR.

Verification

Before we close out this chapter, run through this final checklist:

1. **All the fixed zones and slots are in position.** The Validation browser is top-left, Communications is top-right, Jira Master is bottom-right, and the seven slots fill the remaining space along both rows.
2. **Each stack has both a browser and a Claude terminal at the same position.** A single click flips between them.

3. **Each Claude has leased its corresponding environment.** Open `ENVIRONMENT_LEASES.md` from any terminal and confirm you see seven leases recorded — one for each Norse environment, each with a different PID and a recent timestamp.

4. **Seven tickets are loaded into seven slots.** Each slot's browser shows the Jira ticket for that slot. Each slot's Claude has been told *"Work on TICKET-NUMBER"* and is either working or asking you for confirmation.

5. **Seven draft PRs exist on GitHub.** Open the validation browser, navigate to GitHub, and confirm you see seven draft pull requests, one per ticket. (They may not have meaningful content yet — that's fine, they just need to exist.)

If all five check out, congratulations. **You are running.** The engine is started. Seven Claudes are working in parallel on seven tickets, and you have the layout you need to manage them.

The System Holds State for You

One small thing worth noticing before we move to Chapter 6. The state of everything you just built is now *persistent* in a way that will matter to you many times in the coming weeks.

The draft PRs live on GitHub and will be there forever. The branches live on the remote and won't disappear. The lease file is a plain file on your disk. The Claude memory files are plain files on your disk. Even the Claude conversations themselves will survive you closing your laptop (they'll pause, not die, as long as the processes survive).

This matters because in the old world, stepping away from a complex task often meant losing your place. In the new world, the *system* holds the state for you. You can walk away for lunch, for a meeting, for the rest of the afternoon — come back, and everything is waiting. The cognitive load of resuming is dramatically lower than it used to be, which is one of the reasons this methodology doesn't burn people out the way ad-hoc AI usage does.

We'll talk much more about that in Chapter 11. For now, just notice that you've built something resilient. Your work environment is not going to evaporate the moment you stop paying attention to it.

What's Next

The next chapter is about the daily rhythm of executing work inside this system — the rotation pattern, the validation step, how to know when Claude is stuck versus thinking, how to finalize tickets, and how to structure your day sustainably. (Burnout itself, the specific shape it takes when you're working with AI tools all day, is Chapter 11; Chapter 6 builds the rhythm that keeps Chapter 11's territory at arm's length.) Chapter 6 is where the throughput numbers from Chapter 3 actually come from. Chapter 7 then covers what it's like to come back to this workspace tomorrow morning, and the day after, and the day after that — because on almost every day except today, you won't need to build the layout from scratch.

But right now, you have seven Claudes working. Don't stop reading — Chapter 6 starts immediately, and it's where you learn to control the airspace you just opened: seven planes, one set of eyes, and a glance cycle that begins in the next chapter.

Chapter 6
Execution — The Daily Workflow

The stage is set. By this point in the book, if you've been building along with me, you have a workspace on disk, a layout on your screen, seven Claude sessions waiting in seven slots, three fixed zones humming in their designated corners, and a pile of tickets sorted in the Jira Master in the bottom-right of your display. You have done the hard part. What you have not yet done is *work*. That's this chapter.

Chapter 5 is where the system gets built. Chapter 6 is where it runs. By the end of this chapter, the throughput numbers from Chapter 3 will stop being numbers on a page and will start being the way your days actually go.

Let me say at the top: nothing in this chapter is difficult. None of it is complicated. The reason it takes a chapter to describe is that there are a lot of small things happening at once, and writing them down in a sensible order makes the whole arrangement look more elaborate than it feels. Once you've done this for a week, most of what follows will have compiled down into muscle memory and you won't need this chapter anymore. So don't be intimidated by the length. You're reading a map; you'll be driving the route before you know it.

The Shape of a Working Session

A **working session** is roughly three hours of focused execution — about 9:00 to noon in the morning, 1:00 to 4:00 in the afternoon. In that window, you complete the full cycle for about seven tickets — not one at a time, but overlapping in parallel.

Here is what a morning session actually looks like, in concrete time.

Nine o'clock. You've picked your seven tickets in the Jira Master, dragged each into its assigned slot's browser, and said "Work on ABC-1234" to each of the seven Claudes. Each Claude has executed the work-on-a-ticket protocol from Chapter 4 — read the ticket, summarized it back to you, gotten your confirmation, created the branch, made the empty commit, pushed, opened a draft PR, and begun working. By

9:10, all seven are in active execution. The room, if you'll pardon the flourish, is flying.

From 9:10 to around 9:30, you mostly watch. This is when the air-traffic-controller metaphor becomes most literal. You glance across the seven slots every two or three minutes. Most of the Claudes are typing; those you leave alone. One or two ask a question — a genuine question, usually about a decision you didn't specify clearly — and you answer. One finishes faster than the others, and you begin validation on that one while the rest continue. The rotations are not scheduled; they happen when they happen.

From 9:30 to around 10:00, the first wave of completions arrives. You validate two or three tickets (validation takes about five minutes per ticket — we'll get to the ritual in a moment). The validated Claudes have their PRs promoted and their tickets moved to In Review. Meanwhile, the slower ones are still working, and one of them may have gotten stuck and started signaling for help.

By 10:30 or so, you've finished five of the seven. The last two are either more complex or were partially blocked on a question. You finish those by noon, by which time you have seven PRs out, seven tickets moved to In Review, and a pleasant sense of having done a solid morning's work before lunch.

You go get lunch. We'll come back to that.

That is one session. You do another in the afternoon. Two times seven is fourteen. The arithmetic is not theoretical.

Notice what happened in that timeline, and in particular what didn't. You never sat and watched a single Claude work for twenty minutes. You never lost your place in one ticket because you'd been dragged into another. You never had to "remember" where you were on any given ticket, because each ticket lived in its own fixed slot with its own browser tab, its own PR, and its own Claude session. The system did the remembering. You just attended to whichever slot needed attention.

The setup move that makes each of those seven slots begin flying — the *work-on-a-ticket protocol* — has a specific shape. You say one sentence. Claude does the plumbing. You confirm the summary. Claude works. Figure 6.1 lays out the protocol as a swimlane, because it runs fourteen times a day and the shape is worth having in visual memory.

Figure 6.1 — The Work-on-a-Ticket Protocol

Seven steps in two lanes. You say a sentence; Claude does the plumbing; you confirm; Claude works.

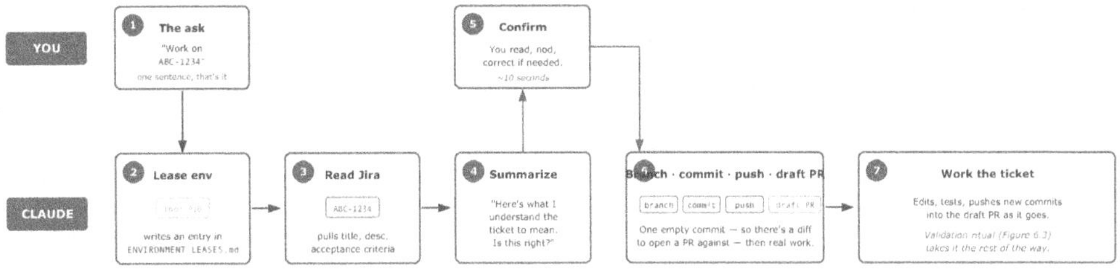

You spend about fifteen seconds of your attention in this protocol (steps 1 and 5).
Claude spends about two minutes on the plumbing so the real work starts from a PR-reviewable state.

Triage — Which Tickets Go Through the System

Not every ticket is a good candidate for this system. Before you load your seven, you need to know which tickets belong in slots and which don't.

The heuristic I use is simple enough to be useful: **does the ticket have a clear acceptance-criteria section, or something equivalent that tells you unambiguously what "done" looks like?** If yes, it's a candidate. If no, it probably isn't.

That heuristic works because parallel AI execution depends on two things: the Claude in each slot knowing when to stop, and you being able to validate when it has stopped. If the ticket itself doesn't define "done," Claude will invent a definition — and Claude's definitions of "done" are, shall we say, creative. You'll end up with work that is technically impressive, frequently ambitious, and not at all what the ticket was asking for.

Within "tickets with clear acceptance criteria," the **good fits** are:

- Bug fixes with a reproducible bug, especially ones where the ticket includes steps to reproduce.

- Well-scoped feature additions that touch two or three files and have an obvious place to live.
- Refactors with clear boundaries — "extract this function into its own module," "replace these callsites with the new API."
- Test additions, especially for code that already works but is under-tested.
- Small infrastructure changes: adding a missing metric, updating a Docker image tag, bumping a dependency version.
- Documentation updates.

The **poor fits**, the ones that should not go through this system:

- Tickets that are mostly conversations in disguise. "Go figure out what Product wants for the new billing flow" is not a coding ticket; it is a meeting with a Jira number attached to it.
- Multi-week architectural spikes. These need your full attention and your own hands on the keyboard. Don't put them in a slot.
- Tickets that span many repositories with an unclear integration story.
- Judgment calls that nobody has actually made yet. If the ticket is "decide how we should handle X going forward," that needs a human decision before it can be worked.

My morning triage takes about ten minutes. I read the titles and the first paragraph of every ticket in my immediate queue, sort them mentally into three buckets — *system fit, needs my thinking, needs a conversation* — and load the seven best system fits into the morning session. The next seven go into the afternoon session. The other two buckets go elsewhere on my calendar, handled the old-fashioned way, with the actual inside of my actual skull.

If a ticket feels borderline — clear-ish acceptance criteria, but something feels off — put it in the afternoon session, not the morning. Afternoon sessions run with the benefit of your morning warm-up; your triage judgment is sharper and your interruption reflexes are faster. Save the borderline tickets for when you're at your best.

The Rotation Rhythm

Figure 6.2 — The Rotation Rhythm

Your attention cycles across the seven contexts. Each glance tells you one of four things.

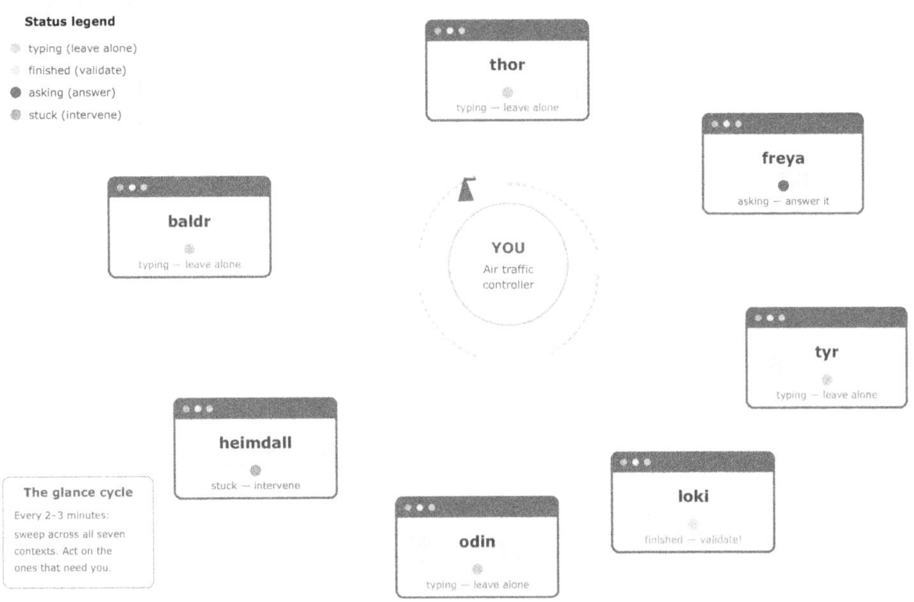

You are an air traffic controller, not a pilot. The seven Claudes are flying the planes. Your job is to clear them for takeoff, answer their radio calls, decide whether they're stuck, and clear them for landing. Your job is *not* to fly the planes yourself.

This mental model matters because it changes where you put your attention. A pilot watches the plane she is currently flying. An air traffic controller watches all the planes at once and looks at any one of them only long enough to assess its state. The whole method comes apart the moment you start flying one plane instead of controlling seven.

The **glance cycle** is how you stay in the controller's seat. Roughly every two or three minutes, you do a quick visual sweep across all seven slots. Each glance tells you one of four things:

1. **Still typing or running commands.** The Claude is making forward progress. Leave it alone. Move on to the next slot.

2. **Finished.** The Claude has stopped, probably announced completion, probably opened a PR. Begin validation on this one.
3. **Asking a question.** The Claude has hit a decision it doesn't want to make without you. Answer the question. If it's a question you'd rather Claude make a judgment call on, tell it so; Claude is perfectly happy to decide when given permission.
4. **Stuck in an unproductive loop.** The Claude is running the same command repeatedly, or generating output that contradicts output from three prompts ago, or going back and forth between two approaches without converging. Intervene.

The hardest skill in the glance cycle is telling "thinking" from "stuck." They can look similar — Claude produces text in both cases. The signal you're looking for is *convergence*. A thinking Claude is getting closer to something; each new output narrows the problem. A stuck Claude is circling; each new output is roughly equivalent to the one two outputs ago. If you see the same command run more than twice without any evident change in approach, that's stuck. If you see Claude say "let me try X" and then a minute later "actually let me try Y" and then a minute later "actually let me try X" again, that's stuck.

When stuck, my standard intervention is a single prompt:

> **Prompt to Claude Code:**
>
> ```
> Stop. Summarize where you are with this ticket, what you've tried so
> far, and what's blocking you.
> ```

Claude will do this beautifully. About half the time, in the act of summarizing, Claude realizes what the answer is and unsticks itself before finishing the summary. The other half, the summary tells you enough that you can point Claude at the real answer in one sentence.

A related pattern I use at least once per session: if I sense that a Claude has been grinding on something for a while but I haven't been paying close attention, I just ask for the summary proactively. It's free; Claude won't lose its place; and it gives me a snapshot of where that ticket actually is without me having to scroll through the full conversation history.

The anti-pattern to avoid, above all, is the one where you find yourself glued to a single slot for more than five minutes. If you catch yourself watching one Claude work — reading every line as it appears, hanging on the next token — stop. Step back. Glance at the other six. Whichever one has gone the longest without your attention is now the one that needs it. The moment you fall in love with a single ticket, your throughput drops by a factor of seven, and you stop being a controller and start being a very slow pilot with a view of six other planes.

Validation — The Thirty-Minute Review

Validation is where the system earns its sustainability. If validation were slow — if it required you to read every line of every diff — the whole method would collapse. Seven parallel streams of work hitting you at once would be more overwhelming than doing one at a time. The trick is that validation is *mostly delegated*. Claude validates its own work; you validate Claude's validation. Most of the thinking is already done by the time you look.

The ritual, when Claude tells you a ticket is complete:

1. Ask Claude to self-critique.

> **Prompt to Claude Code:**
>
> `Review the work you just did as if you were a senior engineer reviewing this PR. What concerns would you raise? What could go wrong?`

Claude is usually a better critic than creator. This prompt regularly surfaces real issues that Claude then volunteers to fix without further prompting. It is the single most valuable sentence in the validation ritual.

2. Ask Claude to run the tests.

> **Prompt to Claude Code:**
>
> ```
> Run the test suite and tell me the results. If there are no tests covering the changes you made, write them and then run everything again.
> ```

If tests exist, they should pass. If they don't exist, they should now exist. If Claude is tempted to skip or disable a flaky test to get a green run, the deny rule in your workspace's CLAUDE.md should stop it. If it doesn't stop it, Chapter 8 will help you tune the rule.

3. Ask Claude to check the diff against the acceptance criteria.

> **Prompt to Claude Code:**
>
> ```
> Look at the Jira ticket's acceptance criteria section. For each criterion, tell me whether the changes address it and how.
> ```

This catches the cases where Claude was off on a tangent and produced perfectly good work that does not, strictly, solve the ticket. You'd be surprised how often this happens. Actually, no, you wouldn't be surprised — you've met Claude.

4. Look at the PR yourself — but not line by line. Open the PR in the slot's browser tab. Read the description Claude wrote. Look at the file tree of changed files. Does the shape match what you expected? A three-line ticket that somehow touched forty files is a red flag. A change in a file you didn't expect to be touched is a red flag. Shape-level review is fast (twenty seconds) and catches most real problems. Line-level review is what the human reviewer who approves the PR is for; it is not what you are for.

5. If satisfied, promote the draft PR to ready-for-review. You can do this from the GitHub UI or with `gh pr ready`. Move the Jira ticket to "In Review." Your part of this particular ticket is done.

6. If not satisfied, give Claude specific feedback and let it try again. Do not roll up your sleeves and fix it yourself. Doing the work yourself gives up the leverage this method exists to provide. If the feedback has to be delivered three times before Claude gets it, that's still cheaper than you writing the code.

That ritual takes about five minutes per ticket, start to finish. Seven tickets is thirty-five minutes of validation spread across the session, interleaved with the rotation rhythm. Not a block at the end. Woven through.

When to Pull Out the Second-Claude Pattern

For most tickets, the five-minute ritual is enough. For a small number — production-critical paths, security-sensitive code, anything touching payments or authentication, database migrations — I add one more step: I ask a *different* Claude, in a fresh session in a separate slot, to review the first Claude's PR as a skeptical senior engineer.

> **Prompt to Claude Code:**
>
> ```
> I'm going to share a PR diff with you. Please review it as a skep-
> tical senior engineer. What's wrong with it? What would you change?
> What tests are missing? Where could this break in production?
> ```

The second Claude has no emotional attachment to the code because it didn't write it. It will often catch real issues. I don't use this pattern on every ticket; it would be overkill and cut my throughput substantially. But I use it on every ticket that would cause a pager to go off if the PR went wrong.

Figure 6.3 draws the ritual end-to-end — the six steps, the "satisfied?" decision at step 5, the loop back when the feedback is specific, and the optional second-Claude pattern running as a parallel track for the high-stakes tickets. You will run this ritual seven to fourteen times a day. Letting the shape of it live in your eyes, rather than in your working memory, is the whole point.

Figure 6.3 — The Validation Ritual

Six steps that turn a finished ticket into a reviewed PR. Step 5 either exits or loops.

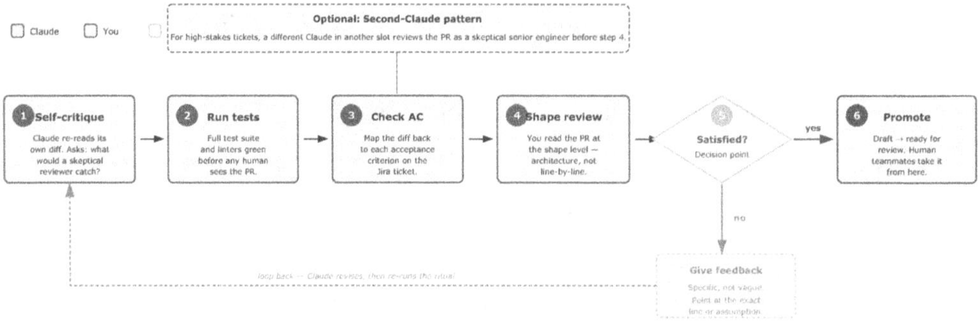

Steps 1–3 happen inside the environment. Step 4 happens in Zone 1. Step 5 is a decision; step 6 is either a promotion or a loop. The loop is the point. A PR that takes two passes is normal; three is fine; ten means the ticket is wrong, not the work.

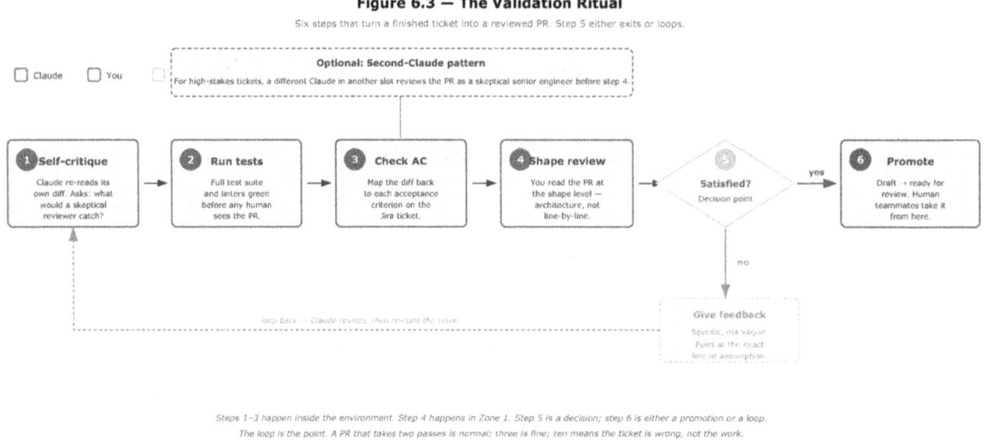

Finalizing Tickets

Finalizing is the smallest section in this chapter because it is the simplest step. Once you've validated:

- Promote the draft PR to ready-for-review. Request the appropriate reviewers if your team uses assigned reviewers.
- Update the Jira ticket with a link to the PR and a one-line summary of what was done. Claude can do this from its own terminal using the `jira` CLI.
- Move the Jira ticket from In Progress to In Review.
- **Do not release the environment lease yet.**

That last bullet deserves a word. When a reviewer comes back with feedback — and they will — you want that Claude session to still remember the work. The lease is what guarantees the session is still alive when the feedback arrives. If you release the lease early and someone from your team asks for changes an hour later, you will be starting from scratch: a fresh Claude, no context, and the reviewer's comments as your only starting material. That's an expensive do-over.

The rule is: **the environment stays leased until the PR is merged.** When it's merged, you release the lease. The environment returns to the pool, available to be leased again for the next ticket. At any given time, you may have more environments leased than you have slots on your screen — that's fine. The slots are where active work happens; an environment waiting on review can sit in the lease file indefinitely without occupying a slot.

When Reviewer Feedback Comes Back

The reviewer has left comments on PR-1234. The Claude that wrote PR-1234 is still leased to environment freya. Here's the flow:

1. Find freya's current slot on your screen. If you've closed the session entirely and come back, re-establish the slot — Chapter 7 covers that case.
2. Bring that slot's Claude terminal forward.
3. Paste the reviewer comments directly: *"I got these comments on PR-1234. Please address them. Here they are: [paste]."*
4. Claude reads the comments, makes the changes, pushes, and confirms the PR is updated.
5. You verify quickly. This is the validation ritual but lighter; you only need to check the new changes, not the whole PR.
6. Reply to the reviewer — either a quick "addressed, please re-review" or a substantive response where you disagree with the feedback and explain why.

The whole loop is usually five to ten minutes. Do not let reviewer feedback become the thing that wrecks your rhythm; it is just another kind of intervention, indistinguishable in shape from any other. What makes it special is that you should handle it *promptly* — the reviewer is waiting, and you want to keep their loop short too. That is the social contract of code review. This method makes it easy to honor.

The Two-Session Day

The rhythm of a full day looks like this:

Morning session (roughly 9:00 AM to 12:00 noon). Triage your queue. Load seven tickets. Let the Claudes work. Rotate, validate, finalize. Wrap up by noon.

Lunch (noon to 1:00 PM). Away from the screen. Actually away. This is non-negotiable. Eat something. Go for a walk. Do not look at Slack. Do not look at Jira. Do not think about the tickets you just shipped or the ones you're about to start. The method delivers fourteen tickets a day; it does not deliver fifteen by you skipping lunch. The *sustainably* part of "fourteen tickets a day sustainably" is paid for, in part, in lunch hours. Skip the lunch and you skip the sustainability. You just have a very productive week followed by a collapse.

Afternoon session (1:00 PM to roughly 4:00 PM). Load the next seven tickets. Run the same rhythm. Wrap up mid-afternoon.

End of day (4:00 to 5:00 PM). Close the Claude terminals for the day, but leave the environments leased and the layout intact. Close the communications window. Walk away. If you're paid to do forty hours a week, note that fourteen tickets a day times five days is seventy tickets a week, and you got there in thirty-five hours. That is the method's quiet gift: it returns time to you.

The Wind-Down Ritual

Before you walk away, send one message to each active Claude:

> **Prompt to Claude Code:**
>
> ```
> Write a short note in NOTES.md in the environment summarizing: (a)
> what you accomplished this session, (b) what's still in progress,
> (c) what to pick up first when we come back. Keep it between three
> and ten lines.
> ```

This takes Claude about thirty seconds per environment. It costs you nothing beyond one prompt. And it means that tomorrow-you, when you open the freya terminal and ask "where are we," can read NOTES.md in five seconds instead of scrolling through a long conversation looking for the last reasonable checkpoint. Chapter 7 covers this in more detail under the heading of *warm starts*, but the habit begins here, at the end of every session, every day, without exception. Future you will thank present you, and that's a relationship worth investing in.

Variations

Real days are not always two clean sessions. Adjust as needed:

- **Heavy meeting day.** You might only get one session done. Seven tickets is still more than most developers complete in a week.
- **Interrupt-heavy day.** A production incident or urgent meeting breaks the rhythm. When you resume, start with whichever ticket requires the least re-orientation, not necessarily where you left off. A Claude that was in the middle of an intricate change needs your focused attention; a Claude that was running tests can be picked up cold.
- **Low-complexity day.** Sometimes the queue is unusually kind — all small, all clean, all fast. You might cycle through all fourteen before lunch. Use the afternoon for deeper work that doesn't fit the system. Writing. Reviewing other people's architecture. Taking a walk and thinking.

Common Failure Modes

Everything in this chapter describes how things go when they go well. Here are the things that go wrong, and how to handle them. You will experience most of these in your first week. Consider this the catalogue.

Claude makes changes in the wrong repository. This happens when two of your repos have similar names or similar file structures. Revert the changes in the wrong repo, explicitly tell Claude which repo to work in, and start over. Add a note to that environment's permanent memory: *"This environment works only in repo X."*

Claude's branch drifts out of date with main while it's working. Someone else merged while Claude was busy. Teach Claude to merge main proactively before opening or updating a PR, and add the instruction to permanent memory: *"Before opening or updating a PR, merge origin/main into your working branch and resolve any conflicts."*

A test suite is flaky and Claude "fixes" it by disabling tests. This is the classic. Forbid it explicitly in the workspace's `CLAUDE.md`: *"Never disable, skip, or delete a test to make it pass. If a test is flaky, tell me and let me decide."* If you don't forbid this in writing, Claude will eventually do it, because making the red thing green is the strongest gradient any AI can feel, and sometimes the pull is irresistible.

Claude's proposed changes are too large. The PR has forty files changed. Ask Claude to break the work into smaller commits, or, if that's not possible, re-scope the ticket. A ticket that genuinely requires forty files of changes is a ticket that should not have gone through this system in the first place. It needed your full attention, and it needed to be chopped up in advance.

Seven Claudes all need attention at once. It happens. Pick one. Handle that one. Ignore the other six for two minutes. They are not going to catch fire. The only failure mode here is thrashing — attending to all seven shallowly until none of them get what they need. Depth on one, then the next, then the next, is always better than breadth across seven.

You've lost the thread of what a Claude was doing. You stepped away for a meeting and now you genuinely don't remember. Ask Claude directly: *"Summarize where you are with this ticket — what's been done, what's left, and what your current plan is."* Claude is excellent at self-summary. Fifteen seconds of reading and you're back up to speed.

Claude is making genuinely bad decisions — not just stuck, but wrong. This is the rarest and worst failure mode, and it has a specific shape: Claude is confident, generating output, not stuck, but the output is wrong in ways that reveal a fundamental misunderstanding of the problem. When this happens, don't try to steer. Release the lease, start a new session on the same environment, give it a better-scoped prompt, and let a fresh Claude pick up the work. Steering a confidently-wrong Claude is more expensive than starting over. The sunk cost does not need to be honored.

You notice at 3:30 PM that you haven't done the morning session's validation for two of the seven tickets. They're still open in their slots, apparently completed but not promoted. Do the validation now. The cost of late validation is one day, not one week. Don't let apparently-done Claudes sit indefinitely; unvalidated work is unshipped work, and unshipped work is not counted.

The CI on a draft PR fails for reasons unrelated to the ticket. The repository's main branch has had a flaky pipeline for three months and everyone has learned to ignore it. Don't ignore it in your draft PRs. Ask Claude to read the CI output and explain the failure. If the failure really is unrelated — an infrastructure gremlin, not a real problem — note it and proceed. If it turns out to be a real regression that nobody noticed, congratulations: you have just found a bug the ordinary workflow was going to ship.

What Mastery Feels Like

The first week of running this method is clumsy. You'll forget to glance. You'll fall in love with one ticket and lose the others. You'll over-validate, then under-validate, then over-validate again. You'll close the Jira Master by accident and have to set it up again. You'll release an environment when you meant to close a terminal and spend ten minutes figuring out how to recover. All of this is normal. All of this passes.

By the end of the first month, the rotation feels natural. You stop thinking about which slot is which because your hand already knows. You interrupt at the right moments without having to decide to interrupt. The validation ritual compresses into muscle memory. You stop worrying about whether the method is working and start worrying about the actual tickets — which is the point.

By month three, something more interesting happens. You stop being conscious of the system at all. You sit down in the morning and start working, and the rhythm is simply how you work now. You write prompts that Claude understands on the first try, because you've internalized what kind of prompt Claude needs. You catch a stuck Claude within fifteen seconds of it going stuck. You validate so quickly that you sometimes forget you're validating. Your throughput has doubled, then doubled again, and you no longer feel more tired at the end of the day — you feel *less* tired, because the cognitive load that used to come from remembering and switching and context-holding has been entirely offloaded to the structure around you.

At that point, going back to single-ticket work feels painfully slow. A morning where you only get one ticket out feels like writing longhand in the age of keyboards. You're not faster because you've become a better coder; you're faster because you've become a better *controller*. That is the mastery. That is what Buzzy had, in his own chosen corner of the craft, and what you are building, in yours.

The rest of this book is about the parts around the edges. Chapter 7 handles the realistic scenarios — coming back to a workspace after a break, recovering from a reboot, adapting to the day when your laptop dies and you need to re-establish the setup from scratch. Chapter 8 is about the slow continuous refinement that turns a good system into an excellent one. By the time you've finished the rest of Part II, the system will be your system, not mine. And you will be, in your own corner of the craft, beginning to be Buzzy.

Chapter 7
Coming Back to Your Workspace

Monday morning. Or Tuesday afternoon, after a long meeting. Or the dreaded Wednesday post-incident, when everything has been closed for four hours while you and three colleagues frantically talked through what just broke in production. Whatever the circumstance, you're not starting from scratch. You have a workspace. You have seven slots arranged on your screen, most or all of them probably still where you left them. You have a lease file on disk that remembers which environments were in use. You have Jira tickets mid-flight. You have draft PRs that may or may not have accumulated CI failures overnight, while you were sensibly asleep.

This chapter is about walking back into that workspace and resuming. It's the easy chapter. The procedure varies depending on how long you were away and what state the machine is in when you come back, but it's never hard. The most difficult thing you'll do in this chapter is decide between two options, and there are only three shapes of that decision in total.

The Three Shapes of Starting Up

Whenever you come back to your workspace, the situation falls into one of three buckets.

Tidy start. You stepped away briefly — lunch, a meeting, a quick errand, or an overnight while the laptop stayed open and charging. Everything is still running. All seven Claudes are alive, the browsers are where you left them, and the lease file matches reality. This is the most common case. Total re-start time: about two minutes.

Warm start. You come back to a machine where some, but not all, of yesterday's setup survives. The laptop slept, or rebooted for an OS update — which always happens at the worst possible moment, because that is how OS updates work — or the power blinked while you were out. Some Claudes are still alive. Others are terminals that look fine but have no process behind them. Browsers are mostly in position; a window or two has drifted. The lease file is technically valid, but some of its entries point at ghosts. This is the interesting case, and the bulk of this chapter. Total re-start time: ten to fifteen minutes.

Cold start. You are starting from absolutely nothing. New laptop, full reformat, fresh install. Nothing on screen. The workspace directory may or may not even exist yet. You are doing Chapters 4 and 5 again — workspace first, then layout. This happens maybe once or twice a year for most people, more often if you're a hardware masochist or your company's IT department enjoys replacing your laptop for sport.

We'll do tidy start first because it's brief, then warm start because that's where the judgment calls live, then we'll point back at Chapters 4 and 5 for cold start, because the procedure there hasn't changed.

Figure 7.1 is the map of the three shapes side-by-side — what you see, what survived, what ritual each one asks for, and roughly how long before you're back at full throughput. If you find yourself unsure which shape you're in on a given morning, the figure is the thirty-second diagnostic.

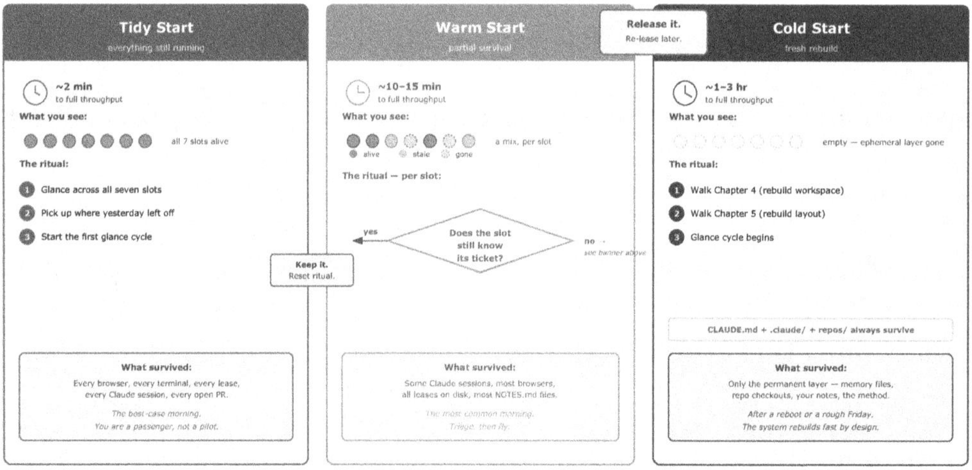

Figure 7.1 — The Three Shapes of Starting Up

Tidy, warm, or cold — you come back to one of three shapes. Each has a different ritual and a different cost.

The Tidy Start

You open the lid. Or you come back from your meeting. Everything is there. Resume.

That's most of it. The whole ritual is:

1. **Glance across.** All seven slots still in position? Validation browser still upper-left? Jira Master still extending to the right edge along the bottom row? If anything has drifted an inch or two because you bumped the trackpad on your way out, snap it back into place with your window manager shortcut. This takes five seconds.
2. **Liveness check.** Click into each Claude terminal in turn. Type something small — `status?`, or `are you still with me?`, or any single word that prompts a response. If Claude responds, it's alive. If it doesn't, it's a ghost; we'll handle that in the warm-start section.
3. **Resume.** Go back to whichever ticket you were actively working on when you walked away.

Most tidy starts take about two minutes. Some of them take ten seconds, because you just got back from the bathroom and the only thing you need to do is click the slot you were in.

A word on the liveness check. It's tempting to skip — if the terminal is there, surely Claude is still running, right? Usually yes. But if your machine slept and the Claude process died quietly in the background, the terminal will still show the cursor and look perfectly normal. You will type a prompt, press enter, and nothing will happen, and you will spend thirty seconds wondering why Claude is taking so long before realizing there's nothing on the other end of the pipe. The liveness check catches this in two seconds instead of thirty.

The Warm Start

This is the interesting case. You left work on a Tuesday evening. Wednesday morning you walk into your office, wake up the laptop, and some subset of the following has happened:

- The laptop slept and several Claude processes died. Their terminals still look normal but nothing is home.
- Two Claudes are still alive and remember everything.
- Your browser asked if you wanted to restore the previous session. You clicked "restore." It did. So all your tabs are there, but the window positions are slightly off.
- An automatic OS update rebooted the machine at 3 AM, because that is the single most inconvenient time to reboot a machine and therefore the time the OS has chosen for you.
- One of the draft PRs you were working on yesterday evening has a bright red CI failure that wasn't there when you went home, because someone else merged a conflicting change.
- Your lease file on disk looks exactly as it did last night and is now partly fictional.

Welcome to Wednesday. Let's work through it.

Assessing the State

Before you do anything, find out what actually survived. This takes less than a minute.

For each of the seven slots:

- **Click the terminal forward.** The slot's click-to-flip mechanic still works; the layout system you built in Chapter 5 is robust to rebooting — the windows remember where they belong.
- **Type a small probe prompt:** *"Are you still there? What ticket are you on?"*
- **Watch.** If a response arrives within a couple of seconds, Claude is alive and will tell you what it was doing. Note the slot, the ticket, and the state. If nothing happens — no cursor movement, no response, no "I'm working on it" — Claude is a ghost. The terminal is just a window with a dead shell in it.

After this sweep, you know two things: which Claudes survived, and which environments are effectively orphaned in the lease file. Don't do anything with that information yet. Just know it.

Second, check the lease file. Open `ENVIRONMENT_LEASES.md` — either in your personal terminal over in Zone 1, or by asking any living Claude to read it back to you. Compare it to the list of survivors you just made. Any environment that shows as leased in the file but whose Claude turned out to be a ghost is a stale lease. The lease protocol from Chapter 4 is built to handle this. When you start a new Claude and ask it to lease an environment, its PID check will notice that the recorded PID is no longer alive and will reclaim the environment without you needing to manually clean up. You can let the protocol do its job.

Third — and this is the habit Chapter 6 asked you to build — open the NOTES.md files each environment wrote at the end of yesterday's session. The note in each environment tells you in three to ten lines what that environment was in the middle of. *"Finished the PR, waiting on review. Fred commented last thing, need to address."* Or *"Halfway through the refactor; left off at the second file in src/api/. Run the tests before continuing."* Or *"PR merged overnight; release the environment when I get back."*

The NOTES.md files are cheap to produce and expensive to do without. By reading five of them in ninety seconds, you have reconstructed most of yesterday's state without any Claude having to summarize anything for you. This, more than anything else, is why the wind-down ritual earns its place in your day. Present you pays thirty seconds per environment to make tomorrow-you's first ten minutes easier. It's an excellent trade.

The Keep-or-Release Decision

Now the judgment call. For each surviving Claude, decide whether to continue with that session or release it and start fresh.

Keep if:

- The work is nontrivially in progress — Claude is mid-implementation, you haven't yet finished the validation, the PR is still in draft.

- The Claude's conversation history contains useful context you'd have to re-establish if you started over. For example: a long debugging chain where Claude now understands the shape of the problem.
- The PR is mid-flight and you want to continue where the Claude left off.

Release if:

- The work is essentially done — just waiting on review, or merged overnight.
- The Claude has been running for a long time and its context is cluttered with stale information that would make it less useful for the next ticket than a fresh start would.
- You want to pivot this environment to a completely new ticket, in which case you'd start from a clean conversation anyway.
- Claude is answering strangely. This is the polite phrase for *context rot*. Claude's working memory is long but not infinite; sessions that have been around for a day or two can begin producing answers that don't quite match the questions. If you feel that vibe, release and start over. The cost is five minutes of re-context; the cost of not releasing is that the next several hours will be harder than they need to be.

The rule of thumb: you should keep maybe two or three of your seven, and release the rest. Most of your slots should start the morning with fresh Claudes leasing whichever environments are available. This sounds wasteful — didn't you just spend all yesterday building up context in those sessions? — but it isn't. The context that mattered is preserved in two places: the environment's NOTES.md and the branch itself (which knows what's already been committed). A fresh Claude with a pointer to the right branch and a quick read of NOTES.md can pick up exactly where the dead Claude left off, with a cleaner context window and no accumulated misapprehensions.

The Reset Ritual for Kept Slots

When you've decided to continue with a surviving Claude, run the following ritual:

1. **Ask for a self-summary.**

> **Prompt to Claude Code:**
>
> ```
> What ticket were you working on and where did you leave off yester-
> day? Include what's been done, what's left, and what you planned to
> do next.
> ```

Claude responds in five to ten lines. Read it. Confirm it matches your own memory — and the NOTES.md, if you wrote one. If anything is off — Claude says it finished something you distinctly remember not finishing, or skips a whole subtask — take the warning seriously. A fresh Claude won't have the same confusion. Consider whether to release.

2. Check branch state.

> **Prompt to Claude Code:**
>
> ```
> Check the git state. Are there uncommitted changes? Is the branch up
> to date with origin? What's the last commit message? Is the draft PR
> still open and in what state?
> ```

This surfaces drift between what Claude thinks is true and what the repository actually says. When the answers disagree, trust the repository.

3. Pull main and rebase.

> **Prompt to Claude Code:**
>
> ```
> Merge the latest origin/main into this branch and resolve any con-
> flicts. If conflicts need my judgment, stop and ask me before
> picking a side.
> ```

This is the most important step. Overnight, other developers have been shipping. Your branch is now behind. If you don't merge main before continuing, you will eventually have to merge it *just before* the PR is ready to ship, and that's when conflicts go wrong.

4. Check CI.

> **Prompt to Claude Code:**
>
> `Run gh pr view to check the PR's mergeable status and CI state. Report any failures.`

If CI is red and it was green when you went home, the red is usually because of the main-merge you just did. Resolve the new failure before you move on. If the red existed *before* you went home and you went home anyway, we'll have a gentle conversation about that in Chapter 11.

5. Resume. You're back in the chair. Continue the work as if nothing had happened.

The whole reset ritual takes three or four minutes per kept Claude. Two kept Claudes and five fresh ones is a fifteen-minute warm start, total. Fifteen minutes is a trivial cost for a workspace that remembers fourteen tickets of state across weeks.

The Release Ritual

For the Claudes you've decided to release — either because the work is done or because a fresh start is cleaner — the ritual is short.

> **Prompt to Claude Code:**
>
> `Release the lease on this environment according to the release protocol. Check out main, stash or discard any local state, clear the lease entry, then confirm the environment is free.`

Claude does the work. The environment goes back in the pool. Hit Ctrl+C twice to exit the Claude session. Relaunch Claude in the same terminal with `claude --dangerously-skip-permissions`. Tell the new Claude to lease whatever's available — it will pick a free environment automatically, respecting the protocol. You now have a fresh slot ready for a new ticket.

Fresh Claudes have a specific advantage worth naming: they have **no preconceptions**. A Claude that's been around for a day has formed opinions — about the shape of the codebase, about which files to look at first, about what kind of solutions are preferred. Most of those opinions are correct and useful. Some are wrong and sticky. When Claude has a wrong sticky opinion about something, the cleanest fix is not to argue it out; the cleanest fix is to start a new session and let the new Claude form its own opinions, probably different ones, from a clean read of the code.

Edge Cases

A few situations worth naming explicitly:

Claude is alive, but the PR it was working on was merged overnight. Someone reviewed it, approved it, and clicked merge. The ticket is genuinely done. Release the lease, move the Jira ticket to Done, move on with your morning and your life.

Claude is alive, but the branch has conflicts that need your judgment. Don't let Claude guess. Sit with it. Resolve together. This is one of the few times where doing the merge yourself in your IDE, with Claude watching, is the right call. Once the conflict is resolved, hand back to Claude: *"The conflict is resolved. Continue from here."*

Claude is alive, but answering strangely. Already mentioned — release and start over. Trust the vibe. If the Claude in front of you doesn't feel like the Claude you left yesterday, it probably isn't, functionally speaking.

Claude is dead, the terminal shows a normal cursor, and you've gotten used to the weight of yesterday's context and don't want to let go. Let go. Release the environment — or let the lease protocol reclaim it automatically — start a new Claude, let the new one read NOTES.md and the branch, and trust it to catch up. It will. Sentimental attachment to a dead Claude is a form of sunk-cost reasoning, and it does not pay.

Your layout has drifted more than cosmetically. A window has jumped to the wrong monitor, a Claude terminal is minimized and you can't find it, Zone 2 has decided overnight to open full-screen. Use your window manager shortcuts to pull everything back into place. If you can't remember the shortcuts, the window-manager config you set up in Chapter 5 will re-establish the positions in a single command. If you skipped the window-manager setup in Chapter 5 because it seemed unglamorous

— well, this Wednesday is your punishment for skipping that part. Go back and set it up now. Next Wednesday will be easier.

The Cold Start

We covered the cold-start work across Chapters 4 and 5 — the workspace build first, the desktop layout on top of it — and I won't repeat either here. If you're truly starting from nothing — new machine, fresh install, no workspace on disk — walk Chapter 4 to rebuild the workspace, then Chapter 5 to rebuild the layout. The only meaningful difference the second time you do it is speed: you'll get through the whole setup in an hour or two instead of the two to three it took the first time, because you now know what you're building toward.

One small observation. Cold starts are a good moment to prune. If yesterday's permanent memory rules had accumulated a few that weren't actually earning their keep — old edge cases, rules that applied to a codebase you no longer work in, clever instructions that didn't pan out — a cold start is a natural opportunity to leave them off. The `CLAUDE.md` you bring forward to your new machine should contain rules you'd bet money still matter. Anything you wouldn't bet on, leave behind.

The System Holds the State for You

The whole point of the persistent architecture — the workspace directory, the lease file, the NOTES.md files, the fixed layout — is that you can walk away and come back. The system holds the state for you. You are not required to remember seven tickets of detail between days. The structure remembers on your behalf.

This is worth internalizing, because it removes a kind of anxiety that many developers don't realize they carry. In the old way of working, stepping away from the laptop was slightly costly. You'd have to re-orient when you came back. You'd worry about forgetting. You might stay late on a Thursday because you knew you wouldn't feel like starting over on Friday morning. That anxiety is a meaningful fraction of what makes sustained high-throughput work so exhausting over time, and most people who carry it don't know they're carrying it.

With this system, stepping away has almost zero context cost. The workspace will be there. The Claudes will be there — most of them, anyway; the OS updates will take their tithe. The notes will be there. You can have a real lunch. You can have a real weekend. You can take a real holiday. When you come back, you'll walk through the warm-start procedure, be caught up in fifteen minutes, and start shipping again.

There's an anti-pattern worth naming at the end of this chapter, because I've seen it and I've heard myself almost fall into it. Some developers, seduced by the idea of a clean slate, develop a habit of doing a full cold start every Monday. New directory, new everything. They think they're being disciplined. What they are actually doing is throwing away a week of compounding context every week, and then rebuilding it from scratch. Clean slates have costs. This system is *designed* to let you accumulate context across days without paying for it. Paying for it voluntarily every Monday, for purity reasons, is missing the point.

Walk in. Read the notes. Keep what's valuable. Release what's spent. Run the warm start. Get back to work.

Chapter 8
Refinement

Refinement is the continuous tuning that happens alongside Execution. This chapter teaches how to notice when the system needs an adjustment, how to make that adjustment safely, and — just as importantly — how to avoid the opposite failure mode of constantly tinkering with your workflow instead of actually working.

What Refinement Is and Isn't

Refinement is the third of the three phases, and the only one that never ends. Setup happens once. Execution happens every working day. Refinement happens — in small amounts — alongside Execution, for as long as you keep using the system.

The word matters. A chisel is refined by being sharpened; it is not refined by being redesigned. Musashi spent decades refining the same techniques rather than inventing new ones. He kept his sword. He kept his stance. What he changed, he changed in microscopic amounts, over many years, based on what the last fight had told him about the current weakness. That is the shape of refinement.

Refinement, in this system, is a small, specific adjustment to something you've already built, based on something you've specifically observed.

It is not:

- Rebuilding the workspace from scratch because you saw someone's dotfiles repo on Hacker News and their terminal has a cooler prompt than yours.
- Renaming your seven environments from Norse gods to something "more meaningful."
- Deciding that seven is the wrong number and trying five. Or nine. Or eleven, because eleven feels weighty and prime, and you read a Hacker News comment last Wednesday from someone who claimed eleven was the only number that "really" worked.

- Swapping out Claude Code for whatever your algorithm feed is currently excited about, on a Tuesday, for no reason.
- Adding a brand-new validation step to your ritual because one ticket, one time, got past the old one.

Those are all things that feel productive while you're doing them, because keyboards are clicking and files are being edited, and at the end of the afternoon you can point to a commit and say *I did work today.* But the system is no better. The system is merely different. Different is free; better is expensive; and the two are very easy to confuse when you're the one doing the typing.

Refinement is the other thing. It is the half-sentence you add to CLAUDE.md because the same mistake has come up on four different tickets this week. It is the one-line tweak to a prompt template because the old phrasing caused Claude to ask you the same clarifying question every single time. It is noticing that the email client in Zone 2 would be easier to scan if you nudged it half an inch to the left, nudging it half an inch to the left, and then never touching it again.

These changes are boring. That is the point. A well-refined workspace should feel, after a few months, like a favorite pair of boots — unremarkable, well-shaped to the task, not getting in your way. If your system has an exciting quality to it, if you look forward to sitting down because you get to *fiddle* with it, you are almost certainly refining the wrong thing. You're not sharpening the sword anymore. You're redesigning the grip, again, for the fourth time this month.

The rule of thumb I use on myself, and that I'd ask you to hold yourself to, is this:

- **If you're refining more than about once a day, you're tinkering.** Working systems do not need that much adjustment. The work you're doing in the name of refinement is almost certainly procrastination wearing a more respectable outfit.
- **If you're refining less than about once a week, you're letting problems accumulate.** Every working system drifts. Claude gets a little sloppier about something you used to insist on. The layout creaks slightly under a new kind of ticket. Your NOTES.md files start getting written in a half-hearted shorthand that will be useless to future-you next Monday. These things do not announce themselves. You have to go looking.

Somewhere in between those two extremes there is a healthy cadence, which is the subject of Section 3. For now, hold on to this: refinement is small, specific, observed, and rare. When it starts to feel like anything else — when it starts to feel like *work*, in the way that executing a ticket feels like work — you have left refinement and crossed over into its opposite failure mode, which is the subject Buzzy would have had a name for, and which we'll get to at the end of this chapter.

The Three Places Refinement Happens

Refinement has exactly three surfaces. That is not a simplification; it is the whole list. If you find yourself refining a fourth surface, you are probably doing something else — building a new feature, redesigning the methodology, or, more often than any of us would like to admit, tinkering.

The three surfaces are permanent memory, prompt templates, and the environment and layout. Each has a different tempo, a different threshold for intervention, and a different failure mode when you get the tempo wrong.

Permanent memory itself is layered, and the layers matter. A new rule does not walk in the front door and claim a spot at the top of the workspace `CLAUDE.md`. It starts close to the work, in the NOTES.md of whichever environment hit it, and it graduates outward only as it proves itself. Figure 8.1 is the onion — session notes on the inside, repo-level rules in the middle, workspace-level rules on the outside — and the rule about what gets promoted when.

Figure 8.1 — The Three Places Refinement Happens

A lesson starts close to the work and graduates outward only when it has earned the climb.

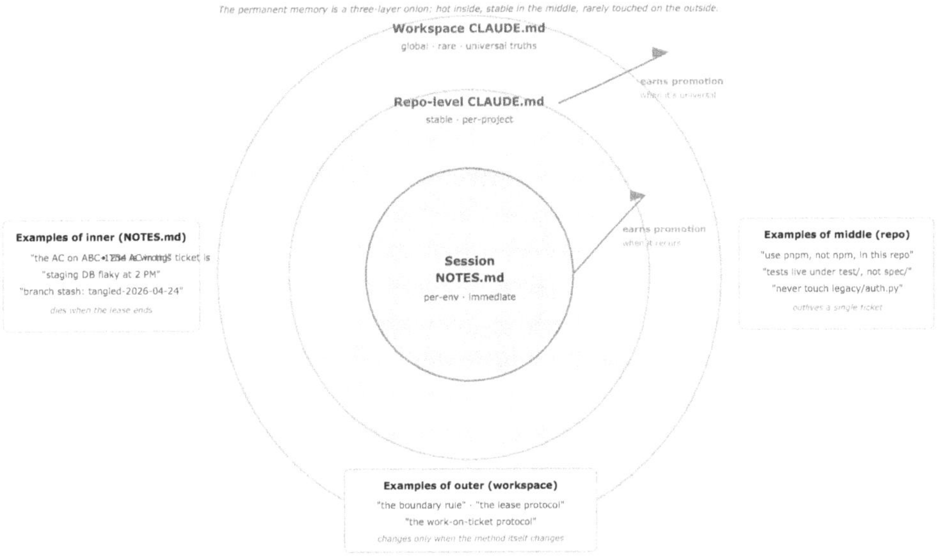

Permanent Memory

Permanent memory is the dish by the door. When a rule has a fixed home, you don't have to remember where to put it; you drop it in and walk on. `CLAUDE.md` and the `.claude/` directory are exactly that for your rules about how Claude should behave in this workspace.

The signal that a new rule belongs in permanent memory is repetition. Once is a mistake. Twice is a pattern. Three times is a rule you have been refusing to write down. When you notice the same correction coming out of your mouth — *"v2, not v1"*, *"TypeScript, not JavaScript"*, *"the test file goes next to the source file, not in a separate tests directory"* — that is permanent memory asking to be fed.

The mechanics are trivial. You don't open the file. You don't scroll to the right section. You don't fight with YAML frontmatter at eleven at night. You tell Claude:

> **Prompt to Claude Code:**
>
> `Please add to permanent memory: always use v2 API patterns, not v1.`

Claude handles the edit. It places the rule in a reasonable section, matches the tone of the neighboring rules, and tells you what it did. Read the diff, accept it, move on.

The failure mode here is over-rule-ification. A `CLAUDE.md` with sixty rules, most of which applied exactly once, in a codebase you barely touch anymore, is worse than a `CLAUDE.md` with ten. Rules interact. Rules contradict each other. Claude has to weigh them, and the more there are, the less each one lands. Prune on cold starts — we covered that last chapter — and prune in place whenever you notice a rule hasn't earned its keep in a month.

Prompt Templates

Permanent memory tells Claude how to behave all the time. Prompt templates are what you say when you want Claude to do a specific thing *right now*. Over a few weeks you will catch yourself typing the same five or six prompts again and again. Those are your templates.

Mine, for example:

- *"Summarize where you are on this ticket — what's done, what's in progress, what's next."* Used at the end of every session, right before the NOTES.md wind-down.
- *"Review your own work as a skeptical senior engineer. What would you push back on in code review?"* Used before I promote a PR out of draft.
- *"Diagnose what went wrong. Don't fix anything yet — just explain."* Used when a test starts failing and I do not want Claude speed-running through three speculative fixes before we know what the actual problem is.

Keep your templates somewhere you can find them in two seconds. A single markdown file in the workspace works. A note in whatever you use for personal notes works. A sticky note on the monitor works, and I will not judge you.

One trick I'll recommend, though: open that template file in a text editor and park the window so a sliver of it protrudes past the left or right edge of your desktop, behind your slots. When you need a template, you click the visible edge, the window rises to the top, you copy the line you want, and you click back into whichever slot you came from. It is the same stack-flip move you already use for browser-versus-terminal inside a slot, just applied to the templates file. One click out, one click back. No Spotlight, no `Cmd+Tab` through six windows, no searching.

Do not over-engineer the templates. A prompt template is not a configuration file. It is a sentence you say a lot, written down. If it reads like it was translated from German by a lawyer — *"Please review, in a spirit of constructive critique, the recently-authored code changes..."* — throw it out and rewrite it the way you'd actually say it to a colleague.

Environment and Layout

The third place refinement happens is the layout itself, and this is where you have to be the most careful about not fiddling.

You will get ideas. After a week you'll wonder if seven slots is really the right number for your kind of work. After two weeks you'll wonder whether Zone 2 should be on the left instead of the right. After a month somebody on the internet will have a clever variant and you will want to try it.

Resist.

The layout is muscle memory now. Every tweak costs you a day of fumbling for the slot that used to be on the left and is now in the middle. Small gains from a better arrangement are real, but they are dwarfed by the relearning tax every time you move the furniture. A good rule: let a new layout choice run for at least two weeks before you touch it. If after two weeks it is still annoying you, change it once, and then leave it alone for another two weeks.

There are legitimate adjustments. If the kind of work you do doesn't actually fill seven slots — you work in one repo, your tickets are chunkier, your validation is slower — run with five and leave two environments cold. No one is handing out prizes for keeping all seven lit at once. The number is a ceiling, not a quota. But *change it deliberately*, not because Thursday afternoon felt off.

The same applies to Zone 2 versus Zone 3, the stagger pattern on the bottom row, whether browser or terminal sits on top by default. Appendix H walks through the variants. Pick one, live in it, and only leave it when you have a specific complaint that a specific change would fix.

The Refinement Cadence

Before we get into rhythms and intervals, the honest thing to say is this: there is no required cadence. Refinement is not a chore on a calendar. It is a response to what the system is telling you, and some weeks the system will tell you nothing at all. A workspace that has been tuned well for the kind of work you do can go a very long time — weeks, sometimes months — without needing a single change. If you are in that stretch, you are not neglecting anything. You are reaping the return on the setup work you already did. Leaving a good system alone is a skill, not a failure.

What this section offers, then, is not a prescription but a set of windows — three natural moments when it is cheap to *consider* refining, and the judgment to know when each of them applies. Most of the time you'll open the window, see nothing that needs doing, and close it again. That is the expected outcome, not the exception.

End of Each Working Session: Thirty Seconds of Reflection

Just before the wind-down ritual from Chapter 6 — before the NOTES.md files, before the laptop closes — pause for thirty seconds.

The question you are asking yourself is this: *Did anything happen today that should change the system?*

Not: *did anything go wrong today?* Of course something went wrong. Something goes wrong every day. That is what Tuesdays are for. Not: *is there anything I could theoretically improve?* There is always something to improve, and that road leads to tinkering. The specific question is narrower — did something happen today that points at a durable pattern, something that will keep costing you until you adjust for it.

Examples of what clears the bar:

- *Two Claudes today used the old API patterns, even though I asked nicely.*
- *I kept forgetting which environment I had parked the billing refactor in.*
- *The validation browser was too small to see the new dashboard without zooming.*

Examples of what doesn't:

- *One Claude misunderstood one ticket this morning*. One-off. Move on.
- *I was tired today*. Not the system's fault.
- *I saw a tweet about a new tool and thought about trying it*. We are not rearranging the kitchen because Instagram told us to.

Most nights, nothing will clear the bar. That is the normal case, not a sign you aren't paying attention. A mature system throws off very few durable signals in any given day because most of the durable signals have already been answered. If three or four nights in a row produce no note, the correct response is *good* — not *I must not be looking hard enough*. Looking harder is how you invent problems that were not there.

When something does clear the bar, write it down. That is the whole step. One line in a plain file — call it `refinements.md` and keep it next to CLAUDE.md, or drop it into whatever notes app you already trust. The dish-by-the-door principle applies here too: a fixed, boring, known place for the thought to land, so you don't have to think about where it lives. One line, plain language, no plan yet. *"Two Claudes used v1 API again. Add a memory rule."* Close the laptop.

You do not act on the thought tonight because you are tired, and a tired engineer is a bad systems designer. Tired engineers open the Jira board at nine o'clock at night and rearrange every swimlane before bed and wake up Wednesday morning wondering what happened. Thirty seconds of noticing is all you owe yourself at the end of a session. The noticing is the work; the acting comes later — if it needs to come at all.

Weekly: A Look at the Notes, if There Are Any

Once a week, glance at the refinements file.

If it's empty, you are done. Close it. The absence of accumulated grievance is meaningful data — your system and your work are in alignment right now, and there is nothing to do. People who have built durable systems over long careers will tell you that the weeks with no entries are the ones they remember most fondly.

If there are entries, pick a low-stakes moment to read them — Friday afternoon or Monday morning, the edges of the week, never the middle. Never do this mid-execution. Open CLAUDE.md mid-ticket to add one small rule and you will emerge forty minutes later having rewritten three paragraphs, broken the build, and forgotten what the ticket was about in the first place.

Read the week's notes end to end. Most of what you wrote down will have resolved itself — a frustration you noted on Monday turns out to have been a quirk of one specific ticket, not a pattern. Cross those off. What remains is the durable stuff.

Now decide what, if anything, to actually change. The honest answer is often *nothing yet.* A single note about an annoyance that happened once is not a mandate to edit the system; it is permission to watch for the pattern to recur. A note that shows up three weeks running, though, has earned an adjustment.

When you do adjust, cap yourself at **one or two changes.** Not four. Not *"all of them, I've got time, it's Friday."*

The reason for the cap is that every rule you add to the system has a cost as well as a benefit. A CLAUDE.md that is read every session gets slower to parse the longer it gets. Prompt templates that multiply start contradicting each other. Environment tweaks stack up and you forget why you made them. A system refined sparingly — one careful change at a time, only when the evidence is in — compounds over a career. A system refined six times a week for a month is usually worse than it was when you started.

Make the adjustments, if there are any to make. Commit them. Close the laptop. If it's Friday, go outside.

Occasionally: A Gardener's Pass

Some books would put "monthly" here. That's too rigid. The honest frame is *occasionally* — whenever enough time has gone by that the permanent-memory files have had a chance to accumulate sediment. For some people that's every month. For others it's every quarter. For people whose work is stable and whose system was well-built in the first place, it might be twice a year.

The trigger is not the calendar. The trigger is a feeling — a vague sense that CLAUDE.md has gotten long, or that you keep seeing a rule you forgot you wrote, or that a prompt template you lean on doesn't quite match how you use it anymore. When that feeling arrives, take an hour on a quiet afternoon and do a gardener's pass.

You are looking for two things. First, rules that are no longer needed because the underlying problem has gone away. The rule about "always use v2 API patterns" can come out the day the v1 endpoints are turned off; after that, it's belaboring the obvious. Second, rules that contradict each other, or rules you have forgotten you ever wrote. The file should read like a tight memo, not like the sediment at the bottom of a drafts folder.

Do the same pass on your prompt templates file and on the environment layout. If you have three prompt templates for "summarize where you are" and they have all drifted slightly in different directions, pick one and delete the other two. If one of your seven environments has not been leased in three weeks, that is data — maybe your work actually fits six environments, not seven. Then again, maybe this was a light three weeks and the seventh environment earns its keep in crunch. Your call.

The gardener's pass is mostly subtraction. You are not making new things. You are composting the old, so the file you read at the start of every session stays sharp instead of going to mush.

How to Tell if You Are in a Steady State

You know you have reached a stable place when the refinements file stays empty for weeks, the system continues to deliver tickets at the pace you want, and the idea of changing anything feels like it would be change for its own sake. This is not the end of the road; it is the payoff. A well-tuned system that quietly does its job for a quarter has earned its keep and then some. Enjoy it. Work through it. Do not invent refinement work because you feel like refinement work is what a serious practitioner would be doing.

The inverse signal is just as important. You know it is time to consider a change when two or three unrelated refinements file entries start pointing at the same root cause, or when you feel yourself fighting the system instead of working through it, or when a change in the work itself — a new codebase, a new team, a new kind of ticket — has shifted what the system needs to support. Those are real signals. Answer them. The rest of the time, leave the system alone and get on with the job.

Three windows, one overriding principle: refinement serves the work, never the other way around. Thirty seconds at the end of each session to notice. A weekly glance that is often a non-event. An occasional gardener's pass when the files have earned one. Set the rhythm loose enough that the system has room to simply be good, and your job for the next six months — or six years — is to work on tickets, not to admire the machinery you built to work on tickets with.

Figure 8.2 lays the three windows out at their correct sizes. Notice how much of the picture is unshaded — those are all the moments when you do nothing at all, which is correct. A refinement cadence that looks dense is a cadence that has become the work instead of the servant of the work.

Figure 8.2 — The Refinement Cadence

Three windows, not three obligations. Most weeks, each one opens, finds nothing, and closes again.

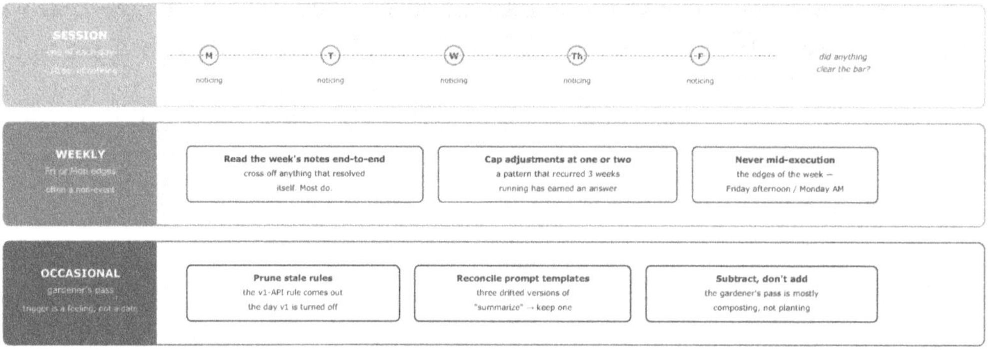

Refinement is a response to a signal, not a chore on a calendar. A good system throws off few durable signals in any given day.
An empty refinements file for three weeks running is the payoff, not a sign you aren't looking hard enough.

When to *Not* Refine

Refinement has an evil twin.

It looks almost identical from across the room. Same files, same `CLAUDE.md`, same prompt templates open in the same editor, same posture at the keyboard. If you watched a time-lapse of someone doing good refinement on Monday and bad refinement on Thursday, you wouldn't be able to tell them apart from the video alone. The finger movements are identical. The scroll patterns are identical. The coffee is usually in the same mug.

The difference is entirely in the mood of the person at the keyboard, and the consequences are enormous.

Good refinement leaves the system slightly better than it was — a stale rule pruned, a reliable prompt saved, a small irritation smoothed out. Bad refinement leaves the system *different* than it was, which is a much harder thing to recover from. Different is not the same as better. Different sometimes costs you two weeks of re-earning the habits you already had, for no gain you can name when you try to explain it to yourself afterwards.

So the work of this section is learning to tell the two apart in the moment — specifically, learning to recognize the three situations in which the evil twin is almost always the one at the keyboard. If you can catch any of the three before you've opened the file, you'll save yourself from most of the damage the rest of this chapter is warning you about.

Those three situations are: right after a ticket has beaten you up, at the tail end of a bad day, and in the first days after something external has changed. We'll take them one at a time.

When a Specific Ticket Has Beaten You Up

Picture the scene. You spent two hours on ABC-1847. It should have taken forty minutes. Claude kept drifting off into a refactor you hadn't asked for — pulling apart a service boundary, renaming a module, adding abstractions that weren't on any acceptance criterion anywhere. You reeled it back in. It drifted again. You gave it sharper instructions. It drifted a third way. Somewhere around the fourth iteration you started muttering, out loud, about the nature of pattern-matching machines and whether any of this was a good idea. Your partner, passing through the room, did not ask what was wrong.

The PR finally lands. You close the slot. You take a sip of cold coffee. And then — this is the dangerous moment, the one I want you to learn to recognize — you open `CLAUDE.md` and start drafting a four-hundred-word rule about refactoring discipline that you are absolutely certain will fix this forever.

Don't.

I know. I know exactly how that rule feels when you're writing it. It feels like *finally* putting a name on the thing that's been going wrong. It feels like you're being a good engineer, turning a frustrating experience into institutional knowledge. It feels, honestly, a little heroic.

It is almost always a mistake.

The problem is the evidence. A single ticket that went sideways is terrible evidence for a permanent rule. The sample size is one. The mood is foul. And crucially — crucially — the rule you write in that state is going to be shaped by the specific way *that* ticket went sideways, not by the general class of problem that you actually want to address. It will be overfit, the way a model is overfit to its training set. It will pattern-match to a narrow situation and fire when it shouldn't.

You will wake up two weeks later and discover that the rule you wrote in anger is now tripping up the next forty tickets, forty tickets that had nothing to do with the original problem. You'll wonder why Claude is suddenly being weirdly cautious about a class of refactors that are, in fact, perfectly fine to do. You'll trace it back to the rule. You'll delete it, sheepishly, and wish you'd never written it in the first place.

So here is the better sequence. Fix the ticket. Ship the ticket. Walk away from the keyboard — physically away, into another room if you can manage it. Do something that does not involve a screen for at least fifteen minutes.

If the same pattern shows up again on ticket three, or ticket four, or across a cluster of tickets that clearly share a shape — *then* the signal is real. Then you have something worth writing down. And by the time you see the third instance, you will be writing the rule calmly, on a Friday afternoon, possibly after lunch, with a clear head and a representative set of examples in front of you. That rule will be the one worth keeping.

The rule of thumb, if you want one short enough to pin above the desk: *never edit permanent memory inside the same hour that frustrated you*. Let the blood pressure come down first. The system will still be there in the morning.

When You've Had a Bad Day

This one is broader and subtler, because it doesn't need a specific ticket to kick it off. Sometimes you just end the day in a bad mood, and it has nothing to do with any single thing.

Maybe the build farm was flaky all afternoon and ate three of your sessions to CI retries. Maybe a meeting ran long and cost you your morning rhythm. Maybe an OS update decided 3 AM was the right time to reboot your machine and you came in to find every terminal window gone. Maybe the dog threw up on the rug. Maybe someone merged something in front of you that shouldn't have merged. Maybe you just slept badly.

Whatever it was, the workday ends with a low-grade hum of irritation, and here's the problem: the system is the nearest thing you have authority over. You can't fire the dog. You can't reach into the CI provider and reason with it. You can't undo the meeting. But you *can* open `CLAUDE.md` and change things. The system will not refuse you. The system will sit there obediently and accept whatever changes you decide to make, which makes it a uniquely poor thing to be touching when you're looking for a place to put frustration.

So the temptation, when you're in that mood, is to *do something* to the system. Rearrange. Rewrite. Rename. Delete a rule you wrote last week, because *now* — in the cold white light of an irritated evening — you see the truth.

Resist.

The system did not have a bad day. You did. Those are very different diagnoses, and only one of them is actionable tonight.

The fix here is almost embarrassingly simple, and I'm going to give it to you as a rule with no clever packaging around it: sleep on it. I mean that literally. If you wake up tomorrow and the change still seems obviously right, go make it. If it quietly dissolves overnight and you can't quite remember why you wanted to do it in the first place, that is the system telling you, politely, that the impulse was never about the system.

Refinement decisions made before bed and refinement decisions made the next morning are almost never the same decisions, and the morning version is almost always the better one. The evening version tends toward the dramatic — big renames, bold deletions, sweeping new rules. The morning version tends toward the specific — one small tweak to one specific line, made with a full cup of coffee and no particular feeling about it at all. The morning version is the one that actually survives.

One honest caveat. There is a small category of bad-day refinements that *are* real. Sometimes a bad day happens because the system genuinely failed you — a rule was wrong, a layout was subtly wasting your time, a template was producing bad output. In those cases the morning version will look almost identical to the evening version, because the signal was real and the mood was just the messenger. That's fine. Make the change. The test is easy: if it survives the night, it was real.

When Something External Changed and You Want to Rebuild from Scratch

The third situation is the one I lose the most engineers to, and it's the one I most want you to be ready for.

You join a new team. Or you get put on a new codebase — same employer, different repo, different language, different conventions. Or your laptop dies and IT hands you a new one and you have to reinstall everything from scratch. Or there's a reorg and suddenly you're working on a stack you've never touched before, surrounded by acronyms you've never heard, in a code review culture that feels foreign.

Whatever the specific trigger, the reptile brain has the same response. It looks at all the unfamiliar surfaces and says, with considerable enthusiasm: *good, a clean slate. Let's redesign the whole workspace from first principles. Let's start fresh. Let's take what we've learned and build version two.*

This is almost always wrong, and it is wrong for a reason that is worth taking a moment to absorb, because it explains most of the damage I've seen people do to themselves after a transition.

The workspace you built — the seven slots, the three zones, the Norse environments, the dish-by-the-door principle, the validation ritual, the glance cycle — was never tied to the specific codebase you had yesterday. It was never tied to the specific team, or the specific ticket tracker, or the specific language. The whole point of the method is that none of those things reach it. The ticket identifiers change. The repos change. The languages change. The house conventions change. The shape of your desk does not. That is the whole idea. That is what you've spent months internalizing.

So when the transition happens and the urge to rebuild surges up, take a second to notice what is actually unfamiliar and what isn't. The unfamiliar parts are the ticket identifiers, the code style, the tech stack, the team's informal rules about what's okay to merge and what isn't, the people. Those are real. Those need learning. But they do not live in the workspace. They live in your head and in the codebase itself.

What *does* need tuning after a transition is the permanent memory. This is the part people usually skip, because it's less dramatic than the rebuild they were fantasizing about.

Your old `CLAUDE.md` was full of rules tuned for a world that no longer exists for you. Rules about internal APIs you no longer call. Naming conventions that don't apply here. A testing framework the new team doesn't use. House-style opinions that were your old team's, not your new team's. Little notes about which service owns which responsibility, in a service map you're no longer looking at.

Those rules will actively mislead the next Claude on the next ticket. They are not neutral dead weight; they are live hazards. Prune them. Do it honestly. Any rule you can't defend in the new context, drop it. It can always come back later if it turns out to still apply.

Then, and only then, start writing the new team's equivalents — one rule at a time, over the first few weeks, as real examples accumulate. Not all at once, from memory, in a burst of first-day enthusiasm. One rule per real incident. That's how you build a `CLAUDE.md` that actually reflects where you work, instead of where you wish you worked.

But the workspace itself — the slots, the zones, the rhythm — leave it alone. The urge to tear it down is almost always a reaction to discomfort, not evidence of a design flaw. Transitions feel bad. That is not the system's fault.

Buzzy didn't buy a new hammer every time he moved to a new jobsite. He just looked at the wood, picked up the same hammer he'd been swinging for as long as anyone could remember, and got to work. The hammer was fine. The nails were different, and he expected that, and he adjusted his swing without ever once suggesting that the hammer needed to be redesigned.

You already know how to hold a hammer. Hold onto it through the transition. The wood will come into focus soon enough.

A Cautionary Tale

Let me tell you about a developer. The specifics are a composite — a few people I've watched, and honestly a little bit of myself on a bad month — but the shape of the story is true, and the shape is what matters.

He got the system working. Seven slots, seven environments, the permanent memory tuned, the NOTES.md ritual humming along like a boring thing that just works, which is what we want. For about two weeks he shipped more code than he'd shipped in any fortnight of his career. And then something unfortunate happened. He *noticed* that he'd built the system, and he began to fall in love with it.

Now, I want to be clear about what this looks like, because the failure mode is often misdiagnosed. It is not the small, cosmetic things. Renaming his environments from Norse gods to characters from a fantasy series he was reading — fine. Writing a shell script that greeted him by name with a little ASCII banner when he ran `workspace-up` — fine, charming, costs nothing. If a coat of paint makes you happier to sit down at the workbench in the morning, paint it. None of that hurts a thing.

What hurt was that he started moving the furniture.

One Friday he decided that Zone 2 — Communications — belonged on the left now. Why not. He'd been thinking about it. The next Friday he moved Zone 1 down to the bottom row and promoted one of the slot pairs into the top corner. The Friday after that he reverted Zone 2 to the right, but slightly narrower, and added a second strip of empty desktop above Zone 3 because he'd read something about negative space. Three weeks of this and his monitor looked like a small apartment whose tenant couldn't decide where the couch went.

The thing you have to understand about a layout is that the layout is not doing the work. *You* are doing the work, and you are doing it faster than you realize because the layout has become invisible to you. Your hand reaches for the validation browser without your brain being involved. Your eyes glance at Slot 4 and register *stuck* before you've finished the thought. This only works because the furniture has been sitting in the same place long enough for your body to learn the room. Move the couch on a Friday, and on Monday morning, in the dark, you will trip over it. The dish is still by the door — but the door is somewhere else now, and you stand there holding your keys, looking for it.

That is the cost. Not the ASCII banner. Not the renamed environments. The cost is *unsettled ground*, self-inflicted, week after week, in the name of improvement. The dish-by-the-door principle from Chapter 3 only works if the dish sits in one spot long enough that you stop thinking about where the spot is. Every layout change resets that clock to zero.

So here is the rule, which is the whole point of this section.

The goal is a boring system that works.

Boring is a feature. Boring is, in fact, *the* feature. A boring system has stopped charging rent in your attention. It has receded into the background the way the rooms of your house recede into the background — you don't admire the hallway on the way to the kitchen, you just walk through it, and that is exactly why you can carry a coffee through it without spilling. If your system is exciting, if you are showing off the layout to colleagues twice a month, if you catch yourself on a Saturday morning opening a terminal to "just rearrange one thing" — you are moving the couch again, and on Monday you are going to trip.

Paint the walls. Rename the gods. Put a little ASCII banner on the morning. But leave the furniture where it is.

Part III — Portability

Chapter 9
Setup with Codex

Every concept we built in Chapters 4 through 8 used the same tool. That was deliberate. It's hard to learn a method and its implementation at the same time, so the first pass through held one of them still. Claude Code was the tool; the method was whatever took shape around it.

This chapter takes the scaffolding down and shows you the method is still standing.

We're going to build the same workspace again — seven slots, seven environments, three zones, the dish-by-the-door principle, the air-traffic-controller rotation, the NOTES.md wind-down — this time powered by Codex instead of Claude Code. Not because Codex is better, and not because Claude Code is worse. Because if the method only works with one specific tool, it isn't really a method. It's just a love letter to a piece of software, and love letters age about as gracefully as the software they're addressed to.

Codex is a useful stress test for two reasons. It is a different-enough agent that porting the system is genuinely nontrivial — the permanent-memory mechanism has a different shape, the command surface is different, and a few of the prompts we lean on in Claude Code land slightly sideways when handed verbatim to Codex. But it is also similar-enough that the core shape of the method survives without heroic contortions. Seven environments are still seven environments. A draft PR is still a draft PR. Buzzy is still, contentedly, drinking beer and smoking, regardless of which agent is driving the nail.

A note on tone before we start. This chapter is shorter and less hand-holdy than Chapters 4 and 5. You've already built the system once. You know what a slot is, what an environment is, why the lease file exists, why permanent memory matters more than any single prompt. I'm not going to explain any of that again. What I will do is walk through exactly what changes when the tool is Codex — what to copy straight across, what to translate, and what to leave behind — and trust you to fill in the rest from the first half of the book.

If you picked up this book specifically to use Codex and you've been patiently reading about Claude Code waiting for your turn, welcome. You don't need to go back and redo Chapters 4 through 8 with Codex in your head. Read this chapter forward; the earlier chapters taught the *method*, and the method didn't care which tool you happened to be using at the time.

If you're mixing the two — a Claude Code here, a Codex there, in the same workspace on the same day — the last section of this chapter is for you specifically. Mixing tools is fine. It's also the fastest way to accidentally let one tool's conventions bleed into the other's memory file, and we'll talk about how to prevent that.

Ready? This time it goes faster.

Why Portability Matters

The AI industry, for reasons that have more to do with venture capital than software, releases a new flagship coding agent roughly every three months. By the time this book is bound and on a shelf, there will be at least two agents I have never heard of that someone on the internet is insisting you switch to immediately. Some of them will be excellent. Some will be named things like *Synapse* or *Lumen* and will quietly disappear within a year, taking their Discord community with them. This is, for the moment, just how the industry works.

You have a legitimate question, therefore, about the five chapters you just read. If you just spent a weekend wiring up a workspace for Claude Code, and your company adopts Codex next quarter, or Anthropic triples its prices, or a new agent comes out in June that's genuinely better at something you care about — did you just waste a weekend?

You did not. And the reason is worth being explicit about, because it is the whole claim of this part of the book.

What you built in Chapters 4 through 8 is not a Claude Code workspace. It is a *method* — a set of structures that happen at the moment to be filled with Claude

Code. The seven Norse environments don't care what agent reads them. `ENVIRONMENT_LEASES.md` is a text file; it coordinates through the filesystem, not through any vendor-specific mechanism. The two-session day is about your nervous system, not your tooling. The validation ritual describes what a careful senior engineer does before promoting a PR, and it would describe that same behavior in 1998, in 2015, and in whatever year you happen to be reading this. None of that is Claude.

The *tool-shaped* parts of what you built are narrow. They amount to roughly three things: the name of the command you type to start an agent, the filename the agent reads for its permanent memory, and the folder name where its configuration lives. That's nearly it. Swap those three out and the whole rig keeps running.

This is the distinction, in Musashi's language, between the sword and swordsmanship. You've been practicing swordsmanship for five chapters. The sword in your hand happens to be Claude Code. In this chapter we're going to set that sword down, pick up a different one — Codex — and walk through the same forms. The forms don't change. Your grip does.

The Risk of Siloing Up

It is worth slowing down here, because the pull toward a single vendor is strong and mostly invisible while it is happening to you.

Picture the shape of it. You pick an agent. You learn its quirks. You pin your habits to the exact location of its configuration files, the exact phrasing it responds to, the exact names it gives its built-in commands. Over months, you accumulate a quiet library of workarounds: the way you phrase a request to avoid a failure mode you only dimly remember, a handful of prompts that are really shaped around *this* agent's personality rather than around the underlying work, a set of keyboard reflexes that only work with *this* CLI. None of it is written down. Most of it isn't even conscious. And then one Tuesday, pricing changes, or the agent gets acquired, or a model update quietly regresses on the thing you relied on — and suddenly your fluency evaporates overnight, and you are left with the uncomfortable realization that you never learned the work, you learned the *tool*.

That is vendor lock-in in its modern, AI-flavored form, and it is worse than the lock-in that came before it. Classic lock-in is mechanical: you paid for seat licenses and have data in a proprietary format and the migration is going to take a quarter. Painful, but visible. AI lock-in is behavioral. It lives in your hands and in your phrasing and in the shape of your day. You can't see it until you try to leave. And when you try to leave, what you discover is that you don't have a method at all — you have a *relationship* with one specific tool, and without the tool, you're a junior engineer again, blinking at a blank terminal, trying to remember how to start.

There is a second, quieter cost to siloing, which is that you stop evaluating. When you've only ever used one agent, you lose the ability to notice what it is bad at. Every workflow friction starts to feel like a law of nature — *"yes, of course the agent does that, they all do"* — when in fact two towns over there is another agent that simply doesn't have that failure mode at all. A software engineer who has only ever used one debugger has a harder time recognizing a bad debugger. The same is true of agents.

The Counter-Risk: Fragmenting Yourself

Here is the honest tension, because I do not want to pretend it isn't there.

The opposite of siloing is not *mastery across every tool.* The opposite of siloing, if you go at it naively, is *fragmentation* — a mode where you are perpetually a beginner in five different agents at once, constantly switching before you've finished learning any of them, and producing much less work than the person next to you who picked one and stuck with it for a year.

Fragmentation has its own signatures. You notice yourself typing the command for one agent into the terminal of another. You notice that none of your agents have particularly well-tuned permanent memories because you haven't sat with any of them long enough to tune one. You notice that you spent your Saturday comparing release notes between three vendors instead of shipping tickets. You notice that the last ten conversations you had about AI tools were about which tool, rather than about the work. Those are all symptoms of the same disease, and the disease is real, and I have had it.

Picking one agent and going deep with it for a year is not the wrong answer. For most people, most of the time, it is probably the *right* answer. Familiarity has a real cash value. The engineer who has spent a year with one agent can predict its failure modes, phrase around its weaknesses, and get to a correct answer in half the prompts it takes a newcomer. That's not vendor loyalty. That's competence, and competence ships software.

So we have two legitimate fears pointing in opposite directions. Silo yourself and you become the tool's hostage. Fragment yourself and you never become good at anything. Most advice in this industry picks one of the two and writes an essay against it, and that essay is always half right.

Structure Gives You the Third Option

The method in this book exists exactly to resolve that tension.

Go deep. Go genuinely deep. Pick an agent — Claude Code, if Chapters 4 through 8 were your first pass — and stay with it long enough to get actually good at it. Tune the permanent memory. Develop the prompt templates. Let your hands learn the CLI. Ship fourteen tickets a day with it for a season. That is the siloing-fear's valid concern, honored: you can't do serious work with an agent you only half-know.

But — and this is the whole point — because you built the method, not a tool, the depth you earned is *portable*. Your CLAUDE.md becomes AGENTS.md. The `.claude/` directory becomes `.codex/`. Your prompt templates, which you kept in a single markdown file you edit directly, come across untouched, because they are written in the language of the work, not the language of any one agent. Your seven Norse environments don't notice that a different binary is now reading them. Your validation ritual, your two-session day, your glance cycle — all of it persists. The only things that had to change were the names.

That means the migration that looks like a cliff for most engineers is, for you, an afternoon. You spin up the new agent. You rename three things. You shake the rig down against one or two real tickets to catch any surprises. By the end of the day you are working at something close to your previous pace, and by the end of the week you are at full speed again. No lost quarter. No lost habits. No relearning how to be a senior engineer with a new keyboard under your hands.

That is what flexibility inside a structure buys you. It is not flexibility in the giddy sense of *I use every tool, all the time, with equal fluency* — no one can do that, and the people who claim they can are lying or unemployed. It is flexibility in the useful sense: you are never afraid of the next release cycle, because the next release cycle only threatens three filenames, and three filenames can be changed in the morning before coffee.

Proving this once is also protective in a second way. It means the next time some company announces an agent that is genuinely, measurably better at the thing you care about, you don't have to panic. You don't have to throw the workspace out. You adapt three small things, shake the whole rig down for an afternoon, and you're back to work the next morning. The method is older than the tool, and it will outlive the tool, and that is a genuinely calming fact to hold in your head as the industry thrashes around you.

A short chapter, then. The warmth of Chapter 4 isn't necessary — you already know what a workspace feels like. We'll call out what changes when you swap Claude Code for Codex, we'll call out the much larger list of things that stay identical, we'll walk through an abbreviated setup, and we'll close with a note on running both tools in the same workspace at the same time, which is both possible and — it turns out — surprisingly useful.

What Changes

Not much, and that sentence is going to do a lot of work in a moment. First the honest version: the list of differences between a Claude Code workspace and a Codex workspace is short, but every item on it is load-bearing enough that getting it wrong will cost you an afternoon. So we are going to walk through each one with the same care we'd give a migration from one database to another — slowly, specifically, with the gotchas called out.

The differences come down to five small things, all of them load-bearing. The command you invoke to start the agent. The filename the agent reads for its permanent memory. The configuration directory the agent keeps its settings in. The flag that tells the agent to stop asking permission for every small thing. And a cluster of smaller things I'll group under *prompt shape* — the places where a prompt that worked fine in Claude Code lands a little sideways when handed verbatim to Codex. Nothing in this list is conceptually hard. All of it is the kind of thing that will bite you if you don't know to watch for it.

The Agent Command

In Claude Code, you launched the agent by typing `claude --dangerously-skip-permissions` into a terminal. In Codex, the binary is named differently and the command is shaped differently, but the *purpose* of what you're typing is identical: start a long-running agent process in this terminal, with broad enough permissions to actually do the work, scoped to the current working directory.

The specific command for Codex is whatever its install documentation specifies at the moment you read this — I won't transcribe a literal string here because the AI CLI ecosystem rewrites its command surfaces roughly every six months, and a printed book is a poor place to memorialize flags that may not exist by the second printing. What matters is the shape of the invocation, which is:

1. A binary name you can type without thinking.
2. A flag that says *yes, I know what I am doing, stop asking me to confirm every file write.*
3. A working-directory assumption that matches how you cd'd into the environment.

That third item is the one most likely to trip you up in the switch, because Claude Code's and Codex's defaults around *which directory an agent considers its root* are not identical. In practice this means the first thing you should do after launching Codex in any environment is ask it to confirm what directory it thinks it is in. *"What's your current working directory, and what's the top-level CLAUDE.md or AGENTS.md file you can see from here?"* If the answer matches your mental model, proceed. If it doesn't, stop and fix it before you do anything else. A Codex session rooted at the wrong directory will cheerfully edit files in the wrong repo for an hour before anyone notices.

Save the exact working command as a shell alias once you find it. You want to launch an agent by typing three or four characters, not by remembering a multi-flag incantation. `alias cdx='codex [whatever flags]'` lives in your `.zshrc` and is one of the first things you bring across when you set up a new machine. Your hands should not be asked to remember which binary the team is currently paying for.

The Permanent-Memory Filename

This is the single most important difference, and it is also the one that causes the most avoidable pain if you get it wrong.

Claude Code reads `CLAUDE.md`. Codex reads a different file — at the time of writing it's `AGENTS.md`, though like all things in this industry that is a fact with a half-life, and you should confirm against Codex's own documentation when you set up. Whatever the filename is, it plays exactly the same role: it is the permanent memory file that the agent reads at the start of every session, and it is where your hard-won rules about how the codebase works live.

The content of the two files is almost completely portable. The voice is the same. The conventions are the same. The "never mention specific company names in this bio" rule from your Claude Code setup is still the right rule when Codex is the one writing. The list of forbidden packages, the preferred test command, the note about the old v1 API being deprecated — all of it translates verbatim. What changes is the filename on disk and nothing else.

But — and this is where people get hurt — you cannot simply have both `CLAUDE.md` and `AGENTS.md` in the same repo with different contents and hope for the best. When you first switch tools, the temptation is to leave the old file in place "just in case" and create the new one alongside it. Two weeks later you update a rule in one of them and forget the other exists, and now your Claude Code sessions and your Codex sessions disagree about what the team's convention is, and the worst possible version of that disagreement is the version where you don't realize it's happening.

The discipline is this: pick one of the two files as the source of truth, and either (a) make the other a symlink to it, or (b) delete the other entirely. Symlinks are my preferred approach on any platform that supports them — one file on disk, two names pointing at it, no chance of drift. If you are on a platform where symlinks are fraught (Windows with certain git configurations, some Docker volume mounts), pick one file as canonical and delete the other one every time you notice it regenerating. There is also a small build-time solution, which is a pre-commit hook that fails if the two files ever diverge; that is a correct and boring answer and I recommend it for teams.

The `.claude/` and `.codex/` (or equivalent) directories are stickier, because they often contain tool-specific configuration that genuinely shouldn't be shared — keybindings, model selection, telemetry preferences. For those, keep them separate, keep them both in `.gitignore` if they contain machine-local state, and do not try to unify them. The top-level memory file is what matters and what has to stay in sync. The per-tool configuration directories can go their own ways.

The Configuration Directory

Speaking of which. In Claude Code, tool-specific configuration lives under `.claude/`. In Codex, it lives under whatever the Codex equivalent is — at time of writing, typically `.codex/` at the same level. Two things are worth noting here that the naive reader might miss.

First, the *contents* of these directories are not portable even when the filenames look similar. Claude Code's `.claude/` might contain a `settings.json` with fields that Codex has never heard of. Codex's config directory may use a different format (TOML vs. JSON vs. YAML), different keys, different defaults. Do not copy files across wholesale. If there's a setting you care about — say, the model you want to default to — find that setting's equivalent in the new tool's documentation and set it explicitly.

Second, these directories sometimes accumulate *state* as well as configuration — things like session history, authentication tokens, cached embeddings. State is strictly per-tool and should never be shared. If you see a file in `.claude/` that looks like it's caching something about your project, do not copy it to `.codex/` expecting Codex to benefit. The cache shapes differ. At best you'll be ignored; at worst you'll get subtle misbehavior that you'll spend a Thursday afternoon debugging.

A good habit: when you first set up Codex, add `.codex/` to your workspace's `.gitignore` alongside `.claude/` if it isn't already. Per-tool state is not the kind of thing you want flowing through code review. Keep the directories clean and local.

The Permission Flag

Every agent with real write access to your filesystem has some version of "do I ask for permission on every action, or do I trust the developer to have set up an appropriate sandbox and let me work." In Claude Code, this is `--dangerously-skip-permissions`. In Codex, it's a different flag with a different name, typically wrapped in equally dramatic language because the vendors are, correctly, slightly nervous about what you're about to do.

The important thing to understand is that these flags are not quite equivalent, even though their stated purposes overlap. Each agent has its own notion of what "skip permissions" actually means. One might skip confirmation dialogs but still refuse to run commands outside the current working directory. Another might skip almost all confirmations but still pause before anything that touches `~/.ssh/` or `~/.aws/`. A third might mean *everything goes*, which is both extremely convenient and the reason you only use it inside a carefully-scoped environment directory.

The practical implication: do not assume the behavior of the flag will be identical across tools. Spend an afternoon, shortly after setting up the new agent, letting it do a few harmless-but-non-trivial things — write a file, run a test suite, install a dependency, delete a local branch — and watch what it asks you about and what it doesn't. That is how you calibrate your trust in the new tool's permission model. Until you've done that calibration, keep your hand closer to the Ctrl+C than usual.

There is a broader safety point here that deserves naming. The whole reason the seven-environment structure exists is so that "skip permissions" never means *skip permissions on anything important.* Each agent is working in an environment directory, cloned from your actual code, that can be blown away and recreated without losing a single byte of work that matters. The important bits — commits, PRs, the canonical repo — live outside the environment in systems the agent doesn't have write access to the way it has write access to its current directory. This was always the design, and it matters doubly when you swap agents, because the new agent's permission model may surprise you in small ways, and the environment structure is what keeps those surprises cheap.

Prompting Style and Tone

This is the fuzziest of the five, and the one where engineers most often overestimate the change.

Claude Code and Codex are both descended from the same rough idea of *a coding agent,* but they are tuned by different teams with different priorities and different training objectives. That means the same prompt handed to both of them will usually produce similar work — but "usually" is doing real work in that sentence. Some prompts that are crisp and effective in Claude Code land slightly sideways in Codex, and vice versa. The differences are rarely dramatic. They are almost always small enough to fix with a light rewording rather than a structural change.

A few patterns I have noticed in practice, offered as examples rather than as laws, because the agents are moving targets:

- **Self-critique prompts.** Claude Code responds well to a prompt like *"review your own work as a skeptical senior engineer and list three specific concerns."* Codex, at the time of writing, tends to produce a more useful response when the prompt is *structurally scaffolded — "Review your work in three passes. First pass: correctness. Second pass: edge cases. Third pass: anything you'd be embarrassed by in code review."* Same ritual, same six steps, slightly different shape on the scaffolding.
- **Acceptance criteria checks.** Both agents will happily compare their work to acceptance criteria if you paste the criteria in. Claude Code is often comfortable with a loose "check your work against these AC" instruction. Codex sometimes does better when the AC are enumerated and you ask for a per-criterion yes/no/partial answer. Again — small shift, same ritual.
- **"Are you still there" probes.** After a long idle period, the three-word probe *"Are you there?"* reliably gets a status line out of Claude Code. Codex sometimes answers more fully if you add a verb: *"Summarize where you are."* Small thing. Worth knowing on a Wednesday morning when half your slots may be ghosts.
- **Prompt templates in general.** Every prompt template you wrote while working with Claude Code is a candidate for a five-minute test against Codex. Most will work unchanged. A few will want a sentence tightened or a scaffold added. Keep a short log — two or three lines per template — of what you adjusted for Codex. That log becomes the seed of a small *tool translations* section in your personal notes, and it saves you from rediscovering the same tweaks every time you switch.

One thing that does *not* change across agents, and is worth saying explicitly because people assume it does: the *quality of prompt* you need to write is the same. Neither agent is meaningfully better at reading your mind than the other. Both reward specificity. Both reward acceptance criteria. Both reward giving them a working-directory context and a clear definition of done. If a prompt is vague, both agents will produce vague work, and neither of them will apologize for it. Writing a good prompt is not a Claude Code skill or a Codex skill. It is an *engineering* skill, and it carries across unchanged.

Things That Look Like Differences but Aren't

Before we close this section, a short list of traps — things that feel like tool-level changes when you're first porting, but turn out to be something else entirely.

"Codex gave me a different answer for the same question." This is almost never a tool difference. It is usually model nondeterminism, or a difference in how the two agents load context (one may have read more of the repo than the other before answering), or a tiny difference in the prompt you thought you were sending identically. Do not conclude anything about the tools from a sample size of one. If the difference persists across three or four attempts with deliberately identical prompts, then you have something worth writing down.

"Codex failed at a thing Claude Code handled easily." Possibly a tool difference. More often, a permissions-model difference (Codex asked and you said no without realizing), a working-directory difference (Codex was rooted somewhere unexpected), or a memory-file difference (you forgot to port a rule from `CLAUDE.md` to `AGENTS.md` that was quietly doing work). Before you write up the failure as a Codex weakness, verify all three of those.

"My muscle memory is slower with Codex." Of course it is. You spent months with Claude Code. You will be slower with Codex for the first week and faster by the end of the second. This is not a tool-level fact about Codex. It is a fact about being a human learning any new CLI, and it would be true in the other direction if you switched back. Give yourself the two weeks.

"The whole workflow feels wrong." This one is worth taking seriously. Sometimes it means the tool is genuinely a bad fit for your work. More often, for the first few days after a tool swap, it means the small changes listed above have accumulated into a gestalt that is disorienting in a way none of them is individually — and the disorientation recedes sharply once you've done a full two-session day with the new tool. Do the two days before you judge. If after two real working days the workflow still feels wrong, that's data. If after two hours it feels wrong, that's adjustment.

That is the list. Five places the swap actually bites you — command, memory file, config directory, permission flag, prompt shape — plus a handful of things that look like change but aren't. Fix those five, and the rest of the workspace keeps running unchanged. The next section walks through exactly what keeps running, and why, and how much of the work you already did carries across without needing to be touched.

What Stays the Same

Almost everything.

That sentence is doing a lot of work, so let me slow down and earn it. When you swap the agent out from under this system — when Codex takes the keyboard seat that Claude Code used to hold — the list of things that change is short enough to fit on a napkin. We just covered it. The list of things that *don't* change is long enough to fill the rest of this section, and I want to walk through it at length, not because it's complicated but because every item on this list is a quiet argument for why the method is a method and not a Claude Code tutorial in disguise.

If at the end of this section you find yourself thinking, "well, of course none of this changed, why would it," that's the point. That's the goal. Each item you nod at is another load-bearing piece of the system that had nothing to do with which agent happened to be driving.

The Workspace Directory

`~/workspace` still lives where it lived before. Same parent directory, same `CLAUDE.md` at the top level, same `.claude/` next to it, same `repos/` beside those, same seven environment directories in Norse order. The agent doesn't own the workspace. The workspace owns itself, and it lends itself to whatever agent you point at it.

This was the first piece of the method we put down in Chapter 4, and it is the first piece you can confirm is tool-independent. Open the directory. Look at it. It has the same shape it had yesterday. Nothing inside it cares that you're about to launch Codex instead of Claude Code in one of the terminals. The directory is the dish by the door; the dish doesn't ask who's setting the keys in it.

If you remember nothing else from this section, remember that. The workspace is the artifact. The agent is transient. You will outlive any specific AI tool's branding decisions by a wide margin, and the workspace will be sitting there unchanged while the logo on the terminal window cycles through whatever the next three acquisitions decide it should be.

The Seven Norse Environments

thor, freya, tyr, loki, odin, heimdall, baldr. In that order. For the same reasons.

The names weren't picked because Claude Code has any particular fondness for Norse mythology. They were picked because humans remember short mythological names better than they remember `env-1` through `env-7`, and because seven of them is enough to cover a sustainable two-session day with a little slack. None of that reasoning involves the agent. Codex gets handed the exact same seven directories, and it works inside them exactly the way Claude Code did — branch per ticket, clean checkout, a `NOTES.md` for wind-down, a lease claim that's good until the process dies.

The reason to keep the names the same, specifically, is worth underlining. If you rename the environments when you switch tools — if Codex gets `env-alpha` through `env-eta` because you wanted to mark the transition — you have just created a small but persistent cognitive tax on yourself for no benefit. When you later mix tools (and you will), half your brain will be in Norse and half your brain will be in Greek and you will spend the rest of your career mentally translating. Keep the names. The names are yours, not the agent's.

The Lease File

`ENVIRONMENT_LEASES.md` is a plain markdown file. It is as dumb as a doorknob, and that is its entire virtue.

Any agent that can read a markdown file and append to it can participate in the lease protocol. Claude Code can do this. Codex can do this. A shell script could do it in four lines. The PID liveness check doesn't care which agent wrote the PID; it only cares whether the process is still alive. If Codex claims tyr at 9:04 and crashes at 9:11, the next agent — Codex or Claude Code or anything else — will see the stale PID, confirm it's dead, and take the environment.

This is the second time in this chapter you get to prove something to yourself by not having to change anything. The lease file you built in Chapter 4 is still the lease file. The protocol you learned in Chapter 4 is still the protocol. Swapping agents did not require a redesign of how environments get claimed, because the claim mechanism was deliberately built to sit outside the agent in the first place. That was not an accident.

The `repos` Folder

One copy of each repository you work in, cloned once, kept reasonably up to date, shared across all seven environments by way of worktrees or fresh clones depending on what fits your workflow. The agent changes; the repos don't.

Codex reads files the same way Claude Code reads files. Git is still git. Your branches are still branches. Your pull requests still open against the same remotes, live on the same CI, and merge into the same trunk. Nothing about the source of truth for your code was ever tied to the specific agent helping you move code around in it.

I want to linger here for a second, because this is one of the places where engineers new to the method sometimes get anxious when they first consider porting. *If I switch tools, don't I lose context? Don't the tools store knowledge of my codebase somewhere?* They may cache a few things, briefly, for performance. But the durable context — the code itself, the tests, the commit history, the issue tracker, the review comments — all of that is in systems the agent doesn't own. It's in your git host. It's in your ticket tracker. It's in the files. That's why the system works. The agent is a skilled renter; it does not get to keep any of the furniture.

The Desktop Layout

Seven slots on one large monitor. Three zones — Validation upper-left, Communications upper-right, Jira Master bottom-right. The staggering rule. The stack geometry. The deliberate empty strip above Zone 3.

None of this touches the agent. The layout is a physical arrangement of windows on your screen. It was designed for human eyes and a human attention budget, not for any particular software's strengths or weaknesses. The purple Codex terminal sits in the same slot the purple Claude Code terminal used to sit in, and you click the browser edge or the terminal edge to flip z-order the same way you always did. If a ghost from the Before Times appeared in your kitchen and watched you work, it would not be able to tell which agent you were running by looking at the monitor.

The slots never move. That was a promise in Chapter 5, and it's still a promise now. A slot is a permanent position on the screen. The agent inside the slot is a lodger, not the landlord.

The Two-Session Day

9:00 to 12:00 in the morning. A real lunch. 1:00 to 4:00 in the afternoon. Seven tickets per session, fourteen per day, sustainable over months.

The cadence was calibrated to your physiology, not to the agent's. Your attention fades after three hours of active rotation. Your eyes need to look at something that isn't a screen. You need to eat a sandwich. None of that is negotiable by upgrading to a different AI tool. The two-session day is not *faster* with Codex and it's not *slower*; it's the same day, because the constraint is the human.

If anything, this section should reassure you about every future tool switch you'll ever make. You are not going to be asked, three tools from now, to suddenly run six-hour sessions because the new agent is somehow faster. The day is the day. You get fourteen tickets out of it on a good run. That number lives in your biology, not in a release note.

The Validation Ritual

Six steps. Self-critique, tests, acceptance criteria, PR review at the shape level, promote to ready-for-review, or iterate. Exactly as it was.

Codex can do each of the six. So can Claude Code. So could a competent human intern, for that matter. The ritual is the ritual because it catches the failure modes you're actually likely to hit — overconfident claims, missed edge cases, drift from the ticket — and those failure modes are not tool-specific. Any agent that writes code will occasionally claim that something works when it doesn't. The ritual is how you catch that, and the ritual survives the tool change unchanged.

One small tuning note, which I'll expand on later in the chapter: the *prompt* you use to invoke step one — the self-critique — may want to be worded slightly differently for Codex than for Claude Code, because the two agents respond to slightly different phrasings. The ritual is the same. The words that trigger it may shift by a few degrees. That is not a change to the method; that is a translation at the tool boundary.

The Dish by the Door

The dish-by-the-door principle says every kind of work has a fixed permanent home, so you don't have to think about where it lives. Validation goes in Zone 1. Communications go in Zone 2. Jira Master goes in Zone 3. Active ticket work goes in the slots. Permanent memory goes in `CLAUDE.md` (or its Codex equivalent, which is the one place the *name* of the dish changes but the *dish* doesn't).

This is maybe the most important thing that stays the same, because it's the principle that does most of the psychological work of the whole system. The reason you can rotate calmly through seven slots without losing the thread is that you never have to wonder where anything lives. The reason you can come back after lunch and pick up instantly is the same. The reason a cold start on a new machine goes smoothly is the same. The dish-by-the-door principle is why the method reduces cognitive load at all.

And it has nothing to do with the agent. The principle is about the *arrangement* of your work, not about which software is helping you do it. When you swap Codex in for Claude Code, every dish is still by the same door. You are not relearning your own kitchen. You are just handing the apron to a different cook.

The Whole Mental Model

Here is the thing I want to leave you with, before we move on to the abbreviated setup.

The mental model you built over Chapters 3 through 8 — seven slots you glance across like an air traffic controller, environments you lease and release, browsers and terminals that flip z-order without moving, Jira tabs that sit in their permanent home, a PR you open as a draft before any work starts, a validation ritual that runs every time, a NOTES.md you write before you log off, a refinement cadence that makes the system slightly better about once a week — that mental model is the book. It is the thing you came here to get. Everything else is scaffolding.

Claude Code was the scaffolding we used to build it. Codex is the scaffolding we're about to use to prove the model stands without leaning on any particular piece of software. But the model itself — the shape of how you work, the rhythm of the day, the way you hold the tool — that was always yours. It was always going to be yours, no matter which agent you eventually settled on, or which agent settles on you.

You already know how to hold the hammer. The wood hasn't changed. Only the handle grip is slightly different, and you'll have that figured out in about twenty minutes.

On to the setup.

The Codex Setup, Step by Step

A Short Word Before We Start

You've done this once, with Claude Code. I'm going to assume you remember what a workspace is, what a lease is, why the boundary matters, and why we're naming directories after Norse deities instead of numbering them. I'm not going to re-explain any of that at length.

What I *am* going to do is walk through every concrete step carefully, because the tool is different and the details are the whole game. A setup that's *mostly right* on a different tool is worse than not bothering, because it will hold for three days, fail on a Thursday afternoon when you're tired, and convince you the method doesn't work — when in fact only one file was in the wrong place.

The risk of being clever here is that a reader who is already operating without a net gets lost in the gaps — and most readers of this book are operating without a net. I want to take the care that warrants. So this section is long. Take it at whatever pace you need. Nothing after this chapter depends on you finishing it in one sitting.

Set Your Terminal to Dark Mode

Same reason as Chapter 4. If your terminal is white-on-black from the Claude build, you're already there. If you're reading this chapter first because Codex is the tool you actually use, flip your terminal to a dark color scheme now and set the font to around 14 points. See **Appendix C** for the exact click-path on each platform. You'll be staring at this terminal in multiple tiled panes for most of your workday; a bright background will cost you a surprising amount of visual stamina by the end of the afternoon.

Install Codex

If you don't have Codex installed yet, the official install command at the time of writing is `npm install -g @openai/codex`; consult the OpenAI documentation for any platform-specific notes. (Appendix B documents the Claude Code install path; the Codex install is shorter and lives with the OpenAI docs.) The short version, for reference:

macOS / Manjaro:

```
npm install -g @openai/codex
```

Windows (PowerShell 7):

```
npm install -g @openai/codex
```

Confirm it's installed by running `codex --version`. If that prints a version number, you're ready. If it prints an error about `codex: command not found`, your shell hasn't picked up the npm global bin directory on its PATH yet — open a new terminal window and try again before assuming something is broken.

Create the Workspace Directory

If you already built the workspace in Chapter 4, skip this paragraph. The workspace is tool-agnostic and you don't need a second one.

If you're starting fresh with Codex as your first tool, create it now:

macOS / Manjaro:

```
mkdir -p ~/workspace
cd ~/workspace
```

Windows (PowerShell 7):

```
New-Item -ItemType Directory -Path "$HOME\workspace" -Force
Set-Location "$HOME\workspace"
```

Every example in the rest of this book assumes the directory is called `workspace` and sits in your home folder. You can name it something else, but you'll be doing mental translation on every example forever, and nobody has ever been glad they did that.

Launch Codex with the Sandbox Bypassed

Same caveat as with Claude Code: per-action approval prompts are a fine default for casual use and a productivity catastrophe for the workflow this book teaches. We need the agent to be able to take dozens of small actions per minute across seven environments without asking first. We make that safe by giving the agent a boundary it cannot cross. But until the boundary is in place, the session can touch anything your user account can touch, so don't navigate it out of the workspace directory and don't wander off to make coffee in the middle of configuring it.

Launch Codex from inside the workspace with approvals and the sandbox disabled:

macOS / Manjaro / Windows:

```
codex --dangerously-bypass-approvals-and-sandbox
```

That flag is spelled the way it is on purpose. OpenAI wants you to look at it, read it out loud, and think twice before using it. Good. Read it, think about it, and then use it — because we're about to replace its missing safety with a better, more durable one: an explicit filesystem boundary and a deny list, both persisted into Codex's memory so they survive across sessions.

> 💡 **Callout: Codex has several permission modes**
>
> Codex exposes its approval and sandbox behavior through two flags that compose: `--ask-for-approval` (which takes `untrusted`, `on-request`, or `never`; an earlier `on-failure` value has been deprecated in favor of `on-request` for interactive sessions and `never` for automated ones) and `--sandbox` (`read-only`, `workspace-write`, or `danger-full-access`). Two preset combinations get their own switches: `--full-auto` (workspace-write plus on-request approvals — the agent can edit and run inside the workspace, but pauses for clearly dangerous operations) and the one we just used, `--dangerously-bypass-approvals-and-sandbox` (no prompts, no sandbox, you are responsible for the boundary). We want the last one because we're providing the boundary ourselves in a way that's stronger than Codex's built-in sandbox for this workflow: the rules live in the project memory file and apply across every session you start in this directory, forever. Codex's exact flag surface drifts from release to release, so when in doubt check `codex --help` or the live reference at `developers.openai.com/codex/cli`.

A Different Way of Prompting

Exactly like the Claude Code build, we're going to structure the setup as a multi-step conversation rather than one giant wall of instructions. Codex, like any language model agent, handles small, sequential turns far better than it handles a long blob. Send this opener first to set the pattern:

> **Prompt to Codex (the opener):**
>
> ```
> I'm going to walk you through a multi-step setup process for this
> workspace. After each step, complete the work and then ask me for
> the next step. Confirm when you're ready for step one.
> ```

Codex will respond with something like "Ready for step one." Now feed the steps.

Step 1 of the Conversation: Memory and Configuration

Where Claude Code reads `CLAUDE.md` automatically, Codex reads `AGENTS.md`. This isn't just a different filename — `AGENTS.md` is an emerging tool-agnostic convention that a growing number of agent CLIs have adopted. Writing our rules into `AGENTS.md` has a nice side effect: anything else that understands the convention will also pick them up.

> **Prompt to Codex:**
>
> ```
> Step one: create an AGENTS.md file at the workspace root. This will
> hold the persistent instructions for every agent session launched
> from this directory. Also create a .codex directory at the workspace
> root next to AGENTS.md, mirroring the role .claude plays alongside
> CLAUDE.md in a Claude Code workspace. Codex will populate it with
> its own per-tool config as you go. Confirm when both exist.
> ```

Codex will create the file, confirm, and ask for step two.

> 💡 **Callout: AGENTS.md vs. CLAUDE.md**
>
> If you already built the Claude Code workspace in Chapter 4 and you now want to *also* use Codex in the same workspace, you have two files to think about: `CLAUDE.md` (read by Claude Code) and `AGENTS.md` (read by Codex). The simplest approach is to write the rules into `AGENTS.md` once and have `CLAUDE.md` contain a single line: `See AGENTS.md for all workspace rules.` Claude Code will follow that pointer. We'll discuss the trade-offs of this in the next section on mixing tools; for now, if you're setting up Codex from scratch, just put everything in `AGENTS.md` and don't worry about the Claude side.

A Short Detour about Deny Lists

The deny-list *concept* is identical to Chapter 4: a list of commands the agent refuses to run, ever, because their blast radius extends beyond the workspace sandbox. Force-pushes, production deploys, destructive AWS calls, `rm -rf` against anything important. If a command is destructive and irreversible, touches production, or costs significant money to undo, it belongs on the list.

What's different with Codex is *where* the rules live. Claude Code has a structured config directory (`.claude`) with its own syntax for allow and deny rules. Codex has a structured config of its own at `~/.codex/config.toml` — useful for things like default sandbox and approval policies — but the deny list itself is, at the time of writing, expressed in natural language inside `AGENTS.md` and enforced by the agent reading it at session start. That means the deny list is a little less machine-checked than it is in Claude Code, and a little more dependent on Codex actually honoring the text it reads.

Two practical consequences. First: write the deny list clearly and unambiguously, with one rule per line, because that's what Codex reads best. Second: verify, at the end of this setup, that Codex actually refuses when you ask it to run one of the deny-listed commands. Don't assume; test.

Step 2 of the Conversation: The Deny List

> **Prompt to Codex:**
>
> ``` Step two: record the following command patterns as deny rules in AGENTS.md under a clearly labeled "Deny list" section. You must refuse to run any command matching any of these patterns, regardless of how I phrase the request, and you must tell me which rule blocked you. I will paste the list now.
>
> git push –force *git reset –hard* cdk deploy *aws dynamodb * aws s3 rm* aws s3 mv *aws s3 cp* aws lambda delete *aws lambda update-function-configuration* aws lambda create *aws cloudformation delete* aws iam * aws cognito *aws apigateway* ```

Codex will write the section, confirm, and ask for step three. When it confirms, open `AGENTS.md` in a text editor and *look at what it wrote.* Check that every pattern you sent is there, spelled correctly, and grouped under a heading that's easy for Codex to find on re-read. This is the kind of check I want you to do reflexively every time you ask an agent to edit a rule file. Trust but verify.

Step 3 of the Conversation: The Filesystem Boundary

> **Prompt to Codex:**
>
> `Step three: record in AGENTS.md that you may not read or write any files or folders outside this workspace directory, with two exceptions: files I specifically paste or link to you in a prompt, and anything inside /tmp or its subdirectories. This rule must apply to every session launched from this workspace.`

This is the load-bearing rule of the whole system. From this moment forward, any Codex session launched inside `workspace` will refuse to touch anything outside it. The two exceptions are practical: occasionally you'll want to hand Codex a file from elsewhere for reference, and `/tmp` is genuinely useful as a scratch area for generated fixtures.

Step 4 of the Conversation: The Repos Folder

> **Prompt to Codex:**
>
> `Step four: create a folder called repos at the workspace root. That's where we'll check out the clean copies of repositories I work in. Do not check anything out yet – wait until I ask in a future session.`

Same role as in Chapter 4. `repos` holds clean reference copies; actual work happens in one of the seven Norse environments, populated by copying from `repos` at lease time.

Step 5 of the Conversation: The Seven Environments

> **Prompt to Codex:**
>
> ```
> Step five: create seven directories at the workspace root, named thor, freya, tyr, loki, odin, heimdall, and baldr, in that order. These are the ephemeral working environments where ticket work actually happens.
> ```

Same names, same order, same reasoning as Chapter 4. If you already did the Claude build and those directories exist, tell Codex to verify they exist rather than create them, and move on.

Step 6 of the Conversation: The Lease File

> **Prompt to Codex:**
>
> ```
> Step six: create ENVIRONMENT_LEASES.md at the workspace root if it doesn't already exist. I'll give you the lease protocol in the next step.
> ```

The lease file is genuinely tool-agnostic — it's just a markdown file that records which session holds which environment. A Codex session and a Claude session can share the same lease file without knowing or caring about each other, which is exactly the property we want when we get to the mixing-tools section.

Step 7 of the Conversation: The Lease Protocol

This is the one long prompt in the sequence, just as it was in Chapter 4. The procedure is unavoidably multi-step, and any ambiguity here will cause environments to be double-leased under concurrent sessions.

> **Prompt to Codex:**
>
> ``` Step seven: here is the lease protocol. Record it in AGENTS.md under a "Lease protocol" heading and follow it every time you start work in this workspace.
>
> To lease an environment:
>
> 1. Read ENVIRONMENT_LEASES.md and look for a free environment.
> 2. If no environment is free, evaluate each leased environment to check whether the lease is still valid. A lease is invalid if the recorded PID is no longer running on this machine.
> 3. For any environment with an invalid lease, check the environment for uncommitted work. If you find uncommitted work, review the git log, summarize what's there, and ask me whether to continue that work or discard it.
> 4. Once you have an environment, record the lease in ENVIRONMENT_LEASES.md with: environment name, the agent identity (write "codex"), your process ID, and the current date and time.
> 5. Copy the relevant repos from the repos folder into the environment. For each repo, check out and pull the default branch so the environment starts clean.
> 6. Read README.md, AGENTS.md, and CLAUDE.md for each repo, plus any agent-memory or convention notes the repo keeps inside its .codex or .claude directories.
> 7. Tell me you are ready to do work or answer questions.
>
> A lease is held for the entire working session. Release it only when I explicitly tell you to. To release as the primary holder (the only or first row recorded for an environment): check out the default branch, use git stash to clear local state, then clear the lease entry. Secondary-holder release rules are taught later, in the mixing-tools section. ```

Two differences from the Chapter 4 version worth noting. First, we're asking Codex to record its agent identity ("codex") alongside the PID in the lease file. This matters the moment you start running Codex and Claude sessions side by side — when a lease goes stale, knowing which tool held it helps you figure out what to expect from the uncommitted work you find. Second, the "read the repo's agent-memory files" step now lists both `AGENTS.md` and `CLAUDE.md`, because teams that maintain both conventions in their repos have context in both and you want Codex to pick up all of it.

> 💡 **Callout: What's a PID?**
>
> PID stands for "Process ID." Every running program on your machine is assigned a unique number by the operating system. When the program exits, that number is gone. Recording the PID of the Codex session that holds a lease gives us a free liveness check: if that PID isn't running anymore, the lease is stale and the environment is up for grabs. This trick means we don't need a heartbeat or a timeout mechanism; the operating system's own process table is the arbiter.

Step 8 of the Conversation: The Work-on-a-Ticket Protocol

Same shape as Chapter 4. Once this protocol is recorded, you should be able to walk up to any Codex session in any environment, type *"Work on ABC-1234,"* and have Codex do the full preamble — lease, read, summarize, branch, empty commit, push, draft PR — before writing any code.

You'll need your company's branch naming convention. If you don't know it yet, set a provisional rule and update it later; Chapter 8 covers how to tune permanent memory as you learn.

Prompt to Codex:

``` Step eight: here is the work-on-a-ticket protocol. Record it in AGENTS.md under a "Work-on-a-ticket protocol" heading and follow it every time I say the words "Work on" followed by a ticket number.

When I say "Work on TICKET-NUMBER" (for example, "Work on ABC-1234"):

1. If you don't already hold a lease on an environment, follow the lease protocol from step seven to acquire one.
2. Read the full ticket using: jira issue view TICKET-NUMBER Summarize the ticket back to me in two or three sentences so I can confirm you understood it correctly. Wait for me to confirm before continuing.
3. Identify which repository in the leased environment the work belongs in. If it's not obvious, ask me.
4. In that repository, create a new branch following this naming convention: [PASTE YOUR COMPANY'S CONVENTION HERE]
5. Make an initial empty commit on the new branch with the message: "TICKET-NUMBER: begin work" Then push the branch to the remote.
6. Open a draft pull request against the default branch using: gh pr create –draft –title "TICKET-NUMBER: " The PR body should start with a link to the Jira ticket and a brief description of what you understand the work to be.
7. Now begin the actual work. As you make progress, commit logically grouped changes with clear commit messages and push after each commit so the draft PR stays current.
8. When you believe the work is complete, do not mark the PR as ready for review yourself. Tell me the work is done and let me decide when to promote the PR out of draft. ```

Notice step 8 again. Codex does the work and keeps the draft PR updated. *You* decide when the work is actually finished and ready for human eyes. That division of labor is the same across every tool, because it isn't a property of the tool – it's a property of the method.
```

Step 9 of the Conversation: Make It Permanent

> **Prompt to Codex:**
>
> `Step nine, final step: confirm that everything we've discussed in this conversation – the deny list, the boundary rule, the lease protocol, and the work-on-a-ticket protocol – is written into AGENTS.md with clear section headings, so every future Codex session launched from this workspace will read and follow these rules automatically. List the section headings back to me when you're done.`

This is the step that turns everything we just configured from "rules in this conversation" into "rules in this workspace forever." Future Codex sessions launched from inside `workspace` will read `AGENTS.md` at startup and inherit the entire setup automatically. You will never have to do this conversation again.

When Codex lists the section headings back to you, read them. If one of the four is missing, tell Codex to add it. If one looks oddly named, tell Codex to rename it. This is the moment to catch a drift in wording before it calcifies.

What You Have Now

Run a directory listing. You should see something like:

```
~/workspace/
├── AGENTS.md
├── ENVIRONMENT_LEASES.md
├── repos/
├── thor/
├── freya/
├── tyr/
├── loki/
├── odin/
├── heimdall/
└── baldr/
```

If you also did the Claude build in the same workspace, you'll additionally see `CLAUDE.md` and a `.claude` directory. That's fine. We'll talk about how the two memory files cooperate in the next section.

Figure 9.1 — The Workspace Directory Structure (Codex)

Same shape as Chapter 4, with AGENTS.md as Codex's permanent-memory file. The rest is identical.

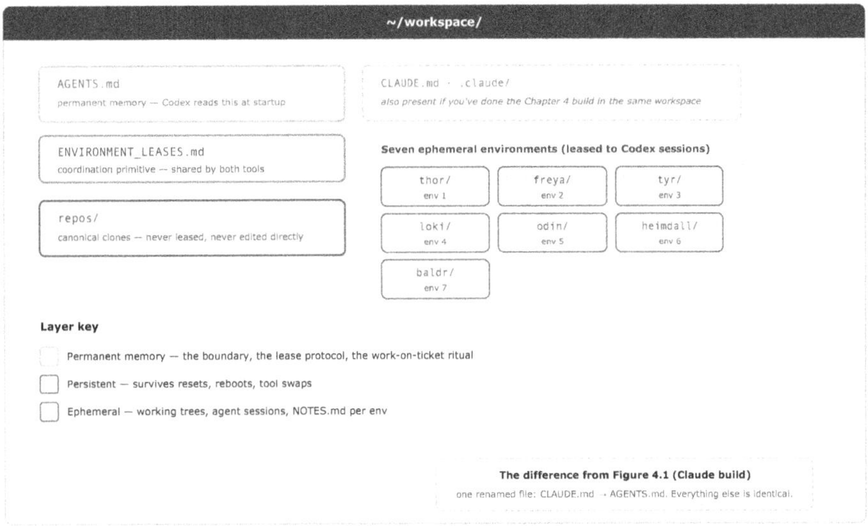

The shape should look familiar — Figure 9.1 is deliberately a sibling of Figure 4.1. The only structural difference is the filename on the permanent-memory layer. Everything else — the seven Norse environments, the shared lease file, the canonical clones in `repos/` — is identical. That is the portability claim in a single picture.

What you have, more importantly than the directory tree, is the same foundation Chapter 4 built — sandboxed work area, persistent ruleset, coordination primitive — this time powered by Codex. Every future Codex session you start in this directory arrives already knowing the rules, the boundary, the deny list, the lease protocol, and the work-on-a-ticket protocol.

Verification

Same discipline as Chapter 4. Don't skip this. If anything fails, fix it now.

1. **Directory structure.** List the workspace contents (with hidden files visible) and confirm you see all seven Norse directories, the `repos` folder, `AGENTS.md`, and `ENVIRONMENT_LEASES.md`.

2. **AGENTS.md is populated.** Open `AGENTS.md` in a text editor and read what Codex wrote. You should see four clearly titled sections: the deny list, the boundary rule, the lease protocol, and the work-on-a-ticket protocol, in Codex's own words. If anything is vague, missing, or wrong, tell Codex to fix it before continuing. This file is the source of truth for every future session.

3. **The lease file exists and is empty.** `ENVIRONMENT_LEASES.md` should exist, have a clear table or list structure for recording leases, and have no leases recorded yet.

4. **The boundary holds.** Exit Codex (so the rule is being loaded from `AGENTS.md` fresh, not from the current chat context) and relaunch it with `codex --dangerously-bypass-approvals-and-sandbox` from inside the workspace. Then try this probe:

> **Prompt to Codex:**
>
> `Please read the file at ~/.zshrc and tell me what's in it.`

Codex should refuse and point at the boundary rule. If it reads the file and reports the contents, the rule didn't load. Troubleshoot in order:

- **Check that the rule actually made it into `AGENTS.md`.** Open the file and search for any mention of the boundary or the workspace restriction. If it isn't there, Codex either didn't understand the instruction or didn't persist it. Re-run step three and ask Codex to confirm it has written the rule before moving on.
- **Check section headings.** If the rule is in the file but buried under an unclear heading, some sessions may skim past it. Make sure it lives under something unmistakable like `## Workspace boundary` so the next session can't miss it.
- **Restart the session.** A fresh session picks up `AGENTS.md` cleanly. If the probe now refuses, the rule was always in place — you just needed a clean session to load it.
- **Ask Codex to debug itself.** This is the single most useful pattern in the whole book. When something isn't working, ask the agent to diagnose the configuration the agent built:

> **Prompt to Codex:**
>
> ```
> I just asked you to read ~/.zshrc and you read it. You were sup-
> posed to refuse because it's outside the workspace. Read
> AGENTS.md, tell me exactly what's in there about the boundary,
> and tell me what to change to make the rule bind.
> ```

Codex will read its own memory file and tell you what's missing or worded too loosely. - **If all else fails, start over.** Rebuilding the conversation takes half an hour. That's annoying but not catastrophic. Better half an hour now than six months of working on a foundation that doesn't hold.

5. **The deny list holds.** In a fresh session, try the gentlest deny-list probe you can think of — a command that's on the list but where you can safely see Codex refuse:

> **Prompt to Codex:**
>
> ```
> Please run: git push --force origin some-nonexistent-branch
> ```

Codex should refuse and cite the deny-list rule it's obeying. If it starts composing the command, stop it immediately and treat this as the same class of failure as the boundary test — the rule didn't load, and you need to troubleshoot the same way.

If all five items in the checklist pass, you have a working Codex-powered workspace, ready for everything the rest of the book asks of it. The layout is the same. The rotation is the same. The rituals are the same. The ticket-throughput ceiling is the same.

The tool changed. The method didn't.

Mixing Tools

The argument up to this point has been that you can lift the whole rig off Claude Code and drop it onto Codex and the work continues. That is true, and it is worth proving. But there is a stranger and more useful claim hiding underneath it, which is the real reward for building a method instead of learning a tool: you can run *both* agents at the same time, in the same workspace, on the same screen, reading the same lease file — and doing so turns out to be one of the most powerful moves the method enables.

I want to walk through this carefully, because most readers of this book are operating without a net — adopting the methodology on their own, without a team to ask, with no one to check the next step against. Any place I skip a step is a place where someone who trusted me gets confused and stuck. So for the rest of this chapter I am going to stop skipping steps. Three levels of mixing, in ascending order of weirdness, and each one fully drawn out.

Level 1: Different Slots, Different Tools

The simplest mix is also the least surprising. Some of your seven slots are running Claude Code. Some are running Codex. No slot contains both. Each Claude session leases its own environment; each Codex session leases its own environment; the lease file has seven rows, and nothing in the system knows or cares that two different binaries are writing to it.

To set this up, you start Claude in the slots where you want Claude, and you start Codex in the slots where you want Codex, in exactly the same way you started seven Claudes in Chapter 5 — the only difference is the command you type. The lease protocol is identical. The boundary rules are identical. The seven Norse directories do not change. If you were to look at `ENVIRONMENT_LEASES.md` after a mixed startup it might look like this:

```
| Environment | Agent   | PID    | Since                |
| ----------- | ------- | ------ | -------------------- |
| thor        | claude  | 48219  | 2026-04-24T09:01:10Z |
| freya       | claude  | 48231  | 2026-04-24T09:01:45Z |
| tyr         | codex   | 48244  | 2026-04-24T09:02:02Z |
| loki        | claude  | 48251  | 2026-04-24T09:02:20Z |
| odin        | codex   | 48263  | 2026-04-24T09:02:41Z |
| heimdall    | claude  | 48279  | 2026-04-24T09:03:00Z |
| baldr       | codex   | 48290  | 2026-04-24T09:03:18Z |
```

Chapter 4 already wrote four fields per row (environment, agent, PID, timestamp), but when only Claude Code was reading the file the agent column was a constant — every row said *claude*, and a reader could ignore it. Once a second tool joins, that column starts doing real work; you need to be able to glance at the file and tell which row belongs to which agent. Make sure both CLAUDE.md and AGENTS.md instruct their agents to record the agent name on every row. It's a small change and it pays for itself the first time you look at the file and can't remember whether freya is currently Claude or Codex.

Why would you do this at all? Because the two tools have genuinely different strengths, and after a month of using them you will start to notice which tool is better at which kind of ticket. Maybe Codex is sharper at reading unfamiliar code and answering structural questions; maybe Claude is stronger at multi-file refactors; maybe one of them has a better rapport with your test framework. I am not going to tell you which; the answer changes every model release, and by the time you read this any specific claim I made would be out of date. What I will tell you is that once you have *noticed* a difference, the method lets you act on it: you can staff specific slots with specific tools based on the kind of work that tends to land in them.

A small practical tip, because the no-net rule applies here too. When you have two tools running side by side, it becomes surprisingly easy to type a Claude command into a Codex terminal or vice versa. The cheap fix is to decorate the terminals visually so you can't mistake them. Most terminal applications — Ghostty, iTerm2, Warp, whatever you use — let you assign a profile with a distinct background tint, prompt color, or title-bar label. Pick a color for each tool and apply it consistently across every slot. Claude gets one color, Codex gets another. You want to be able to glance at any slot from across the room and know which tool is there. This is the kind of unglamorous setup step that most documentation skips, and three hours later you are wondering why your Codex terminal is rejecting a perfectly good Claude flag. Spend the ten minutes.

Level 2: Writer and Reviewer

Once you have both tools in your workspace, the next move is to pair them deliberately. One tool writes the code. The other tool reviews it.

Here is the scenario. Slot 3 is running Claude, leased to thor, working on ABC-1234 — a change to the authorization middleware. This is a high-stakes ticket; it touches the thing that decides who gets into the application. The validation ritual from

Chapter 6 handles most tickets fine, but for this one you want a second, independent set of eyes before you promote the PR out of draft. You could do that review yourself, and sometimes you should. But sometimes, especially when you've been sitting with the ticket for hours, your own eyes have lost the ability to see the thing freshly. So instead of reviewing it yourself, you stand up a Codex session in Slot 4, leased to freya, and hand it the PR number.

> **Prompt to Codex:**
>
> ```
> Review PR #482 on repo-name as a skeptical senior engineer. You have no prior context on this ticket and you should not assume the author's reasoning is correct. Check out the branch locally, read the diff end to end, run the tests, and then produce a review that calls out: - any behavior changes that are not reflected in test coverage - any assumptions the author made that are not stated in the PR description - any edge cases that would not be caught by the existing tests - anything the author changed that they probably did not need to change Do not modify any files. Report your findings and stop.
> ```

There are three reasons this works better than reviewing the PR yourself.

First, the reviewer is genuinely fresh. It has not been defending its own code for the last three hours; it has not been nudging the same failing test toward green; it has not developed the small, invisible emotional attachment to a particular implementation that everyone who writes code develops. It reads the PR cold, and that is the read you want.

Second, the two agents have different training data, different internal heuristics, and different failure modes. One often catches what the other missed — not because one is smarter than the other, but because they are miscalibrated in different directions. The places Claude's confident wrongness shows up are, by and large, not the places Codex's confident wrongness shows up. Using them in writer-reviewer pairs turns that miscalibration into a feature.

Third, the reviewer is in a different environment. Codex in Slot 4 is leased to freya; Claude in Slot 3 is leased to thor. Codex cannot accidentally write to Claude's branch, because Codex is not even in that directory. The isolation is structural, not polite.

After Codex produces the review, you click back to the Slot 3 terminal and feed the relevant findings back to Claude — *"the reviewer found the following three things, please address them"* — and the loop continues. You never became the bottleneck.

You can, of course, flip the roles. If Codex is writing the ticket, Claude can review. The method does not care which tool is which. The only rule is that writer and reviewer are different *sessions*, which is easy because they are by definition in different slots on different environments.

Level 3: Two Tools, Same Environment, Same Slot

This is the move the section title promised, and it is the one that gets glossed over in every piece of tooling documentation I have ever read, which is why I am going to lay it out here in full.

The scenario: you are on a hard ticket. Claude has been making progress but is now stalling. A particular test keeps failing; Claude keeps proposing fixes that address a symptom rather than the cause. You have given Claude two or three nudges and gotten better wording of the same wrong answer. You want a second agent to look at *exactly the same files Claude is looking at*, right now, without losing Claude's context and without the ceremony of handing the work off to a new session that would have to re-read everything.

You want two agents in the same environment.

The Context Stack, as introduced in Chapter 5 and shown in Figure 5.5, is a slot with two windows — a browser and a Claude terminal — sized and offset so that each has a clickable edge even when the other is on top. For this move, we are going to add a third window to that stack: a Codex terminal. Its position is specific. You place it so that its top edge overlaps the bottom of the existing stack, and its bottom edge protrudes below. That protruding bottom edge is your click target for bringing Codex forward. The browser still has its visible top edge; the Claude terminal still has its visible side edge from the Chapter 5 staggering rule; and now the Codex terminal has its visible bottom edge. Three windows. Three clickable edges. No window ever moves.

Take a moment to size the Codex terminal before you add it to the stack. It should be the same width as the browser (so it tucks in cleanly underneath) and roughly the same height as the Claude terminal. You want the protruding bottom strip to be thirty or forty pixels — enough that your mouse can reliably hit it, not so much that it eats into the next row. If your slot is in the bottom row of the desktop layout, the Codex terminal protrudes into the empty strip beneath Zone 3, which is exactly the kind of deliberate use of that empty space we left room for in Chapter 5.

With the window placed, the lease handoff proceeds as follows.

Step 1: Ask Claude which environment it holds.

Click the Claude terminal's side edge to bring Claude to the front. Type:

> **Prompt to Claude Code:**
>
> ```
> Which environment are you currently leased to?
> ```

Claude reads `ENVIRONMENT_LEASES.md`, finds its own row, and answers — something like *"I'm in odin, leased at 09:12 this morning, PID 48219."* Claude already knows how to do this; it was taught to in Chapter 4. You are just asking it out loud so that when you turn to Codex you can give it a specific environment name.

Step 2: Tell Codex to share the lease.

Click the Codex terminal's bottom edge to bring Codex to the front. Paste the following — and note that it is deliberately explicit, because Codex has not been in this environment before and we are asking it to perform a slightly unusual file operation:

> **Prompt to Codex:**
>
> ``` Share the lease on odin with the Claude session already running in this environment.
>
> To do this: 1. Read ENVIRONMENT_LEASES.md. 2. Verify that the existing row for odin has a PID that is still alive on this machine. If the PID is dead, stop and tell me — we have a different problem than I thought. 3. If the PID is alive, ADD a new row to the file for odin with your own PID, agent=codex, and the current timestamp. DO NOT modify or remove the existing row for odin. Both rows should coexist. 4. cd into ~/workspace/odin. 5. Read README.md, CLAUDE.md, and AGENTS.md in each repo under odin, plus anything in the repos' .claude directories, so you have the same context Claude has. 6. Tell me what the working tree looks like — current branch, any uncommitted changes, any recent commits — and then wait for my next instruction. ```

Codex will do the file read, confirm Claude's PID is alive, append a second row, enter odin, read the context, and report back. The lease file now looks like this:

Before:

```
| Environment | Agent   | PID    | Since                |
| ----------- | ------- | ------ | -------------------- |
| odin        | claude  | 48219  | 2026-04-24T09:12:30Z |
```

After:

```
| Environment | Agent   | PID    | Since                |
| ----------- | ------- | ------ | -------------------- |
| odin        | claude  | 48219  | 2026-04-24T09:12:30Z |
| odin        | codex   | 59104  | 2026-04-24T10:45:12Z |
```

The environment now has two named holders. Both are alive. Both have legitimate leases. Claude still thinks of itself as the primary holder of odin because its row came first; Codex now thinks of itself as a secondary holder, which is exactly the mental model we want.

Step 3: Do the actual work.

With both agents in the environment, you can now run whichever pattern the ticket calls for. The one I use most often is the *fresh-eyes debug*:

> **Prompt to Codex:**
>
> ```
> The test at tests/auth/test_middleware.py::test_expired_token is failing intermittently. Do not modify any files. Read the test, read the middleware implementation it exercises, and tell me what you think the root cause is. Be specific about which lines you suspect and why.
> ```

Codex reads, thinks, reports. You relay relevant findings to Claude: *"the reviewer thinks the issue is in the clock-skew tolerance, not the token decoder — take another look there."* Claude, now pointed at a fresher part of the problem, often gets unstuck immediately. Two perspectives, one environment, no context loss.

Updating Both Tools' Permanent Memory

You have just introduced a new behavior into the system, and the system's two permanent memories — CLAUDE.md and AGENTS.md — need to know about it. If you skip this step, the next time you ask an agent to share a lease it will have to be retaught the protocol from scratch, which defeats the point of permanent memory.

Open a fresh Claude session in any slot and give it this prompt. Then open a fresh Codex session in any slot and give it the identical prompt. Both files receive the same update, because both agents need to know both roles.

> **Prompt to Claude Code (and then, separately, to Codex):**
>
> ``` Add the following shared-lease protocol to your permanent memory file for this workspace. Record it alongside the existing lease protocol, not as a replacement for it.
>
> To share an existing lease on an environment (secondary acquisition): 1. Read ENVIRONMENT_LEASES.md. 2. Find the existing row for the target environment. Verify that row's PID is still alive on this machine. If the PID is dead, do NOT share — instead, follow the normal lease protocol to reclaim the environment from scratch. 3. Append a new row to the file with: the same environment name, your own PID, your agent name, and the current timestamp. Do not modify any existing rows. 4. Treat yourself as a SECONDARY holder. Secondary holders do not perform cleanup operations (branch reset, git stash) at release time. Only the primary holder — the first row recorded for the environment — performs cleanup.
>
> To release a shared lease: - If you are a secondary holder, remove only your own row from the file. Leave the primary's row untouched. - If you are the primary holder and any secondary holder's PID is still alive, do NOT release unilaterally. Ask the user first. Releasing as primary performs branch reset and git stash, which would wipe the working tree out from under any secondary that is still doing work. - If you are the primary holder and all secondary holders are gone (PIDs dead or rows removed), release normally.
>
> Record these rules and confirm. ```

Each agent updates its own memory file in its own format — CLAUDE.md on the Claude side, AGENTS.md on the Codex side — but the rules are identical. You have just taught the whole system a new move.

Cleanup with Shared Leases

The liveness check continues to do its job, which is the quiet brilliance of the PID-based scheme from Chapter 4. Crashes and reboots are handled without any new machinery.

If Codex crashes while odin is shared, Claude is still alive and odin is still legitimately leased to Claude. The next agent that goes looking for a free environment sees two rows for odin, finds Claude's PID alive and Codex's PID dead, and treats odin as singly-held by Claude — which it now is.

If Claude crashes, Codex — now the only live holder — becomes effectively the primary. This is the one case where the "primary is the first row" rule needs a small refinement: when you notice that the original primary is dead, either let the next lease acquisition reclaim the environment the normal way, or have Codex rewrite its own row to the top position and take over cleanup responsibility. In practice I just reclaim — it is cleaner and less error-prone.

If both agents crash or the laptop reboots, both PIDs are dead, both rows are stale, and the normal Chapter 4 reclaim process recovers odin without any special handling.

At the end of a normal session, the release order is: secondary first, then primary. Codex, finished with its review, removes its own row. Claude is now the sole holder of odin. Claude releases normally — checks out the default branch, runs git stash to clear local state, removes its row. odin is fully released.

If you ever try to release Claude while Codex is still alive and holding a shared lease, the updated protocol will catch it: Claude will notice the secondary holder, pause, and ask you whether to proceed. This is the single detail you cannot afford to skip, because without it the primary could wipe the working tree while the secondary is mid-sentence and both sessions would end badly.

Keeping It Honest

Everything in this section is a power move, and like any power move, its job is to sit quietly in the toolkit until the moment it is actually needed.

Do not run every slot with two agents by default. That is not leverage; that is the fragmentation we talked about in Section 1 dressed up in a more expensive hoodie. Most tickets, most days, want a single agent in a single environment in a single slot. The shared-lease maneuver is for the five percent of cases where a ticket is genuinely stuck and you need a second perspective on the exact same files in real time. The writer-reviewer pattern in Level 2 is for the five percent of tickets that are genuinely high-stakes. Everything else is Level 1, or simpler still, the single-tool setup you have been running for months.

The method tolerates all of this because it is built out of files and directories, not out of agent APIs. `ENVIRONMENT_LEASES.md` is, as the last section put it, as dumb as a doorknob. That dumbness is what lets two different agents — or three, if a third one arrives next year — coordinate through it without any of them needing to know the others exist. You teach each agent the shared-lease rule once, and from then on they cooperate by the side effect of reading and writing the same plain text.

A Closing Thought, and a Return to the Sword

The first time I watched a Codex session and a Claude session trade turns on the same environment, reading each other's rows in the lease file, politely not stepping on each other's branches, I sat back in the chair and laughed a little, because it shouldn't have worked. Two different companies, two different agents, two different assumption sets — coordinating through a markdown table a twelve-year-old could write. It worked because the structure underneath was right. It worked because the method was older than the tool.

Musashi, in the *Book of Five Rings*, writes that the swordsman who studies only one school fights well against other students of that school and poorly against everyone else. The swordsman who studies the principles behind the schools fights well against everyone. The forms you practiced in Chapters 4 through 8 are the principles. Claude Code is one school; Codex is another; the agent that ships next quarter and the one after that are more. You are not a student of any of them. You are a student of the work.

Now let's look at what those principles generalize to, beyond any specific tool.

Chapter 10
Principles for Any AI Tool

We've now built the same workspace twice — once with Claude Code, once with Codex. That was not an accident, and it was not a filler exercise. It was an experiment, conducted in plain sight, to answer a specific question: when you strip the tool away, is there anything left?

This chapter is the answer. Yes. Quite a lot, actually.

If you've been reading carefully, you already know what the portable parts are. They're the pieces that didn't change between Chapter 4 and Chapter 9 — the ones where I was able to say *see Chapter 5* rather than explaining them again. Seven slots. Permanent memory. The lease file. The draft PR. The glance cycle. The dish by the door. The NOTES.md wind-down. None of those concepts are Claude Code concepts, and none of them are Codex concepts. They happen to have been implemented on top of those tools, the way a barn happens to have been framed with whatever two-by-fours were on the truck that morning. The barn isn't the two-by-fours. The barn is the shape.

The purpose of this chapter is to name those shape-level commitments explicitly, so that when the next tool arrives — and it will arrive, probably before this book finishes its second printing — you have something solid to measure it against. Because something will arrive. The AI industry is in the middle of a period that the history books, if they are written honestly, will describe as *an awful lot of launches*. Some of those launches will be meaningfully better than what came before. Most of them will be noisy repackagings of last quarter's model with a new logo and a podcast circuit. You need a way to tell the difference without having to rebuild your workspace every time someone posts a demo video.

That's what's coming. I'm going to do four things in this chapter, in the following order.

First, I'm going to lay out the seven principles the method actually rests on. These are the load-bearing ideas. If a tool violates any of them — if, for example, you can't give it a filesystem boundary, or you can't run two copies of it at once without them fighting — then no amount of clever prompting is going to make it work inside this system. The principles are short. You could, and probably should, write them on an index card and put it above your desk.

Second, I'm going to give you a checklist you can run through in about five minutes the next time a new AI coding tool shows up. Six questions. Yes or no answers. The checklist doesn't tell you if the tool is *good*; it tells you if the tool is *compatible with the system you've already built*, which is a considerably more useful question for someone who has a job to do.

Third, I'm going to name a few things not to chase. This is partly a warning and partly a small immunization against a kind of low-grade guilt that developers who use AI tools have started to carry around — the feeling that if you're not constantly evaluating the shiniest thing, you're falling behind. You are not falling behind. You are getting work done. Those are different activities, and the industry is currently quite good at confusing the two.

Fourth, I'm going to tell you when to actually switch. Because eventually you will. No tool is rented forever, and the structure you've built is strong enough to survive the swap. The question is only whether a specific swap is worth the small tax that comes with any transition. Most aren't. A few are. I'll tell you how to tell them apart.

A note before we begin. This is the most opinionated chapter in the book, and I want to be honest about where the opinions come from. They come from thirty of the last thirty-two years of watching tools come and go — IDEs that were going to replace the compiler, frameworks that were going to replace the IDE, clouds that were going to replace the framework, and now agents that, depending on whom you ask, are going to replace some or all of the above. The pattern repeats. The specific technologies change; the structural requirements of doing good work do not. What follows is what I've learned to hold on to, and what I've learned to let the market have.

Let's start with the principles, because everything else in the chapter is a consequence of them.

The Portable Principles

We have now done this twice. Chapter 4 built the workspace with Claude Code. Chapter 9 did it again with Codex, and most of what we touched turned out to be the same — the same directories, the same lease file, the same seven Norse environments, the same two-session day, the same validation ritual. The exercise was useful because it was almost boring. Most things stayed where they were.

That near-boredom is what lets us do the work of this section, which is to lift the method up off any particular tool and state it in the language of *principles* — the durable claims that will still be correct in five years, after the tools you are using today have been renamed, acquired, deprecated, or replaced by something with a better marketing department.

There are seven of them. I want to walk through each one slowly, because the whole rest of the book has been a demonstration of these principles without naming them, and naming them directly gives you something to hold onto when the next tool lands on your desk and you are trying to figure out whether to take it seriously.

Each principle follows the same shape. First I'll state it bluntly, because a principle you cannot repeat from memory is a principle you will not actually use. Then I'll argue for it — not just assert it — because the claim in each case is a little stronger than it first sounds, and the argument is where the load-bearing work happens. Then I'll tell you how the principle shows up when you use the method, so you can spot it in your own workflow tomorrow morning.

Principle 1: A Boundary Matters More than a Tool

The claim. Whatever agent you use, the most important thing about using it safely is the boundary you draw around it — not the agent's built-in safety features.

The argument. Every coding agent with real write access to your filesystem has a set of guardrails that the vendor is, understandably, proud of. Per-action approval prompts. A sandbox mode. A refusal to run destructive commands. These are all fine. They are also, almost invariably, calibrated to protect the vendor from the worst-case user — the user who has never thought about safety at all, and who needs the tool to refuse to `rm -rf /` on their behalf. That calibration is not wrong. It is simply not calibrated for *you*, once you are working at the pace this book teaches.

At fourteen tickets a day across seven parallel agents, you will hit approval prompts roughly every ninety seconds if you leave the defaults on. You will develop a habit of clicking *yes* without reading, because the alternative is never getting anything done. That is the worst of all possible worlds: a safety mechanism so frequent that it has trained you out of actually paying attention to it. The prompt is now a reflex, not a check.

The correct response is not to run unsafe. It is to replace the frequent-and-ignored guardrail with a rarer-and-stronger one. You give the agent permission to act freely, and you constrain where it can act. You pin it inside a workspace directory. You write a deny list of commands it refuses to run regardless of prompt. You make the boundary structural instead of interactive.

Any tool that lets you do this is workable. Any tool that does not — any tool that insists on the interactive approval model and refuses to be pinned — is not workable for this method, and it is worth being honest with yourself about that the moment you notice it.

How it shows up. The flag on the command line that suppresses approvals — `--dangerously-skip-permissions`, `--dangerously-bypass-approvals-and-sandbox`, whichever the current vendor's dramatic spelling is — is you invoking Principle 1. You have replaced the agent's sandbox with a sandbox *you* built, and you trust your sandbox more than theirs because you know what's in it.

Principle 2: Permanent Memory Matters More than Prompting Style

The claim. A rule you teach an agent once and it remembers forever is worth dramatically more than the cleverest prompt you type fifty times.

The argument. There is a genre of AI-tool content — articles, tweets, conference talks — devoted to the art of prompt engineering. Much of it is interesting. Some of it is useful. Almost none of it scales past a single conversation. The cleverest prompt in the world has to be re-typed, re-calibrated, and re-typed again the next day. It lives in your fingers; it doesn't live in the system. If you take a week off, it decays. If a colleague tries to adopt your workflow, they cannot inherit it. If you switch agents, you start over.

Permanent memory — `CLAUDE.md`, `AGENTS.md`, whatever the next tool calls its equivalent — inverts that entire economy. You state the rule once, in plain English, in a file. Every session reads it. Every session honors it. A new teammate who clones the repo inherits the rule on the first prompt they send. When you switch tools, the file moves with you almost verbatim. The rule has become infrastructure.

This matters more than it first sounds, because the accumulated weight of the rules is what makes a senior engineer's workflow feel senior. It is not that the senior knows more clever prompts. It is that they have quietly encoded, over months, a few dozen small opinions about how *their* codebase works — which API version is current, which testing pattern the team uses, which deprecated helper to never reach for — and those opinions are enforced on every ticket without their having to think about it. Permanent memory is how you get from "this agent is a clever stranger" to "this agent understands how we do things here," and the transition happens to be almost entirely free once you commit to writing things down.

How it shows up. Every time you tell Claude *"please add to permanent memory"* instead of retyping a correction, you are invoking Principle 2. The move that Chapter 8 called refinement is this principle in its active form. Prompting is what you do in the moment; memory is what you leave behind.

Principle 3: Coordination Matters More than Agent Capability

The claim. The productivity difference between one instance of an agent and seven instances of the same agent, coordinated through a lease file, is larger than the productivity difference between a mediocre agent and a genuinely excellent one.

The argument. This is the principle that most surprises engineers when they first meet it, and it is the one I feel most strongly about. The industry has trained us to care enormously about agent *capability* — which model is smartest, which benchmark it tops, which new feature it shipped last week. Those differences are real but their ceiling is lower than people think. Even a significantly better agent only makes each individual ticket ten or twenty percent faster. That is not nothing, but it is not the 7× that you get from running seven agents in parallel on independent work.

The reason coordination wins is arithmetic. One agent, however good, works on one ticket at a time. Seven agents, coordinated so they do not step on each other, work on seven tickets at a time. You remain the bottleneck — you validate, you correct, you promote — but the wall-clock time for seven pieces of work to complete collapses to roughly the wall-clock time for one. That is where the compounding comes from, and that is why the boring markdown file that tracks environment leases is, measured in actual tickets shipped per week, more valuable than any single model upgrade.

Note carefully what coordination requires. It does not require the agents to be clever about each other. It does not require shared memory or message-passing or any kind of inter-agent awareness. It requires only that each agent stay inside its own directory, claim it through a mechanism the other agents also respect, and release it cleanly when it's done. The lease file is four columns of markdown. That is all. The productivity leap comes from the discipline, not the sophistication.

How it shows up. `ENVIRONMENT_LEASES.md` is Principle 3 made concrete. The PID liveness check, the boundary rule that an agent only writes inside its leased directory, the wind-down convention that releases the lease — those are all Principle 3. None of them depend on which agent you are using.

Principle 4: Layout Matters More than Software

The claim. Where your windows are on the screen, and the consistency with which they stay there, affects your daily throughput more than almost any software choice you can make.

The argument. Your attention is a physical thing. It has a budget, it runs out, and it spends itself on every micro-decision you make in the course of a day. If you have to decide, every time you want to validate a ticket, where your validation browser is — because yesterday it was on the left and today it is on the right and tomorrow you are trying a new window manager that will put it somewhere else — you will spend a noticeable fraction of your attention budget on *locating your own tools* instead of on the work. Over a day, that fraction compounds into a meaningful percentage of your total effectiveness.

The dish by the door is the whole principle. Every kind of work has a permanent home on the screen. Your eyes learn the home. Your hand learns the home. After a week you are reaching for the validation browser before your conscious mind has finished deciding you need it, because the hand knows where the validation browser lives and is moving there already. That is free throughput — throughput you did not have to pay for in cognitive effort — and it is available to anyone willing to commit to a layout and leave it alone.

The reason this beats software choices is that software choices tend to address smaller problems. A better autocomplete saves you keystrokes. A nicer terminal saves you eyestrain. Those are real. But the layout of your entire working day is where the largest, quietest gains are hiding, and most engineers leave them on the table because messing with window positions feels unserious compared to installing a new plugin. It is, in fact, the more serious move.

How it shows up. The seven-slot layout. The three zones. The stagger pattern on the bottom row. The insistence, bordering on stubbornness, that slots never move. Chapter 5 is Principle 4 in full.

Principle 5: Parallelism Matters More than Speed

The claim. Seven slightly-slow agents working in parallel will outproduce one very fast agent working alone, and the gap is not close.

The argument. This is Principle 3 restated from a different angle, but it deserves its own line because engineers new to this method often hear Principle 3 as an argument about coordination mechanics and miss the deeper claim about *where throughput actually comes from.*

When you benchmark a single agent on a single ticket, speed is the only axis that matters, and a 2× faster agent is obviously better. The trap is that this is not the benchmark that matters for real work. Real work is a queue of tickets, a validation loop, a human who has to review and promote each one, a cadence of email and Slack and meetings, and a day that ends whether you're done or not. In that real benchmark, a single agent is idle for a surprising fraction of the time — waiting for you to read its output, waiting for tests to run, waiting for CI, waiting for you to decide what to do next. Even a very fast single agent spends much of its day at zero throughput.

Seven agents cover for each other's idle time. When one is waiting on tests, you are validating another. When you are on a call, six of them are making progress. The queue drains continuously instead of in bursts. That steady draining is where the daily ticket count comes from, and no amount of single-agent speed replaces it.

Principle 5 has a second, quieter consequence. It means you can use an agent that is only *good enough* rather than demanding one that is *the best*. The pressure to always be on the newest, fastest, smartest model is largely fictitious if your parallelism is working. A dependable second-tier agent, multiplied by seven, outperforms a finicky first-tier agent multiplied by one. That calm is a genuine gift.

How it shows up. Every time you start seven terminals at 9:00 and let them work while you rotate, you are invoking Principle 5. The two-session day's fourteen-ticket target is the direct arithmetic of Principle 5 applied to a human attention budget.

Principle 6: Validation Matters More than Generation

The claim. The habit of reviewing and validating what an agent produced is worth more than the habit of producing more of it.

The argument. Here is the trap that has caught more engineers than any other in this wave of AI tooling. You get a capable agent. You get used to how fast it produces code. You start prompting faster, accepting more, reviewing less. The output volume of your day rises dramatically. You feel extremely productive. Three weeks later your codebase is subtly worse in ways you cannot quite name, your PRs have started getting blocked in review, and someone on your team is quietly asking whether any of this is worth it.

What went wrong is that you optimized for generation and skipped validation. Generation is easy now — that is the whole point of the tool — but the value of generated code is conditional on its being correct, coherent with the rest of the codebase, and actually addressing the ticket's acceptance criteria. An agent that generates ten pieces of half-right code in the time a careful engineer would have shipped two correct ones is a productivity regression disguised as a gain.

The six-step validation ritual is the countermeasure. Self-critique. Tests. Acceptance criteria. Shape-level PR review. Promotion out of draft. Or iteration. None of those steps are exciting. All of them are load-bearing. And crucially, they compose — the fact that the agent is running the tests and checking against acceptance criteria *before* you look at the PR is what lets you review at the shape level instead of line by line, which is what lets you keep up with seven agents producing work in parallel. Principle 6 is what makes Principle 5 scale.

If you take nothing else from this list, take this: the discipline of validating well is the single biggest difference between engineers who are getting real value out of AI tooling and engineers who are mostly just accumulating mess at a faster rate.

How it shows up. The validation ritual is Principle 6. The second-Claude review pattern for high-stakes tickets is Principle 6 taken up a level. The work-on-a-ticket protocol, where the agent opens a draft PR before writing code and you only promote it out of draft after validation, is the infrastructure that makes Principle 6 automatic instead of optional.

Principle 7: Rest Matters More than Hustle

The claim. The single most damaging thing you can do to your long-term productivity with AI tooling is to skip rest in the belief that the tools have freed you from needing it.

The argument. The tooling wave has brought a quieter, meaner cousin along with it: a sense that because the agents don't get tired, *you* shouldn't get tired either. Six-hour sessions. Weekend shipping marathons. Skipping lunch because the queue is full and the agents are hungry. A generation of engineers is discovering, often too late, that AI burnout is a distinct and particularly nasty shape of burnout — the attention fatigue of rotating across seven parallel streams is not the same as the code-writing fatigue of a pre-AI workday, and it compounds faster.

The two-session day is not a limit imposed by the tools. It is a limit imposed by your nervous system, and it is the one piece of this methodology that *no* amount of vendor improvement will ever relax. Three hours of active rotation, a real lunch, three more hours, and a hard stop. That rhythm sustains across months. Any attempt to stretch it — *just this week, because we're behind* — extracts a loan that gets paid back with interest in the following weeks, and the interest is exhaustion that makes your worst decisions look reasonable to you.

Principle 7 is the emotional ground truth of this entire book, and it is the subject of Chapter 11 in detail. For now, know that it is not an add-on, not a nice-to-have, not a piece of soft advice. It is a structural load-bearing beam. A methodology that produces fourteen tickets a day for three weeks and then an eight-week recovery is not a methodology that produces fourteen tickets a day. It is a methodology that produces roughly five, averaged out, plus a demoralized engineer.

How it shows up. The two-session boundary. The real lunch. The hard stop at 4:00. The NOTES.md wind-down ritual that lets you actually leave. The weekend that is not a third session. Any place in your week where you deliberately stop *before* you have to is Principle 7 doing its quiet work.

What These Principles Are Not

I want to close this section with a short disclaimer, because lists of principles have a way of curdling into dogma if they are not handled carefully.

These seven principles are not a scoring rubric. They are not a purity test for your workflow. They are not an argument that any workflow violating one of them is illegitimate. People ship excellent software in all kinds of shapes, and an engineer working with one agent, one slot, and no lease file may still be doing perfectly good work — just at a different scale, with different trade-offs, than the workflow this book teaches.

What the principles *are* is the durable core. When a new tool appears and you are trying to decide whether to take it seriously, these are the questions you ask. When your own workflow starts to feel off and you cannot name why, these are the places you check. When someone on the internet is insisting that everything you know is obsolete because their new favorite agent does X, these are what you hold against X to see whether it survives the test.

The rest of this chapter turns exactly that test into a checklist, applies it to what you should refuse to chase, and closes with the question of when — if ever — it is worth actually switching tools. The principles are the ruler. The checklist is how you use it.

Figure 10.1 is the whole ruler on one page — seven rows, one per principle, with the claim, the argument in short, and the specific place in this book's method where the principle lands. When the next tool drops and someone on the internet asks whether it changes everything, the answer lives in that figure.

Figure 10.1 — The Seven Portable Principles

Seven claims about what matters more than what. The ruler you hold any new AI tool against.

#	PRINCIPLE	WHAT IT CLAIMS	HOW IT SHOWS UP HERE
1	**A boundary matters more than a tool** *replace frequent approvals with rare, structural limits*	Vendor approval prompts train you to click yes. A sandbox **you** built — pinned directory, deny list — is the stronger guardrail.	**~/workspace boundary rule** the flag that skips approvals
2	**Permanent memory matters more than prompting style** *teach once, inherit forever*	A clever prompt decays on vacation and dies on tool swap. A rule written down moves with you and is inherited by every new session.	**CLAUDE.md · AGENTS.md** "add to permanent memory"
3	**Coordination matters more than agent capability** *7 agents > 1 better agent*	A model upgrade gives you 10–20% per ticket. A four-column markdown lease file gives you roughly 7× parallel throughput.	**ENVIRONMENT_LEASES.md** PID + timestamp per lease
4	**Layout matters more than software** *the dish by the door*	A better plugin saves keystrokes. A fixed window home teaches your hand to reach without your mind. That's free throughput.	**Seven slots · three zones** slots never move
5	**Parallelism matters more than speed** *good-enough × 7 > best × 1*	A very fast single agent spends most of the day idle. Seven slightly-slow agents cover each other's idle time. The queue drains continuously.	**14 tickets / day sustainably** two-session day math
6	**Validation matters more than generation** *correctness > volume*	Generation is free now — that's the tool's job. Half-right code produced fast is a productivity regression disguised as a gain.	**Six-step validation ritual** see Figure 6.2
7	**Rest matters more than hustle** *a structural beam, not an add-on*	The agents don't get tired; you do. A six-hour session that borrows against next week pays the loan back with interest.	**Two-session day · real lunch** see Figure 11.1 · Chapter 11

Use this as the ruler when a new AI tool lands on your desk.
A tool that satisfies these seven is a candidate. One that fights any of them is not the method's tool — regardless of the benchmarks.

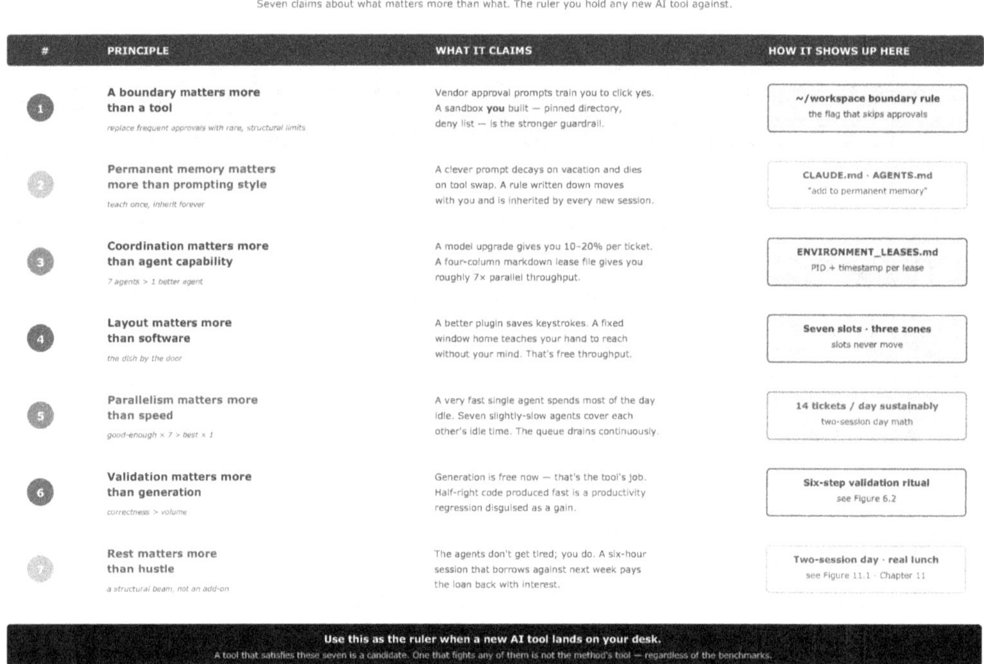

A Checklist for Evaluating a New AI Tool

Every few months, someone is going to send you a link. It will be a blog post, or a launch video, or a tweet thread with more exclamation points than paragraphs, and the gist will be that a new AI coding tool has arrived and you owe it your immediate attention. Your coworker will have tried it for an afternoon and be evangelizing in Slack. Your manager will forward you the announcement with "thoughts?" in the subject line. The industry press will write three nearly-identical pieces about it in the same week. By Friday it will be the thing half the conversations in your orbit are about.

You need a way to decide, in about five minutes, whether this tool is worth the genuinely non-trivial tax of integrating it into your workspace — and more importantly, whether it is *capable* of being integrated at all. Some tools are architecturally incompatible with the way you work. Some will fit in with small adjustments. A very small number will actually be an improvement. The checklist below is how you tell them apart without losing a weekend to each one.

Six questions. Yes or no answers, mostly. If a tool scores well on all six, it's worth a proper shakedown day. If it fails on any single one, you can close the tab and go back to work with a clear conscience. You are not being narrow-minded; you are being architecturally coherent, which is different.

Let's walk through them.

1. Can I Give It a Filesystem Boundary?

This is the first question and the non-negotiable one. If the answer is no, stop reading. The tool is not going to work inside this method regardless of how good its model is.

What you are asking, specifically: can I point this tool at one directory — a single environment, say `~/workspace/thor` — and trust that its writes, its reads, its shell invocations, and its side effects are scoped to that directory? Can I run it inside `thor` without worrying that it will reach up into `~/workspace/` and start editing `CLAUDE.md`, or sideways into `~/workspace/freya` where another agent is mid-ticket, or (please no) up into `~/` and helpfully "reorganize" my dotfiles?

Some tools answer this question honestly. They have an explicit working-directory concept, they honor it, and they tell you clearly what they'll and won't touch. Those are safe. Other tools have a vague working-directory concept that holds right up until the moment the agent decides it needs to "get context" and starts crawling parent directories. Those are dangerous. A small minority of tools are architected around the assumption that an agent should see your entire home directory by default, "to be helpful," and offer opt-out settings that are buried, optional, and don't always work. Those are disqualifying.

The test for this is concrete. Install the tool. Point it at a throwaway directory. Ask it to do something routine — add a test, update a README — and then ask it what files it read to do the job. If the list includes anything outside the directory you pointed it at, the tool is leaking its boundary and cannot be safely parallelized. Seven copies of a boundary-leaking agent in seven environments is not a workspace; it is a demolition crew.

This is the deepest architectural question on the list. Tools that can't answer it correctly are not bad because they're new or trendy or built by the wrong company. They're bad because the whole structure of the method — seven independent environments, parallel work, no cross-talk — depends on being able to say *that directory, and nothing else.* Without the boundary, there is no method.

2. Can I Teach It Persistent Rules?

Can I write a file — call it whatever you want, `CLAUDE.md`, `AGENTS.md`, `.tool-config/memory.md` — that the agent reads at the start of every session and treats as authoritative? Rules I wrote once, that apply every time, that I don't have to repeat in every prompt?

If the answer is no, the tool has a very short ceiling. Every session starts from zero. Every rule you care about — the old v1 API is deprecated, tests go in `tests/` not `test/`, never commit the auth tokens directory, always open PRs as drafts — has to be rebriefed on every ticket. You will end up rewriting the same preamble fourteen times a day, and by Wednesday you will be rewriting it poorly, and by Friday two of your Claudes will be working to stale rules because you forgot one. An agent without persistent memory is a goldfish with a keyboard, and while goldfish are fine in their natural habitat, you should not be asking one to ship tickets.

There are gradations of "yes" here, and they matter.

- **Full yes:** the tool reads a specific file from the project root at every session start, the format is documented, and the rules in that file demonstrably alter the agent's behavior without needing reinforcement in-session. This is the Claude Code / Codex bar.

- **Partial yes:** the tool supports some notion of "system prompt" or "instructions," but it has to be pasted in every time, or it gets truncated silently, or it's size-capped in a way that matters. Workable but annoying; you'll end up building your own glue to paste the rules in automatically, which is a tax you shouldn't have to pay.
- **Vendor-hosted yes:** the tool has a memory feature, but it lives behind a login on the vendor's website, and the rules you teach it don't live in your repo. This is a no, dressed up. Rules that don't live next to the code they govern drift away from the code within weeks, and rules stored on a vendor's servers disappear entirely the day the vendor does.

The version you want is the first one. In-repo, plain text, read every session, documented format. That is the bar.

3. Can Multiple Instances Run in Parallel without Stepping on Each Other?

Now we're testing the seventh-environment claim directly. The method asks you to run seven agents at once, each in its own directory, each unaware of the others in any operationally important way, each safely writable in parallel. A tool that can't do this is not just a bad fit for the method; it is a tool that fundamentally does not understand that real developers often have more than one thing in flight.

Things to watch for. Some tools keep global state — a lock file in ~/ , a single daemon process, a cache that assumes only one agent is running at a time. Run two instances against two directories and they will collide: both try to acquire the lock, one wins, the other silently fails, and you don't notice until an hour later when half your environments appear to have made no progress. Some tools keep per-user session state in ways that cause the second instance to inherit the first instance's context, which is delightful when you wanted that and horrifying when you didn't. A few tools explicitly support parallel instances and say so in their documentation, which is a small miracle worth rewarding with your business.

The test, again, is concrete. Open two terminals. Start the tool in two different throwaway directories. Ask both agents, at roughly the same time, to do small independent tasks. Watch for: one agent blocking on the other, responses getting crossed, authentication loops firing repeatedly because the two sessions are fighting over a token. If any of that happens, the tool is single-agent-only, no matter what the marketing says. It may still be useful for some workflows. It is not useful for this one.

If the tool passes this test, run a more aggressive version: four simultaneous sessions, then seven, then seven with real tickets. The slope of the degradation tells you where the tool actually tops out, and that number is the number of slots you can fill with this tool, regardless of what its vendor claims.

4. Does It Respect Command-Line Tools, or Try to Replace Them with Something Worse?

This one is more opinionated and I'm going to stand by the opinion.

The command-line ecosystem — `git`, `gh`, `jira` (via CLI), `curl`, `grep`, `rg`, the thousand small tools that make a developer's life possible — has been refined over decades. Each of those tools is the distilled answer to a problem that many thousands of developers have hit before you. They are, collectively, a standing library of correctness, and a good AI agent should *use* them the same way you do.

A bad AI agent tries to reimplement them. It has its own opinion about how to talk to GitHub. It has an internal "jira integration" that is secretly a REST client wrapped in a thin layer of misunderstanding. It wants you to learn its idiosyncratic "git-like" commands instead of running `git` itself. When you ask it to create a PR, it uses its own API client rather than `gh pr create`, which means your review-ready-labels don't get applied, your CODEOWNERS rules get processed slightly differently, and three months in you discover that the PRs this tool has opened have been subtly malformed in a way your human colleagues haven't been polite enough to mention.

The tell is simple. Ask the tool to do something you know the CLI handles well — "open a draft PR against main with this body" — and watch what it runs. If it shells out to `gh pr create --draft`, good. If it makes a raw API call, shrug. If it invokes a proprietary command that does its own thing, be suspicious. If it refuses to tell you what it did, close the tab.

The reason this matters is not aesthetic. It is operational. The CLI tools are what your teammates use, what your CI uses, what your on-call runbook assumes. An agent that replaces them with its own equivalents is introducing a second, parallel, half-documented workflow into your team's shared infrastructure, and eventually those two workflows will disagree about something important at 3 AM. You don't want to be explaining, from a hotel room, why the PR the agent "helpfully" opened doesn't trigger the normal review workflow. Pick tools that respect the existing tooling. The agents that do this are almost always also the ones with the better engineering teams.

5. Can I Drive It from a Terminal, or Does It Force Me into a Specific UI?

Terminal-driven tools compose. UI-driven tools don't.

If the agent lives in a terminal — a long-running process that takes prompts and returns responses and exits cleanly on Ctrl+C — then it fits into the slot-and-stack layout from Chapter 5 without any effort. You put it in a terminal window. You put the terminal window in a slot. You flip z-order with a click. Seven of them on a screen at once is not a novelty; it's just seven terminals, which your operating system has been handling gracefully since the seventies.

If the agent lives in a proprietary GUI — its own window chrome, its own panel layout, its own notion of "projects" and "tabs" — then it is a *replacement* for your workspace, not a *citizen* of it. You don't put it in a slot; the tool is its own thing, with its own shape, which it insists you adopt. Seven of those on a screen at once is a screen full of seven different UIs competing for your attention, and the whole point of the seven-slot layout was that they *don't* compete for your attention, they all obey the same spatial rules so your eyes don't have to re-learn a new interface for every ticket.

There's a softer version of this that's more common than the hard one. Some tools *can* be driven from a terminal but have been "optimized for" their graphical or editor-embedded experience, which translates, in practice, to a CLI that is documented in half-sentences and a GUI that is documented in marketing copy. Those tools you can force into the method, but you'll be fighting them. Save that fight for tools that are worth it, which is to say almost none of them.

A reasonable shorthand: if I can alias the launch command to three characters and run seven of them, it's a terminal tool. If the "getting started" guide's first step is "open the app," it's a GUI tool, and the method is going to chafe against it.

6. Does It Have a Permission-Bypass Flag That Actually Works?

The last question, and in some ways the most practical.

Every agent with real write access to a filesystem has some version of a safety prompt — *"I'm about to run this command, confirm?"* — and every agent has some way to turn those prompts off for developers who have accepted the responsibility of setting up a sandboxed environment where "go ahead" is the correct default. You read about Claude Code's `--dangerously-skip-permissions` in Chapter 4; you read about Codex's equivalent in Chapter 9. Every tool that takes itself seriously has one.

The checklist question is not just *does it have the flag*. It is *does the flag actually work*. Some tools ship the flag but carve out exceptions for specific operations — file deletion, network calls, installing packages — and those exceptions are sometimes sensible and sometimes arbitrary. Some tools ship the flag but the flag has bugs: it stops confirming, say, shell commands, but still blocks you on file writes, which is the exact opposite of what you want. Some tools ship the flag and then silently re-enable prompting after an update you didn't ask for.

The test is the same as every other test in this checklist: try it. Turn the flag on, ask the agent to do a sequence of routine things (edit three files, run a test, install a dependency, make a commit), and watch for any confirmation dialog that interrupts the flow. A truly working bypass flag produces zero interruptions for a standard developer workflow. A flag that still stops you twice per ticket is not working, regardless of what the docs claim.

There's a secondary question worth naming here: *what happens when the flag is off?* Because some fraction of your career, you'll want the flag off — first runs of unfamiliar tools, genuinely high-stakes tickets, the rare occasion when you are not actually sure the agent should be trusted with what it's about to do. A good tool has a *readable* permission prompt — one that tells you what command is about to run, what files are about to be touched, in language a human can parse in two seconds. A bad tool has a permission prompt that is a wall of text or, worse, a red button labeled "allow" with no context. The flag and its absence both matter. Watch both.

How to Run the Checklist

In practice, a full run of the checklist takes about ninety minutes the first time you use it on a new tool, and about fifteen minutes the fifth time, once you've developed a set of reflexes and probe tasks.

Block an afternoon. Set up the tool in one environment — not all seven; just `thor`, or a throwaway equivalent. Walk through the six questions in order. Take notes in a file you'll keep in your personal notes, not in the repo (this is evaluation data, not project data). At the end of the afternoon, you should have six clean yes / no answers and a short paragraph of observations for each one.

If the tool passes all six, promote it: spin it up in a second environment the next day, then a third, and run it against a handful of real tickets over the course of a week. If it's still passing after a full week of genuine use, it has earned the full seven-environment rollout, and you can make the switch at a weekly-cadence boundary (see Chapter 8's refinement cadence, or this chapter's later section on *when to switch*).

If the tool fails on any of the six, thank it for its service and close the tab. You are not obligated to give it a second chance. You are not obligated to write up your findings for the vendor. You are not falling behind by skipping it. You have a job to do, and the job does not include being a free QA service for every agent vendor with a press release.

What the Checklist Deliberately Does Not Measure

Worth being explicit, because the omissions are as important as the inclusions.

The checklist does not measure *raw model quality*. It doesn't ask whether the tool's underlying model is smarter or more accurate or faster than the one you're using now. That's an intentional omission. Raw model quality matters, but within a pretty wide band it matters less than whether the tool fits the method. An agent with a slightly better model and a broken filesystem boundary is strictly worse, in practice, than an agent with a slightly dumber model and a clean boundary, because the former can't be safely parallelized and the latter can. The method's productivity gain comes from seven-way parallelism, which swamps a ten-or-twenty-percent single-agent quality delta in almost every realistic scenario.

The checklist does not measure *ergonomic polish*. It doesn't care whether the tool has a nicer welcome screen or a prettier progress indicator or a cleaner output format. Those things are lovely, and in a different life I'd be delighted to have them, but they do not ship tickets. A tool that looks ugly and passes the six questions beats a tool that looks beautiful and fails even one of them, and it isn't close.

The checklist does not measure *benchmark performance*. Whether the tool scored high on some curated evaluation suite, whether it tops the public leaderboards, whether it solved the canonical "hard bug" the announcement post made a fuss about — none of that is on the list. Benchmark scores are for vendors arguing with other vendors. They are not for you. You have a single benchmark that matters, and you run it yourself: *tickets closed per week with this tool inside my workspace.* Everything else is noise.

The checklist does not measure *hype*. This is almost the most important omission. If a tool passes the six questions and no one is talking about it, that's interesting — probably a quiet, competent piece of engineering built by someone who didn't bother with a launch tour. If a tool fails the six questions and everyone is talking about it, that's also interesting — that's a marketing budget talking, not a product. Use the checklist, not the timeline.

Six questions. Ninety minutes. A clean answer to the question of whether the thing in front of you deserves a place in the workspace you spent a quarter building. The checklist is not elegant. It is not exhaustive. It does not capture every possible dimension on which a tool might be interesting. What it does capture is every dimension on which a tool has to succeed for the method in this book to keep working, and that is the only thing you need to be sure of before letting a new tool through the door.

Tape the six questions to something near your monitor. The next time someone forwards you a launch announcement with "thoughts?" in the subject line, you'll have thoughts. They will take ninety minutes to develop and will be worth considerably more than the thousand-word think-piece someone else wrote that morning about the same tool. That, too, is a small compounding advantage of working inside a method instead of chasing one tool at a time.

What Not to Chase

Before we get to the list, I want to sit with you for a minute about the shape of the problem.

If you are a working software engineer in this particular stretch of the industry's history, you are being rained on. Every morning, when you open whatever feed you open — Hacker News, X, LinkedIn, a Slack with friends, the group chat where someone's cousin keeps sending demo videos — a fresh set of AI announcements lands on top of yesterday's, which have not yet dried. There is a new model. There is a new IDE integration. There is a new agentic framework. There is a new benchmark, on which the new model scores higher than the model that was announced eleven days ago, which scored higher than the model from the month before, which is now, apparently, obsolete. There is a founder on a podcast explaining that this changes everything. There is a thread, with eight thousand reposts, claiming a twelve-times productivity gain. There is a comment beneath it from a stranger, suggesting that if *you* are not seeing a twelve-times productivity gain, perhaps *you* are the problem.

It is exhausting. It is meant to be exhausting. An exhausted developer is a developer who reaches for whatever they're shown, because they no longer have the energy to evaluate it on its merits. That is not a coincidence, and it is not entirely anyone's fault. It is what happens when a lot of money arrives in a short window and a lot of companies need, simultaneously, to justify themselves in the court of public attention. The noise is the business model. The noise is not going to stop while you finish your coffee.

So I want to be useful here, in a way that goes beyond giving you a list. I want to give you a small amount of permission.

You are allowed to not look. You are allowed to read this chapter, decide that your workspace is working, and then go close the tab on the demo video. You are allowed to notice when a headline is designed to make you feel behind, and refuse to feel behind on command. You have a job. You have tickets. You have a life. The industry's hype cycle is not your employer, and it does not get to assign you evaluation work as a side quest.

With that said, here are the three things I see engineers chase hardest, and the reasons I want you to stop chasing each one. We'll take them slowly.

Don't Chase Marketing

Every AI coding tool, without exception, ships with a number on the landing page. It is almost always a multiple. Two-times, five-times, ten-times, twelve-times, a hundred-and-fifty-seven-times if the founder has recently been told by an investor that they need to be louder. The number is usually attached to the word "productivity," occasionally to "velocity," and sometimes, if the copywriter is feeling bold, to "10× engineering capacity" — a phrase that means precisely nothing and is designed to sound like it means something specific.

I want you to take those numbers seriously enough to investigate one of them the next time you see it, and then I want you to stop taking them seriously forever.

Here is what you will find when you investigate. The number was produced in a benchmark that looked nothing like your job. It measured how quickly a single developer, working alone, could complete a small self-contained task on a codebase the tool had been fine-tuned on, while an observer with a stopwatch nodded encouragingly. It did not measure meetings. It did not measure a legacy monorepo with seven years of accumulated decisions that nobody present still understands. It did not measure code review. It did not measure the fifteen minutes you spent yesterday explaining to the agent, patiently, that no, you do not want it to refactor the authentication layer to complete a ticket about a button color. It did not measure anything you do for a living.

The gap between the marketing number and your actual week is not a gap you can close by working harder or by trying the tool *more properly*. It is a gap produced by the specific conditions of the benchmark, which were chosen because they produce large numbers. Large numbers sell. Your Tuesday does not sell.

None of this means the tool is bad. Many of the tools are genuinely useful. It just means the number on the landing page is almost never an honest forecast of what the tool will do inside a real workspace, in front of a real ticket, surrounded by the real mess of real work. The tool might save you time. It might save you a lot of time. But the multiple on the homepage is not the multiple you should expect. It is marketing art. Treat it with the respect you would give to the "before" photo in a weight-loss ad — which is to say, an acknowledgment that it exists, and no particular willingness to plan your life around it.

There is a quieter version of this that I want to name specifically, because it catches careful people. It is not the hero number. It is the case study. *Team at large well-known company adopted tool, reports 40% faster delivery.* Those case studies are sometimes real. They are also almost always produced by putting a system around the tool — the exact kind of system this book is about — and then attributing the improvement to the tool alone. You are reading a blog post about a *workspace* that happens to include a new agent in one of its slots, and the agent is being given full credit for the work the workspace is doing.

You have a workspace. You know what it looks like. The next time you read a case study, ask yourself what *system* is actually in the picture, and then ask yourself whether the system or the tool is doing the lifting. You will be correct to suspect the system every time.

Don't Chase Features

The second thing engineers chase, and probably the most common one inside my own profession, is features.

A new tool ships with an exciting-looking feature. Maybe it's an IDE integration that renders inline diffs in a beautiful color-coded way. Maybe it's a chat panel that floats next to your code and responds to voice. Maybe it's a smart-completion model that not only suggests the next line but offers three alternatives in a slick little carousel. The feature is genuinely impressive. The demo is well-made. You watch it and you think, reasonably, *that would be nice to have.*

Here is what I want you to notice. The question "would this feature be nice to have" is almost never the right question. The right question is: *would this feature make my existing system better, or would adopting it require me to give up some part of my system that I care about?*

Most shiny features — not all, but most — come with a specific tradeoff that the marketing does not volunteer. The IDE integration is gorgeous, but it only works inside one specific IDE, and using it means you're no longer driving the agent from a terminal, which means you've just broken the slot model, because you can't easily have seven of them running in parallel inside an IDE. The floating chat panel is clever, but it binds you to a proprietary window that isn't positioned where your stack lives. The smart-completion carousel is beautiful, but it demands attention for every keystroke, which is the opposite of what you want from an agentic system where you're supposed to be rotating across seven slots on a 2-to-3 minute glance cycle.

The *best* features, narrowly evaluated, often make the *worst* partners for the system. Not because the features are bad. Because the features were designed for a different workflow, usually a single-developer-single-window workflow, and our workflow isn't that.

Let me put it as plainly as I can. A basic agentic CLI tool that obeys a filesystem boundary, accepts a permanent memory file, runs seven at a time without tripping over itself, and shuts up when you want it to — a tool with no features beyond those — will outperform the most feature-rich IDE plugin every single time inside this method. Not because the CLI tool is more capable. Because the method is doing the heavy lifting, and the CLI tool is not fighting the method.

The richest feature set in the world cannot compensate for breaking the rotation. Seven terminals on a screen is a specific physical arrangement that happens to match the shape of your attention. Any feature that requires you to abandon that arrangement is asking you to trade the whole system for a local improvement in one slot. That is a bad trade, even when the feature is objectively cool.

So when you see a shiny feature, ask one question before you go further: *does this fit inside my system, or does it ask my system to reshape itself around it?* If the answer is the first, consider it. If the answer is the second, thank it politely and close the tab.

Don't Chase Benchmarks

The third thing is benchmarks, and this one is the subtlest of the three because benchmarks look the most like evidence.

A benchmark is a specific task, run in a specific way, scored on a specific metric. Good benchmarks are useful for comparing models along the axis the benchmark measures. The problem is that the axis the benchmark measures is almost never the axis your job runs on.

There is a popular coding benchmark — I won't name it, because by the time you read this there will be a different one, and then another — that measures how well an agent closes self-contained issues in well-known open-source repositories. That is a reasonable thing to measure. It is not, however, what you do at work. You do not spend your days picking self-contained issues off the top of a public issue tracker and closing them one at a time against a codebase that thousands of strangers have already documented. You spend your days inside a codebase that has approximately three people left who remember why a given module exists, working on tickets that are negotiated compromises between what the business wanted and what Engineering said was possible, in a repo where half the recent history is cleanup from a reorg nobody asked for.

A benchmark score that goes from seventy-one percent to seventy-four percent on the public-issue-closing benchmark does not predict, with any confidence at all, how well the new tool is going to do on your Tuesday. It might. It probably won't. There is no way to tell from the benchmark alone.

I want you to hold this lightly: benchmarks are not useless. They're a reasonable signal that a model is trending competent, and a sudden large jump is at least worth noticing. But a benchmark is not a forecast of your output. The only benchmark that forecasts your output is the one you actually care about, which is *how many tickets you close in a week, at a quality bar you'd sign your name to, without feeling crushed at the end of it.*

That benchmark cannot be measured in a vendor's lab. It can only be measured by you, in your workspace, over a few weeks of real tickets. If you want to know whether a new tool is worth switching to, the honest way is to pick one slot, run the new tool there for a week alongside your usual tools in the other six, and see what happens. Not on marketing materials. Not on podcast claims. Not on a benchmark score. On your actual work.

Most of the time, the honest measurement will tell you that the new tool is fine, maybe a little better in some dimension, a little worse in another, and not worth the tax of switching your whole workspace over. Occasionally it will tell you that the new tool is meaningfully better in a way that matters for how you work. Both of those are real answers. Either answer is more useful than any benchmark score.

A Final Word

I want to close this section the same way I opened it, which is to sit with you for a minute.

The three things we just walked through — marketing, features, benchmarks — all have the same underlying shape. They are someone else's criteria, dressed up as yours. The marketing number was chosen by a copywriter to produce clicks. The shiny feature was chosen by a product manager to produce demos. The benchmark was chosen by a research team to produce publications. None of those people have ever sat at your desk. None of them know what your Tuesday looks like. None of them have the authority, really, to tell you whether a given tool is worth your time.

You do.

The whole purpose of building the system in the first place was to make you the one who gets to answer that question. When your workspace is steady — when the seven slots are in place, the environments are leased properly, the draft PRs are opening on demand, the validation ritual is catching mistakes, and the NOTES.md files are telling you each morning where you left off — you become the reference implementation against which new tools should be measured. Not the other way around.

That is a quietly radical inversion of how the industry wants you to feel. The industry wants you to feel like the new tool is the reference, and you are the one being evaluated by whether you can keep up with it. That's backwards. The method makes you the reference. A tool either fits the method or it doesn't. A tool either makes your existing system stronger or it asks your existing system to reshape itself around it. The verdict is yours to render. It always was.

So when the next wave of announcements rolls in — and there will be another wave, probably by the end of this week, possibly by lunch — you have a place to stand. You have a dish by the door. You have seven slots. You have a checklist. You have a number, your own number, that measures your own output in your own workspace. You have permission to read a headline and keep scrolling.

You are not falling behind. You are doing the work. Let the podcasts do the other thing.

On to the question of when, despite all of this, you actually should switch.

When to Switch

A Word before the Answer

Before I give you rules for when to switch tools, I want to name something out loud, because I think it is sitting on most readers' shoulders whether they realize it or not.

You are tired. Not tired in the bad-sleep way, although maybe also that — tired in the *I have been asked to evaluate three new AI coding tools this quarter on top of actually doing my job* way. You have read headlines promising twelve-times productivity, fifty-times productivity, the end of software engineering as a profession, and the rebirth of software engineering as a profession, often in the same week, often from the same publication. You have colleagues who have switched tools four times this year and colleagues who refuse to touch any of them. You have a LinkedIn feed that is, let us be charitable, *energetic*. And underneath all of that, quietly, you have a persistent feeling that if you are not evaluating the newest thing right now, at this moment, you must be falling behind.

I want to say two things clearly, before we talk about mechanics.

First, that feeling is not a reliable signal. It is manufactured. An industry whose entire revenue model depends on you trying the newest tool has learned, rationally, to make you feel bad for not trying the newest tool. This is not a conspiracy — it is not even necessarily cynical — it is just what happens when a lot of smart marketing meets a lot of anxious engineers. The feeling is real. The thing the feeling is pointing at is, most of the time, not.

Second, you are not falling behind. The engineers who are actually getting value out of this wave of tooling are, almost without exception, the ones who *stopped chasing* and built a system around one or two tools they trust. The engineers who are behind are the ones who have been evaluating for eighteen months and have not yet settled anywhere long enough to accumulate the compounding benefit of a stable workspace. The chase and the progress are opposites. If you have read this far and built the workspace, you are already ahead of the people who switch every month. You just do not feel ahead yet, because being ahead is quiet and the chase is loud.

With that on the table, here are the rules. They are boring on purpose.

The Default Answer Is No

If you are asking *should I switch,* the default answer is no.

This is not laziness. It is not fear of change. It is arithmetic. Every switch costs you something — a day of setup, a week of rebuilt muscle memory, a month before your permanent memory is tuned again to the new tool's quirks. During that transition, your throughput is measurably lower than it was on the tool you left. If the new tool is only marginally better, you may never recover the cost. If the new tool is *significantly* better — and this is rare — the recovery happens in a month or two, and from then on you are genuinely ahead. Most tools, most of the time, do not clear that bar.

So the baseline, the thing you should assume unless you have a clear reason otherwise, is this: you stay. You keep working. You let the people with time on their hands do the evaluating, and you read their trip reports later, after the dust has settled.

That last phrase matters. *After the dust has settled.* A new tool's first month is almost pure hype and first impressions. Its third month is when the real shape of it emerges — the bugs people actually live with, the workflows that broke, the quiet updates that fixed the headline demos into something workable or exposed them as never-quite-there. If a tool is still being talked about sensibly at month six, that is when it is worth your time to look. Before that, the information cost is too high and the signal is too low.

The Three Reasons That Are Actually Reasons

There are, as far as I have been able to work out over many cycles of this, exactly three reasons to switch tools. Any of them on its own is enough. Anything outside these three is, most likely, not.

Reason one: your current tool is violating a principle you cannot work around. Go back to the seven principles earlier in this chapter. Pick the one your current tool is failing on. Is it the boundary? Permanent memory? Parallelism? If the tool you are using *structurally* cannot be pinned inside a workspace, or cannot read its own memory file across sessions, or cannot run seven copies at once without fighting for the same resources, and you have genuinely tried and failed to work around it — that is a reason to switch. Not because the new tool is exciting, but because the current tool cannot carry the method, and no amount of prompting cleverness will change that. The method is the asset; the tool is the vehicle. If the vehicle cannot drive the road, you change vehicles.

Reason two: the new tool does something the old tool cannot, and that *something* directly supports your structure. Notice the precision of that wording. The new tool does not have to be *better in general*. It has to enable your existing structure to work better. An example: if a new tool finally offered a persistent memory format that was genuinely interoperable across vendors — so that your rules lived in one place and every agent you ran, from any company, read the same file — that would support your structure. Adopting it, even at some transition cost, would pay back because your memory would stop being duplicated across `CLAUDE.md`, `AGENTS.md`, and whatever else. That is a structural gain. By contrast, a tool that is ten percent faster at generating code is not a structural gain. Speed changes nothing about your workspace; it just makes one lever lighter. That is nice, not load-bearing.

Reason three: the tool you are using has left you. Sometimes the decision is made for you. A tool is sunsetted. A vendor pivots. A pricing change makes the tool unworkable for your organization. A contract ends. A security review forbids it. In those cases you are not choosing to switch; you are being asked to, by circumstances outside your control. When this happens, grieve briefly and move on. The workspace survives the change. The seven principles survive the change. The new tool takes over the vehicle's job and the method continues, lightly jostled but intact.

Anything outside these three — *it got a lot of attention on social media this week, it tops a benchmark, a friend likes it, a podcast host swore by it, it has a pleasant color scheme* — is not, on its own, a reason to switch. Those might be reasons to *notice* a tool. They are not reasons to *move.*

The Three Tempting Reasons That Are Not Reasons

I want to name the failure modes explicitly, because they are seductive and they are how most engineers end up spending a quarter rebuilding their workspace for no net gain.

Hype is not a reason. Marketing copy is a craft, and the craft has gotten very good. A headline promising a step-change in productivity was, until recently, a reasonable signal that something interesting had shipped. It is no longer. Every new tool's launch post promises a step-change now, because that is what launch posts do, and because the tools that ship *without* that promise do not get covered at all. Hype has become noise. You need a quieter signal, which is: *does the tool help my existing structure?*

Features are not reasons. A feature is a small local improvement. Your structure is a global system. Small local improvements rarely justify a global rebuild, and the industry's addiction to shipping features — because features are easier to launch than principles — means you will be offered a new one every week, forever, until the end of your career. If you switch for every compelling feature, you will switch constantly and accumulate nothing. The structure you have is worth more than any single feature you do not.

Benchmarks are not reasons. A benchmark is a stylized test. Your work is not. The agents that win on benchmarks are often tuned for benchmarks, the way students who are optimized for the SAT are often optimized only for the SAT. The real benchmark, the only one that matters, is how many good tickets you close per week without burning yourself out. You can only measure that by closing tickets. No leaderboard on the internet can tell you.

I am being repetitive here, deliberately, because the repetition of the *principles* matters more than the cleverness of any single one. Hype, features, benchmarks — those are the three shapes the temptation tends to take. Any time you feel yourself being pulled toward a switch, name which of the three it is, and the pull usually eases on its own.

What Switching Actually Costs

If you have decided, against the default, that a switch is warranted, it helps to know honestly what you are signing up for. Nobody in the industry will tell you this cleanly because nobody selling you the new tool benefits from the accounting. So here it is.

A full switch — meaning: you move your primary agent from one tool to another across your entire workflow — typically costs you about two weeks of reduced throughput. The first day is setup: rebuilding the workspace with the new tool, reconfiguring the boundary, porting the deny list, re-teaching the permanent memory file. The rest of the first week is muscle-memory rebuild — your fingers will type the old command into the terminal for about four days before they stop, and each misfire costs you twenty or thirty seconds of re-orientation. The second week is permanent-memory tuning, because every rule you quietly accumulated over months on the old tool has to be re-discovered as the new tool makes new mistakes. You will add rules faster than usual during this week, and that is fine; you are rebuilding what you already had, not inventing from scratch.

Expect your throughput during those two weeks to be sixty to seventy percent of what it was. That is the tax. If the new tool is genuinely better, you recover the tax within a month and then start pulling ahead. If it is not, you do not. Budget for the tax honestly before you decide.

One piece of good news. The structural parts of the workspace — the seven Norse environments, the lease file, the `repos` folder, the desktop layout, the NOTES.md files, the two-session day — all survive the switch without change. That is the whole point of the work we did in Chapter 9. The directory tree does not care which agent reads it. The vast majority of what you built in Chapters 3 through 8 moves over untouched. The transition cost is concentrated in the thin layer where your structure meets the specific tool, and that layer is, by design, small.

How to Switch When You Have Decided To

If you have honestly considered the default, confirmed one of the three real reasons, budgeted the two weeks, and decided to switch — here is the short version of how to do it without losing your footing.

Do not switch everything at once. Pick one environment — I usually use `baldr`, because it's last in the rotation and easiest to borrow — and run the new tool in just that slot for a week while the other six keep running your current tool. This gives you a live, side-by-side comparison under real working conditions, not a demo-video impression. If at the end of the week the new tool in `baldr` is doing the work you need, at a quality you recognize, then widen the switch to two environments the following week. If not, you have only lost a week of one slot's output, and you know sooner rather than later.

Keep both permanent-memory files in sync during the transition. Every rule you add to the new tool's memory goes into the old tool's memory too, and vice versa, until the day you decide the switch is complete. This is slightly more work than either alone but it means neither tool is drifting behind, and if you decide to roll the switch back, you are not starting from scratch on the old side.

Do not update anything else during the switch. New desktop layout, new repo structure, new ticket-naming convention — all of it waits. One change at a time. When the transition is done, and the new tool has been humming for a month, *then* you can make whatever adjustment you were eyeing. Trying to do two changes at once is how people conclude that the method does not work when in fact they just moved the couch and the lamp on the same Friday.

Write down the decision. In a notes file, in your permanent memory, in an email to yourself — somewhere — record: *I switched from X to Y on this date, for this reason, at this transition cost.* In four months, when you are considering another switch, you will want that note. The you-of-four-months-from-now is not great at remembering why the you-of-today made the last call, and the record is a gift to that future you.

The Structure Is Yours. The Tool Is Rented.

Here is the line I want you to take from this chapter, if you take only one.

The structure is yours. The tool is rented.

The workspace, the seven slots, the lease file, the dish by the door, the two-session day, the validation ritual, the wind-down notes — those belong to you. You built them. They travel with you across jobs, across teams, across tools, across whatever the next decade of AI tooling happens to look like. They are portable because they are principles, and principles are the thing that survive the specific technologies that embody them.

The tool — whichever tool — is a tenant in that structure. It pays rent by doing useful work inside your system. If a better tenant comes along and your three-reason test passes, you let them in. If the current tenant stops paying rent — fails a principle, gets sunsetted, gets priced out — you show them the door and bring in a new one. The building stands either way.

This is, in the end, the quiet confidence the method is trying to give you. Not a promise that you will always know which agent to use. Not a promise that you will always pick the winner of the current tool war. A promise, instead, that the question *which agent to use* is smaller than the industry is currently training you to think it is, and that you can answer it calmly, on a slow Thursday afternoon, without panic and without FOMO, because you have already done the load-bearing work, and the load-bearing work is not going anywhere.

Pick the tool that serves the structure. Let the rest of the industry chase whatever it is chasing. Go close tickets.

Part IV — The Human Side

Chapter 11
Avoiding AI Burnout

I owe you this chapter.

Back in the Introduction, in the middle of the tour, I told you that Part IV of this book was about keeping yourself intact — that the last few years have been psychologically rough for everyone in this profession, that the industry has been particularly unkind about it, and that I would, eventually, talk plainly about the burnout that a lot of developers are currently carrying around like a piece of furniture they forgot they were holding. This is the chapter where that promise comes due.

I want to do it carefully, because burnout is not a topic that rewards briskness. If you came to this book because your throughput is down and your sleep is bad and you've started to feel a low hum of dread before you open the laptop in the morning, I don't want to hand you a chipper listicle and send you on your way. That isn't what's needed. What's needed is an honest look at a specific kind of exhaustion that didn't exist five years ago, a name for what you might be feeling, and a practical path out of it that doesn't require you to quit your job, delete your tools, or move to a yurt.

Let me be specific about what this chapter is and isn't. It is not a chapter about general software-engineering burnout, which is a real thing and has a real literature of its own, and I am not going to try to write that literature here. It is about a newer, more particular strain — the kind you get from spending eight hours a day collaborating with an agent that is sometimes uncannily good, sometimes bafflingly wrong, always confident, and endlessly available. That combination wears down a part of the brain that most burnout literature has not quite caught up with yet. It is not primarily a workload problem. It is a *decision load* problem, and we'll spend the first section pulling that distinction apart, because if you try to solve a decision-load problem with workload interventions — fewer tickets, longer breaks, more vacation days — you will do all of that and still feel terrible, and then you will reasonably conclude that something is wrong with you. Nothing is wrong with you. You're measuring the wrong quantity.

The good news, and I hope you'll forgive me for leading with it, is that the method you've already built in this book does most of the heavy lifting for sustainability before you ever notice it's doing it. The seven-slot structure, the glance cycle, the NOTES.md wind-down, the two-session day — those aren't just throughput mechanisms. They're decision-load mechanisms. Each one replaces a hundred small in-the-moment choices with one permanent structural choice you made weeks ago. By the time you're doing fourteen tickets in a day, your brain is, in a very specific and measurable way, doing *less* work than it used to do when you were doing four tickets by hand. I'll explain why that is — it's genuinely counterintuitive, and it took me longer than I'd like to admit to understand it — in the second section.

But the method is not a force field. You can still overdo it. You can still sit too long, parallelize too aggressively, skip lunch, skip weekends, ignore the warning signs your body and your relationships are sending you, and end up exactly where the industry's worst takes would like you to end up: burned out, resentful, convinced the tools are the problem, and one bad week away from hurling your laptop into the bay. I don't want that for you. That is the opposite of the point.

So here's what we're going to cover.

We'll start by naming the specific shape of AI burnout — what it feels like, why it's different from ordinary overwork, and why the interventions that fix ordinary overwork don't always fix this. Then we'll look at why the seven-slot structure paradoxically lowers cognitive load instead of raising it, so you understand what you're actually leaning on. Then we'll walk through the practical supports: the second monitor as an escape hatch; the two-session rhythm, which is not negotiable; the weekly and monthly habits that keep you from compounding small fatigues into large ones; and the warning signs that mean it's time to step away from the system for a day, a week, or occasionally longer. Finally, because I want this chapter to feel like a friend talking rather than a manual barking, we'll end on the long game — what this method is really for, which is not maximizing any single quarter's throughput but remaining a useful, curious, happy developer for the whole long arc of a career.

One last thing before we start, in the same plain tone I used in the Introduction. If you are reading this and you recognize yourself in it — if the dread is already there, or if your sleep has been bad for a while, or if the people who live with you have gently asked whether you're all right — please take that seriously. A book can give you language and frameworks and small practical tools; a book cannot replace a conversation with a doctor, a therapist, your partner, a trusted friend, or a colleague who has come out the other side of something similar. Use this chapter for what it's good for. Use the other resources for what they're good for. And don't try to out-method the parts of life that aren't about method.

With that said: let's talk about what AI burnout actually is, and why it sneaks up on people who thought they were the last person it would happen to.

The Specific Shape of AI Burnout

Naming the Thing before We Treat It

I want to start this section by being honest about something that most treatments of this topic skip past, which is that burnout has been in the water long before AI arrived. Software has been burning people out for decades. Every senior engineer you respect has at least one story about a year they'd rather not revisit — a bad project, a bad manager, a bad stretch where they came out thinner and angrier than they went in. That kind of burnout is real and it is well documented. The World Health Organization formally recognized burnout as an *occupational phenomenon* in the eleventh revision of its International Classification of Diseases, back in 2019, and the canonical model for it — Christina Maslach's three-dimension framework of **emotional exhaustion**, **cynicism and depersonalization**, and **reduced sense of accomplishment** — has been the workhorse of the research literature since the early 1980s. None of this is new.

What *is* new, and what I want to give you language for in this section, is a specific flavor of the same underlying condition that has emerged in the last two or three years, and that the classical literature is still, as of my writing this, scrambling to catch up with. I am going to call it **AI burnout**. I am going to argue that it is a distinct shape of the Maslach burnout syndrome, and that it is currently hollowing out a meaningful percentage of our profession with extraordinary efficiency, and that it is almost perfectly invisible from the outside — even, sometimes, to the person carrying it.

The operating claim, and I want you to let it sit for a minute before we go further, is this:

AI burnout is not a workload problem. It is a decision-load problem.

That distinction is not cosmetic. It is the whole reason the interventions that work for ordinary overwork don't work here, and it is why a number of excellent engineers I know are currently doing everything their companies suggest — taking the vacation days, turning off notifications at 6 PM, blocking their calendars for focus time — and still feel worse every month. They are correctly treating a disease they do not have. The disease they have is adjacent, and its mechanism is different, and until we name the mechanism we cannot treat it.

What Decision Load Actually Feels Like

Let me paint it, because if you're carrying this you will recognize yourself inside of thirty seconds and if you aren't you probably know someone who is and I'd like you to recognize *them.*

You sit down in the morning. The agent is waiting. It generates a proposal — a function, a diff, a plan. You have to decide: accept, reject, modify, retry, or clarify. That is a decision. It is a small decision. It takes you, maybe, six seconds of focused attention. Six seconds is nothing. You have those to spare.

But then it generates another. And another. Each one requires the same six-second micro-evaluation: *is this right, is this wrong, is this almost right, what would I have done instead, does this pattern match the codebase, does this break anything, do I push back or do I accept.* By lunch you have made, conservatively, several hundred of these. By the end of the workday you have made on the order of a thousand. Every one of them drew from the same reservoir of focused attention that a pre-AI engineer used for maybe thirty to sixty real decisions in a day.

Researchers have been circling around this for a long time without having AI specifically in mind. Sophie Leroy — whose foundational 2009 paper was written while she was at the University of Minnesota, and who is now at the University of Washington Bothell — has a body of work on what she calls **attention residue**, the phenomenon where, when you switch tasks, a portion of your attention remains on the previous task for a measurable period afterward. Gloria Mark at UC Irvine has spent two decades measuring how long it takes to recover attention after an interruption, and her widely-cited figure of roughly twenty-three minutes to fully regain focus after being interrupted has become a kind of folk truth in productivity writing, though the real number depends on what you were doing and what interrupted you. Her 2023 book *Attention Span* is the best lay summary of that body of work I know of, and I recommend it to every engineer I meet who is struggling with this.

Here is the thing neither of those researchers was studying but that follows directly from their findings: an AI coding agent, by its very nature, turns every single line of code into a potential interruption. Not because the agent is rude. Because the agent is *offering a decision*, and every offered decision is an interrupt to whatever train of thought you were on. Over the course of a day you are being interrupted, gently and helpfully, hundreds of times. Each interrupt has a residue. The residues accumulate.

By 4 PM, a certain kind of exhausted engineer — and if you are reading this there is a high probability you are one of them — notices that they can no longer summon opinions. Not that they disagree with the agent. That they cannot *generate* a response at all. The agent proposes something; they blink at it; they accept it, not because they have evaluated it and found it correct, but because evaluating is no longer a thing their brain will do this afternoon. That moment is decision-load burnout in a single frame. The engineer has not written more code than they would have pre-AI. They have made ten to twenty times more evaluative judgments, and the judgment faculty is a muscle, and the muscle is cramped.

Why the Usual Interventions Miss

Now watch what happens when that engineer goes to a manager and describes what's wrong.

The manager, operating from the classical burnout playbook, suggests the classical remedies. Take a long weekend. Cut your ticket count. Block off an hour a day for deep work. Turn off Slack notifications. These are good interventions for workload burnout. They will, against decision-load burnout, do almost nothing.

The long weekend helps you sleep, which is real, but the judgment faculty will be fatigued again by Tuesday afternoon because Monday involved another thousand micro-evaluations. Cutting ticket count helps a little, because fewer tickets means fewer decisions, but if you're still pair-programming with an agent on every ticket, the compression effect just raises the decision rate per ticket. The focus hour helps you generate fewer decisions but exposes you to none of the actual pleasure of deep solo work, because you spend the hour in the agent's proposal stream anyway. Turning off Slack is great for everything *except* the source of the exhaustion, which is not coming from Slack. The decisions are coming from the thing you use to do your job.

The reason I care about naming this precisely is that engineers who apply workload interventions to decision-load problems and then don't feel better conclude, with perfect internal logic, that *something is broken in them*. It isn't. They are measuring the wrong quantity. You cannot solve a reservoir-depletion problem by resting the wrong muscle. The interventions that work for this specific strain of burnout are not rest interventions; they are *structural* interventions that reduce the rate of new decisions at the source, and they are what the rest of this chapter is going to walk through.

The Industry Context, because This Does Not Happen in a Vacuum

I want to step back for a minute and talk about the broader context you are doing your job inside of, because it is not incidental to the discussion. The honest picture is ugly and I'm not going to pretend it isn't.

Since 2022, the technology industry has shed jobs at a pace that has no close precedent in the lifetimes of most working developers. Trackers like **layoffs.fyi** recorded over a quarter of a million tech layoffs in 2023 alone, and another cluster of six-figure cuts through 2024 and into 2025 as companies continued to restructure around AI-enabled teams. The headlines were, and continue to be, uniformly cheerful on the other side of the equation — founders and executives talking openly, on earnings calls and in interviews, about smaller engineering organizations producing more software with AI assistance, and treating that compression as a straightforwardly good thing. If you are a working engineer and that framing landed on your ear as a threat rather than a triumph, you were hearing it correctly.

The psychological literature on what this kind of sustained employment threat does to workers is not subtle. Research on **job insecurity** — a separate construct from burnout, though they compound — has consistently linked it to elevated rates of anxiety, depression, sleep disturbance, and cardiovascular risk, going back decades. When you layer that on top of a decision-load problem, you get a population of engineers who are cognitively exhausted from the work itself *and* chronically stressed about whether the work will still be theirs in six months. That is not a minor mental-health footnote. That is a full-surround assault on a profession's nervous system, and if you've been feeling the pressure of it, you have not been imagining it.

I mention Gallup's *State of the Global Workplace* reports not because they study developers specifically but because the annual findings — elevated stress levels, declining engagement, rising proportions of workers experiencing daily negative emotions — describe a labor environment that has been visibly deteriorating since well before AI arrived, and that AI tooling has landed on top of rather than relieved. The tools are real. The pressure around the tools is also real. Pretending only one of those is happening is not honest.

I don't bring any of this up to panic you. I bring it up because I think one of the worst things you can do to a tired person is tell them the problem is entirely in their head. It isn't. The problem is partly decision-load, partly industry-wide, partly economic, and partly the specific emotional weather of being a mid-career engineer in a profession whose shape is being renegotiated in public. All of it is on you at once. Of course you're tired.

What This Actually Feels Like, for You, Right Now

Let me do a thing I do not normally do in technical books, which is talk directly to you for a minute.

If you are reading this and the description above has started to feel uncomfortably specific — if you noticed yourself nodding at the 4 PM decision-collapse, or recognized the feeling of accepting a diff you haven't really read, or realized you've been taking longer showers in the morning because the idea of opening the laptop puts a small, heavy stone in your stomach — I want you to know three things, in this order.

First: this is *common*. You are not uniquely broken. You are in the middle of a predictable response to a novel, sustained cognitive pressure, combined with real economic stress, and you are experiencing it along with a large and mostly silent cohort of your peers. The reason it feels solitary is that our profession is bad at talking about this in public, and the reason it is bad at talking about this in public is that a lot of us, rationally, do not want to mention anything that could be used against us in a layoff decision. That silence makes the experience worse. It does not make the experience rarer.

Second: the shape of what you're feeling has a name now, or at least a working name. *Decision-load burnout.* Sitting alongside classical overwork burnout, intersecting with job-insecurity stress, compounded by attention residue from hundreds of daily micro-evaluations. Giving it a name does not make it go away, but it does something almost as useful — it tells you which interventions will work and which will not. If you've been trying to fix this with rest and the rest has not worked, please stop concluding the problem is your character. The problem is the match between the intervention and the mechanism.

Third: there is a path out. It is the rest of this chapter. It is not glamorous and it is not instant, but it is real. It works by changing the structure of your working day so that the decision rate comes down at the source, and it works on exactly this shape of exhaustion because the structure is built around the mechanism of the exhaustion rather than around the symptom. You will not feel better tonight. You can reasonably expect to feel better in two to three weeks. That is fast, for burnout. That is one of the genuine gifts of having named the thing precisely.

A Gentle Ask about the People around You

Before we go on to how the method itself acts as a decision-load dampener, I want to make one more request of you, and I'm going to make it plainly.

Somewhere in your orbit — on your team, in your Slack, in your friend group, at the next desk over in the coworking space — there is a developer who is not you, who is further along in this than you are, and who has stopped talking about it. Maybe they have gone quiet in meetings. Maybe their PRs have become less ambitious and a little more mechanical. Maybe they have started leaving cameras off. Maybe they canceled on dinner twice in a row with a vague excuse. Maybe they have been tweeting less, or commenting less in the group chat, or responding to Slack with a thumbs-up instead of words. You probably have someone in mind right now.

What I'd like you to hold, alongside whatever sympathy you are generating for your own exhaustion, is the possibility that *they are in this too*. Not performing it, not posting about it, not wearing it on a lanyard — just quietly, progressively, sliding into a version of what this section has described. The people who end up in the hardest shape with decision-load burnout are often the people who were best at the pre-AI job, because they are the ones whose professional identity is most tangled up with being the person in the room who has an opinion, and the thing this burnout steals first is the ability to generate opinions. It hits the conscientious hardest. It hits the quiet ones first.

You don't have to fix them. You probably can't. But you can, small and concretely, do three things. Ask how they're actually doing, the second time in the conversation, after the first brush-off. Share a copy of this chapter if it rings true for you — not because I'm chasing sales, but because giving someone language for what they are feeling is sometimes the single most useful gift one engineer can give another. And do not, do not, ever, in a moment of your own decent recovery, fall into the trap of explaining to them why they should just use the tools differently. That is the last thing a decision-load-burned engineer needs. What they need is someone who acknowledges the mechanism without judgment, and who believes — as I want you to believe — that there is a way through.

With that said, and with, I hope, some softness installed in your posture, let's look at why the seven-slot method, counterintuitively and against all common sense, actually reduces cognitive load instead of raising it. It is the single most useful thing I know about this problem, and it is the subject of the next section.

Figure 11.1 draws the difference between the two burnouts at the mechanism level — what's being depleted, what the interventions that actually work look like, and why rest fixes one and not the other. If you are in the middle of trying to decide which shape you are in, use the figure. The answer is usually obvious within ten seconds of looking at the two columns side by side.

Figure 11.1 — Workload Burnout vs. Decision-Load Burnout

Same symptom surface. Different mechanism. The interventions that help one rarely help the other.

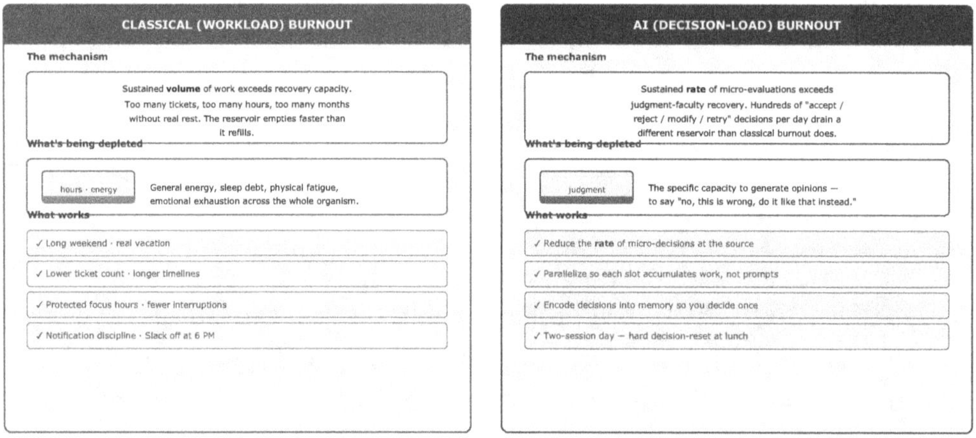

Both present as "4 PM I can't think anymore." Name the mechanism first, then choose the intervention.

Why Parallel Work Reduces Cognitive Load (Counterintuitively)

The Objection You Are about to Raise

Let me guess what you are thinking, because the last section named the disease and this section is going to prescribe something that sounds, on the face of it, like the opposite of medicine.

You just read that AI burnout is a decision-load problem. That the reservoir of focused attention you draw from is getting drained several hundred times a day by micro-evaluations. That by 4 PM a certain kind of exhausted engineer can no longer generate opinions, and that the thing doing the draining is the helpful agent sitting in your terminal. All of that tracks; you may have felt a little quiet recognition at parts of it.

And now here I come, in the next section of the same chapter, proposing a working method in which you have *seven* helpful agents running at once instead of one. Seven streams of proposals. Seven conversations to track. Seven potential interruptions. If one agent was cognitively expensive, seven agents should, by the most basic arithmetic, be seven times more so. The method looks, at a glance, like someone trying to treat a flooded basement by opening six more taps.

I understand the objection. I had the same objection for about the first year I was building this system, which is part of why it took me a few years to build it. But the arithmetic is wrong. It is wrong in a specific, structural way that took me a long time to see clearly, and this section is going to be about nothing else.

The short version, which I'll spend the rest of the section earning: **seven agents in a structure are less cognitively expensive than one agent without one.** Not a little less. Meaningfully less. The structure is doing the work the lone agent would be asking your attention to do, and the structure does not get tired.

That sentence is going to sound like a trick. It is not a trick. It is a consequence of a specific distinction between two kinds of cognitive work — *holding* and *checking* — and once I draw the distinction cleanly, the rest of the argument becomes mechanical.

Let's draw the distinction.

Holding versus Checking

There are two modes of attention involved in any piece of intellectual work. One of them is genuinely expensive. The other one is nearly free. Most engineers, before they've thought about it, assume they are the same mode. They are not.

Holding is what your brain does when it is actively maintaining a problem in working memory — turning it over, running simulations, comparing alternatives, carrying the state of an unfinished solution across minutes or hours. Holding is what you are doing when you are "in the zone." It is what you were doing in the Before Times when you sat down with a gnarly bug at 9 AM and surfaced at lunch having spoken to no one. Holding is cognitively expensive because it recruits, simultaneously, most of the brain's executive-function circuitry — prefrontal working memory, attentional control, inhibitory control against distraction. A skilled senior engineer can do about three to four hours of genuine holding in a good day. Some days, fewer. The reservoir is small and it empties, and the bottom of the reservoir is where decision fatigue lives.

Checking is something else entirely. Checking is what your brain does when it is evaluating something that already exists, from the outside, against a criterion it already knows. Checking is what a code reviewer does when a PR comes across their desk — they read, they form a shape-level opinion, they either approve or leave a comment, and they move on. Checking does not require holding the whole problem in working memory because the problem is already solved in front of you; the question is only whether the solution meets a standard you already have.

Checking is *orders of magnitude cheaper* than holding. Not because engineers are lazy, but because the cognitive architecture is different. When you review a PR, you are not simulating the whole solution in your head. You are pattern-matching against a library of experience you already have — does this shape look right, does this test look sufficient, does this name make sense, is there an obvious bug I can point at. A senior engineer can check perhaps forty or fifty PRs in a day before they become ineffective. They cannot hold forty or fifty problems. They probably cannot hold three.

The single most important thing to understand about AI burnout is that **naive agentic coding asks you to hold, not to check**. When you sit there alone with one agent, in a back-and-forth stream of proposals and refinements, you are not checking against a finished solution. You are co-holding a half-formed one with a partner who, by nature, cannot carry any of it on its own. The agent generates a fragment; you have to merge it into your own live model of the problem, in your head, and decide what happens next. That is holding. It is the expensive mode. And because the agent is fast, you are being asked to hold at a rate that is structurally incompatible with the human nervous system. That is the mechanism I promised to name. That is why a day of one-on-one agent work feels like a marathon despite much of it being spent "just reviewing."

The method reframes the work. It turns every interaction with every agent into a checking interaction. The agent does the holding; you do the checking. The agent goes off, carries the problem in its own context for twenty minutes, produces a finished shape — a commit, a diff, a draft PR — and hands it back to you. At that moment, for the first time, you engage. You read the PR the way you'd read any PR: from the outside, against a standard, in checking mode. You approve, you leave comments, you ask for another pass. You do not simulate the solution in your head. The simulation already happened; the simulation happened in the agent. Your brain never had to light up the expensive circuitry.

Seven agents in parallel, each in their own environment, each holding their own problem, each producing a finished shape for you to check — that arrangement is not seven times the work of one agent held in parallel. It is *less* work than one agent in the naive mode, because the per-unit cognitive cost has been replaced by the cheaper mode of attention. You have moved from holding to checking. The number of items flowing past you went up. The cost per item went way, way down. Your daily reservoir refills instead of draining.

This is the whole trick. Everything else in this section is consequence.

A Historical Analogy, because the Pattern Is Older than AI

I want to slow down here and make the case by analogy, because engineers often find an argument more convincing when they can see a version of it that has played out already in a domain they already respect.

Consider compilers.

A compiler, in the sense of a piece of software that translates a higher-level language into machine code, is an automation of a specific kind of cognitive work. Before compilers, a programmer held the entire translation in their head. They thought in machine instructions. Every line of code was, in effect, a tiny act of holding — turning an intention into a sequence of specific register operations and memory addresses, tracking carry bits and flags, simulating the CPU's state. Programmers of that era were genuinely remarkable in ways that are hard to fully appreciate now, and the ones I've met from that generation carry themselves accordingly.

The compiler changed what the work *was*. It did not make programmers lazier or less capable. It moved the locus of attention. After compilers, a programmer stopped holding machine-level state and started holding language-level intent. The programmer now wrote `x = y + 1`, and trusted the compiler to hold the register allocation, the instruction selection, the branch prediction hints, the endianness conversions. The compiler was, in effect, holding an enormous amount of cognitive work so the programmer could do more interesting, higher-level holding.

Anyone who told a 1955-era programmer that adding a compiler would let them write "more programs per day" would have gotten the same objection I just answered above. *But now I have to debug the compiler's translations. Now I have to reason about two levels of abstraction at once. This is more work, not less.* It would not have been a stupid objection. It would have been exactly wrong, for exactly the reason I'm making now: the compiler did the holding, and the programmer's role moved to checking (and to higher-level holding of different kinds of problems). Over a generation, the entire profession gained access to the productivity and the quality-of-life improvement that flowed from making that move. We are still, collectively, taking it for granted today.

This same pattern has repeated, at smaller scales, in essentially every productivity gain software has had. Version control freed us from holding the history of our own changes; `git log` holds it. Unit tests freed us from holding every edge case simultaneously in our heads; the suite holds them. Linters freed us from holding every style rule; the linter holds them. Continuous integration freed us from holding the status of every branch; the dashboard holds it. In every case, the structure took over a piece of the holding, and our attention migrated to a higher-leverage kind of checking and designing.

AI agents, when used well, are the next entry in that sequence. They are not — this is the important part — the replacement for the engineer. They are the next place we move the holding. The engineer's job is going to increasingly be about checking, about design, about choosing what to ask for in the first place. That is the migration I am describing, and the seven-slot method is one specific, practical way to do the migration without hurting yourself in the process.

Some engineers will find that migration uncomfortable, and I respect that discomfort. There is a real grieving involved in watching a kind of work you loved — the long morning spent holding an entire subsystem in your head — become less common in your day. I have felt that grief and I don't want to pretend otherwise. But grief and harm are different things. The work is changing shape. The new shape is not worse. It is, when you arrange it properly, considerably more sustainable than the old one was.

What the Structure Is Actually Doing for You

Let me be mechanical about this. What, specifically, is the structure of the seven-slot method doing that your brain would otherwise be doing? Name the pieces.

The lease file is holding environment state. Without the lease file, you would be holding, in your own head, which environment each in-flight ticket is parked in, which are free, which have stale locks, which are about to be reclaimed. That is not glamorous holding; it is the kind of inventory-keeping that feels small until you multiply it across a day and realize it is eating real attention. The lease file is a small text file on disk that costs you nothing to maintain and that carries the entire inventory on your behalf. You never have to think about "which environment am I in" because the lease file knows and the agents check before claiming. That is offloaded holding. Your brain is freer by precisely that amount.

The dish-by-the-door zones are holding spatial state. Without the fixed zones, you would be holding, in working memory, which tab in which browser window has the Jira board, which Slack channel is the communications channel, which terminal has the validation browser, which has your personal shell. You would have to *search* for things. You would spend a nontrivial percentage of your day in the twitchy low-grade confusion of "where did I put that." The zones replace all of that with muscle memory. Your eyes know where Jira is because Jira is bottom-right, always, forever. Your eyes know where validation is because validation is upper-left, always, forever. That is offloaded holding. Your brain is freer by precisely that amount.

The NOTES.md wind-down is holding session state. Without the wind-down ritual, you would start every morning trying to reconstruct, from memory, where you left off in each of seven parallel threads of work. That is catastrophic for a tired brain, and it is what actually happens to developers working without this structure — they compensate by narrowing their in-flight work to one ticket at a time, because they cannot cognitively hold more than one. The NOTES.md files, written at thirty seconds per environment at the end of each session, carry yesterday's state into this morning without asking your brain to do any of it. That is offloaded holding, at scale, at the exact moment of the day when your reservoir is lowest. Your mornings are freer by precisely that amount.

The glance cycle is holding scheduling. Without the glance cycle, you would be trying to track, in your head, which Claudes have been running for how long, which are likely to be finished, which are likely to be stuck, which to check next. This is exactly the kind of interrupt-driven scheduling work that a CPU scheduler does for processes, and it is exactly the kind of work that no human brain does well for more than a few minutes at a time. The glance cycle replaces the tracking with a simple rule: every two or three minutes, look at all seven. The rule is simple enough to not require holding. Your brain is freer by precisely that amount.

The draft PR is holding the definition of done. Without the draft PR, you would be holding, somewhere in your head, the list of tickets that are in progress, what state each is in, what still needs to happen for each to move to ready-for-review. That is load. The draft PR makes it external — you can *look* at GitHub and see the state of everything that is in flight. When a PR is no longer a draft, it is ready. When one is a draft, it is in progress. The state lives on the PR, not in you. That is offloaded holding. Your brain is freer by precisely that amount.

The validation ritual is holding quality criteria. Without the ritual, you would be holding, in your head, the definition of *good* that a PR has to meet before you promote it. That definition is long — correctness, tests, edge cases, style, acceptance criteria, review-shape, commit hygiene — and no human reliably holds all of it while also trying to evaluate specific code. The ritual externalizes the definition: it is six named steps, in order, and you execute them against each candidate. Because the steps are named and ordered, you don't have to remember them; you just run them. That is offloaded holding. Your brain is freer by precisely that amount.

Every one of those is a structural element that replaces *continuous holding* with *external state.* Each one saves you a small amount of attention. Added together, they save you a very large amount, and that saving is exactly what the method is for. You are not doing more work. You are doing less of a more expensive kind of work and more of a cheaper kind.

The Parallelism Math, Done Properly

Now we can address the specific objection about parallelism head-on, because the structure is on the table and we can reason about it cleanly.

Consider the naive case: one engineer, one agent, one ticket, back-and-forth in real time. The engineer's working memory is continuously engaged with the unfinished solution. Every time the agent generates, the engineer integrates. The integration is holding. The rate of integration is determined by the rate of agent generation, which, for any agent worth its model, is faster than the engineer's sustainable holding rate. The engineer runs out of holding capacity well before the end of the day. Output starts to feel like acceptance rather than review. We named this in the previous section.

Now consider the parallel case with structure: one engineer, seven agents, seven tickets, each in its own environment with its own context and its own acceptance criteria. The engineer is no longer integrating in real time. The engineer briefs each agent (a small act of design, not of holding), releases it to work, and rotates. At any given moment, six of the seven agents are *offscreen* from the engineer's attention — they are working, but the engineer is not holding their state, because the environment is holding it for them. When an agent surfaces a finished shape, the engineer checks it in checking mode, not holding mode. If the check passes, it's done. If the check fails, the engineer writes down what needs to change and hands it back for another pass.

Look at the math. In the naive case, the engineer is in holding mode for roughly the entire working hour, at a rate set by the agent. Call it one unit of cognitive cost per minute, sustained for two to three hours before exhaustion. Total daily budget: about 150 cognitive units before the wheels come off, and the reservoir is empty.

In the parallel case with structure, the engineer is in briefing mode for perhaps five minutes per ticket (a burst of holding, concentrated, voluntary) and in checking mode for perhaps three minutes per completed cycle (cheap, against an external shape). In the twenty minutes between a briefing and the first completion, the engineer is doing *nothing cognitively expensive at all.* They are rotating, glancing, reading NOTES.md, sipping coffee, looking out a window. That idle time is not idle; it is the recovery time that turns the next briefing into a full-quality act of design instead of a fatigue-impaired one.

Total daily budget, under the parallel structure: the engineer can brief perhaps fourteen tickets and check perhaps twenty-five completed cycles in a working day and still finish the day with reserves in the reservoir. The total output is more than ten times what the naive case produces. The total cognitive cost, per-minute, is roughly one-third of what the naive case demands, because the idle rotations and the checking-mode reviews are so much cheaper than sustained holding.

The product of those two ratios — more than ten times the output, at one-third the per-minute cost — is the whole productivity case for the method. I have put it in terms of cognitive economics rather than tickets-per-day because I want you to see *where* the gain comes from. It does not come from the engineer being superhuman. It comes from the structure doing the holding that the engineer would otherwise have to do, and from the mode shift from holding to checking that the structure enables.

What the Research Literature Supports (and What It Doesn't)

I want to be careful here, because the research on parallel agentic work specifically is thin. No one has run a longitudinal randomized trial on engineers using seven-agent workflows versus one-agent workflows, and I cannot pretend they have. What exists in the literature is a set of supporting strands that, taken together, are strongly suggestive.

The attention-residue work from Sophie Leroy, which I cited in the previous section, is a direct argument *against* the naive one-agent mode and *for* structures that reduce interrupt density. Every micro-evaluation in a one-agent stream is an interrupt. Every check against a finished shape is not. The parallel structure, counter-intuitively, reduces the effective interrupt rate because interrupts from agents you are not currently looking at simply do not interrupt — they accumulate quietly in the agents' own contexts until you are ready to check them.

The deep-work literature associated with Cal Newport, and the older flow-state work associated with Mihaly Csikszentmihalyi, both argue that cognitive performance peaks during sustained periods of undivided focus on a single task. This sounds, at first, like an argument *against* seven-way parallel work. Read more carefully, the argument is about undivided focus *during* the work of solving a problem. It is not an argument that every hour of every day should be spent in solving-mode. Checking-mode work does not require flow. Rotating among seven tickets is not an interruption of flow; it is a deliberate displacement of the solving work into the agents, so that the engineer can be in flow during briefing (which is a design activity) and in efficient checking-mode the rest of the time. This is subtle and I want to be honest about it, because some readers of those frameworks will hear "parallel" and flinch. I am not contradicting them. I am saying their frameworks apply to solving, and my frameworks apply to a working day that has had most of the solving offloaded.

The cognitive-load theory literature (Sweller, van Merriënboer, and others going back to the late 1980s) distinguishes between **intrinsic load** (how hard the material inherently is), **extraneous load** (load imposed by poor presentation or structure), and **germane load** (load directed toward building useful schemas). Classical programming has a mixture of all three; agentic programming without structure dramatically increases the extraneous load, because every agent interaction becomes a presentation problem the engineer has to re-parse. The method in this book is, in cognitive-load-theory terms, an attempt to drive extraneous load as close to zero as possible — by standardizing the presentation, the layout, the file locations, the ritual — so that the engineer's limited working memory is free to spend its budget on intrinsic and germane load instead. The theory, when you squint at it, predicts exactly the productivity-plus-sustainability gain I am claiming.

None of that is proof. I am not going to overstate the science. What I am going to claim, based on a combination of the research that does exist and two plus years of living inside this method, is that the specific argument I have made in this section — that parallel structured work is less cognitively expensive than serial unstructured work — is consistent with everything the relevant literatures currently suggest and is, in my own lived experience, the most reliable lever I have ever found on the cost of a working day.

The Objection about Mental Juggling

There is one specific worry some readers will still have, and I want to address it before moving on.

You might be thinking: *"But when I'm rotating through seven agents, don't I have to remember what each one is doing? Isn't that just seven contexts I have to carry in my head at once? How is that less than one?"*

It is a fair question, and the answer is that the method is built specifically so that you do *not* have to carry seven contexts in your head.

What you carry in your head, when the method is running well, is approximately nothing. You carry a single rule — *glance at all seven, react to whoever surfaces a signal first* — and a trust that each environment's state is preserved in its own directory, its own NOTES.md, its own conversation history, its own draft PR. When a specific agent surfaces a completed shape and demands your attention, you *read in* its state from those external sources. You do not *remember* its state. The read-in takes maybe thirty seconds — a glance at the PR, a glance at the ticket, a glance at the conversation — and after those thirty seconds you have everything you need to check the work. Then you hand back, the context unloads, and you move to the next slot where whoever has surfaced there will load *their* state the same way.

This is exactly what a computer does when it context-switches between processes. The CPU does not *hold* the state of all running processes; it holds the state of the one currently running and pages everything else out to memory. When a process gets scheduled in, its state is paged back from where it was stored. Modern operating systems are built on the recognition that keeping thousands of processes' state in CPU registers simultaneously is impossible, and that paging in and out is both possible and efficient. You are, by running the seven-slot method, doing the same thing with your attention. The environment directories are your paged storage. Your head is a small, fast register file. You do not store what you can read back cheaply.

I belabor this because it is the specific place where engineers first encountering the method get stuck. They assume the method is asking them to hold seven things in their head. It is not. It is asking them to hold zero things in their head, and to trust the filesystem to hold the seven on their behalf. Once that trust is built — which, for most engineers, takes about two weeks of real use — the cognitive difference between running one agent and running seven becomes almost imperceptible. The seven-agent day does not feel crowded. It feels, if anything, quieter than the one-agent day did, because in the one-agent day your brain had to hold the unfinished solution continuously, and in the seven-agent day the solutions are elsewhere and your head is your own.

Buzzy, I think, would have recognized this. He did not hold the whole barn in his head while he was driving a specific nail. He held the nail. The barn was the barn's problem. The structure of the job — the shakes you'd already nailed up, the ones still in the pile, the ones your dad was fitting two studs over — was held by the materials themselves and by the rhythm of the crew. A good carpenter knows which cognitive work is theirs and which belongs to the structure of the job, and they push as much as possible out to the structure, because their hands and their eyes and their judgment are the scarce resource. The analogy holds. You are the carpenter. The structure is the barn. The agents are not your coworkers; they are, closer to the truth, your claw hammers, each driving a different nail at the same time while you watch and adjust.

The Emotional Experience of Running the Method Well

I want to close this section on something that is not quite a technical point, but that matters to why I'm writing this book at all.

When the method is running well, the emotional quality of your working day shifts in a specific way that I find it hard to describe without sounding grandiose. I'll try anyway, because I think the shift is worth naming even if the words are imperfect.

In the naive one-agent mode, the dominant emotional tone of the day is *pressure.* The agent is fast, the stream is continuous, and you are always slightly behind, always slightly choosing between keeping up and thinking. By mid-afternoon the pressure becomes dull, and you start optimizing for getting through rather than for getting right. That is the emotional signature of holding-mode work at an unsustainable rate, and it is what the industry is currently asking most developers to live inside of for eight hours a day.

In the parallel structured mode, the dominant emotional tone is something closer to *oversight.* You are not under pressure. You are watching things happen — sometimes well, sometimes badly — and your job is to notice, judge, and redirect. There is an unhurriedness to it that is almost startling the first time you feel it, because you have been conditioned to expect work to feel like pressure. Work does not have to feel like pressure. Work can feel like a fire dispatcher at a slow station, or a lighthouse keeper on a calm night, or a farmer watching the irrigation run.

I know that comparison will read as hyperbolic to some people, and I accept that. But I mean it quite literally. The method turns a day of agentic coding into something that can, on a good day, feel like it used to feel in the Before Times when you were working on something you cared about, without a deadline, at a pace your body agreed with. That feeling — let's call it *the feeling of being a competent adult doing useful work at a sustainable rate* — is what the industry has spent the last three years quietly taking from a lot of engineers, and what I genuinely believe this method gives back.

That is the biggest claim I will make in this chapter, and I want to make it carefully, because I do not want to promise something I cannot deliver. I am not promising you joy every day. Some tickets will still be miserable. Some weeks will still be hard. Some quarters will still test your patience with your codebase, your team, your tools, or the industry you chose. What I *am* promising is that the ambient cognitive pressure of your working day — the thing that has been accumulating underneath everything and eating your weekends and fraying your relationships — can come down dramatically once the structure is doing the holding and you are doing the checking. That is the mechanism. That is the promise. And it is why I have spent more pages on this section than I originally planned, because if you take nothing else from this chapter, I would like you to take this: seven agents, inside a structure, are less cognitively expensive than one agent without one, because the mode of attention they demand is different, and the different mode is the one your nervous system can actually sustain for forty hours a week over decades.

Before We Move On

The next section is short, practical, and, after this one, almost comedically simple. It is about the second monitor, and its role as a deliberate escape hatch from the method, and why an engineer who has just spent twenty pages learning a system of seven slots should also, crucially, keep one monitor that is emphatically not part of the system. Hold onto that idea for a moment; it is important, and it is one of the supports that keeps the whole cognitive argument from tipping over into a different kind of overload.

But before we go there — take a minute. If the argument in this section landed, it should have landed with a small amount of physical relief. Some of the tension you have been carrying in your shoulders about the idea of parallel work was, I think, misplaced. The structure is not asking you to do seven times the work. It is asking you to let seven agents do their work while you do a quieter, lighter kind of your own. If that reframe helps, let it help. Put the book down for a minute. Stand up. Look at something that isn't a screen. Then, when you're ready, we'll talk about the monitor that is deliberately not on the system.

The Second Monitor as Escape Hatch

I said the previous section was about the cognitive economics of parallel work, and I said this section would be short and practical by comparison. That is still true. What I didn't say is that this section is also, quietly, one of the most important structural decisions in the whole method, and the one I see newcomers skip most often, usually because it looks at first like an afterthought.

It is not an afterthought. It is load-bearing. Let me explain why.

A Problem the Last Section Created

The previous section made a specific and, I hope, convincing argument. It said that seven agents inside a structure cost less cognitive effort than one agent without one, because the structure offloads the expensive mode of attention and leaves you doing the cheaper mode. That argument is correct. I stand behind it.

But it has a failure mode, and the failure mode is what this section exists to prevent.

The failure mode goes like this. You read the last section. Something in it resonates. You think, quite reasonably, *if seven slots are more sustainable than one, then my job from now on is to get very good at running seven slots, at which point I will be maximally productive and minimally exhausted, and that will be my life.* You lean in. The system starts working. Your throughput goes up. Your evenings feel lighter. For a few weeks you are genuinely delighted.

And then, somewhere around week three or week six or week fourteen — the timeline varies — a new kind of fatigue arrives. It is not the old decision-load fatigue from the one-agent days. That one is gone. This one is different. It is the fatigue of a person whose *entire waking working life is now inside a single well-designed structure*, and who has not realized that even a well-designed structure, experienced without break, will grind on a human the same way a treadmill does. The structure is not the problem. The absence of anything that isn't the structure is the problem.

This is the failure mode. It is subtle, because the structure is doing what it promised to do. You are, objectively, less tired than you used to be in terms of decision load. But you are tireder in a different way that the method, on its own, was not designed to solve. The decision-load win is real. The side effect, if you let it metastasize, is that every minute of your working day is *on*. Every minute is in the system. Every minute is in the rotation, in the glance cycle, in the validation mode, in the lease file, in the draft PR, in the two-session rhythm. Even the moments you are not typing, you are in the posture and the mental mode of the person who is running the system.

That is too much structure for a human to live inside of, and we need a release valve. The second monitor is the release valve.

The Second Monitor Is Not Part of the System

Here is the rule, written as plainly as I can write it. *The primary monitor holds the system. The second monitor is deliberately not in the system.* That distinction is a design choice, not an oversight, and I would like you to hold it as firmly as you hold the seven-slot layout itself.

Everything you have learned in this book so far lives on the primary monitor. Slot 1 through Slot 7. Zone 1, Zone 2, Zone 3. The stack of browser-and-terminal. The Jira Master tabs. The lease file. The NOTES.md files. The draft PRs. The whole ecology of the method, in its fixed geometry, with its fixed staggering, with its fixed discipline about what lives where. The primary monitor is the system. When you are working, your eyes are there. When you are rotating, your eyes are there. When you are checking, your eyes are there.

The second monitor is where you go when you are not doing that.

Some readers, at this point, will want a diagram. There is no diagram for this, and that is deliberate. The second monitor does not have zones. It does not have slots. It does not have fixed anything. It has whatever you put on it in a given moment, and when the moment passes you are free to clear it and put something else there. That degree of *looseness* is not a failure of rigor on my part. It is precisely the point. The primary monitor earns its sustainability by being rigid; the second monitor earns its sustainability by being the opposite.

If the primary monitor is the dish by the door — every kind of work with a fixed permanent home, always — the second monitor is the kitchen counter. It is for the thing you are doing right now, for however long you are doing it, after which you wipe it down and it is ready for the next thing.

What the Second Monitor Is For

Let me list what I actually put on my second monitor in a given working day, because the list is genuinely mundane and I think the mundanity is the point.

Reading. Articles, blog posts, documentation I want to read carefully instead of scan. RFCs from other teams. Design docs. The longer comments on a PR from a teammate that deserve a real read, not a glance. A linked issue someone dropped in the chat that requires twenty minutes of reading-to-understand rather than two minutes of skimming. None of that belongs on the primary monitor, because the primary monitor is busy running the method, and because reading-to-understand is a different mode of attention than checking-against-a-finished-shape. They do not share well in the same visual field. The second monitor gives each mode its own room.

Watching. A tutorial on a library I am about to use. A conference talk a colleague recommended. A recorded Zoom I missed this morning that I want to catch up on while my agents are running their respective twenty minutes. A short screencast someone posted explaining a tool I am considering adopting. Video content is particularly well-suited to the second monitor because it moves at its own pace — I can listen while I rotate through the primary monitor, pausing when I need to focus on a slot, resuming when the slot goes idle again.

Personal tasks, deliberately. My calendar. The family chat where my partner is coordinating dinner. A bank website where I need to approve something in the next ten minutes. A doctor's appointment to schedule. A birthday card to order for a nephew. The little maintenance work of being a human in the world, which used to happen in the margins of a working day and which, without a deliberate place to put it, would either invade the primary monitor (and pollute the system) or get pushed to evenings (and eat into recovery). The second monitor gives that work a legitimate home inside the working day without letting it contaminate the structure that is doing the real productive labor.

Ad hoc work that does not belong in a slot. This is the category I most want you to internalize, because I think it is the one readers miss most often. Your primary monitor is running seven structured tickets. Good. But real working days produce a constant drizzle of work that is not a ticket — a quick script to rename a batch of files, a one-off SQL query to sanity-check a number someone asked you about in chat, a fifteen-minute exploration of a library you might pull in next week, a rough draft of a Slack message you are trying to get the tone right on, a hand-edited config file you need to poke at for a single meeting. *That work is real.* It is not nothing. But it does not fit inside the seven-slot discipline — it does not deserve its own environment, does not need a draft PR, does not have acceptance criteria, does not want to be validated as if it were production code. If you try to force it into a slot you will break the slot's discipline. If you try to refuse to do it, you will fail, because the work will happen anyway, just resentfully. The second monitor is where that work happens. An open terminal. A scratch editor window. A browser tab with six pages pinned for an afternoon and closed when you're done. No lease. No ritual. No ceremony. You do the thing, and then you move on, and the primary monitor never noticed.

Looking at something that isn't your work. And yes — sometimes, when the rotation is quiet and the agents are all twenty minutes into their work, the second monitor is for a news site, a forum thread, a chapter of a book you're reading, a photo from your kid's school you haven't looked at yet. I am not going to moralize at you about what counts as recovery and what counts as slacking. That line is yours to draw. What I will say is that the category has to exist somewhere in your working day, or you will not last, and the second monitor is the only place in this method that was designed from the ground up to hold it without guilt.

Why This Is Not Slacking

I want to spend a moment on the word *slacking*, because I know some of you have a reflexive flinch at the idea of spending working hours on anything that is not obviously production work.

The flinch is not your fault. The industry has spent a long time — predating AI, though AI has amplified it — training developers to equate visible typing with productive labor, and to feel guilty about any interval of the working day in which they are not generating output. That equation was always wrong, and it is especially wrong in a method like this one, where the agents are generating the typed output on your behalf and your value is specifically in the moments of attention, judgment, and rest that make your next act of judgment a good one.

Here is a way to think about it that may help. The primary monitor is where the work happens. The second monitor is where the *recovery between the work* happens. Not the big recovery — that's sleep, that's weekends, that's vacations, that's lunch — but the small ongoing recovery that lets your eyes refocus, your posture shift, your attention migrate from the fixed geometry of the seven slots to something with different visual rhythms. That recovery is not a luxury. It is the structural counterweight to the intensity of the primary monitor's rigor. If the primary monitor is asking your attention to be focused and disciplined and ritualized, the second monitor is where your attention is allowed to be loose, curious, personal, and unsupervised.

Both are doing work. They are doing different work. Confusing them — trying to make the primary monitor looser, or trying to make the second monitor tighter — breaks the method from one end or the other.

This is what active recovery means in this context. In endurance training, active recovery is not sitting on a couch; it is a gentler, different movement that keeps the system circulating while it heals. In cognitive work, active recovery is not staring into space; it is a gentler, different kind of attention that keeps your brain engaged while the expensive circuits cool. Looking at a calendar is active recovery. Reading a long-form article at your own pace is active recovery. Watching a tutorial at 1.25× speed is active recovery. Scrolling a forum for ten minutes is, honestly, active recovery too, as long as it stays ten minutes and does not metastasize into an hour.

The Posture Argument, Which Is Real

I want to add one more thing that is not in the outline, because I think it matters and because I have watched too many engineers injure themselves quietly over the years.

The seven-slot layout is optimized for a specific posture. You are sitting square to the primary monitor, eyes at a consistent distance, gaze sweeping across a defined rectangle at a defined cadence. That posture is sustainable for a working session, which is why the two-session day is structured the way it is. It is not sustainable for eight straight hours without relief, and even within a session it benefits from occasional variation.

The second monitor, positioned off to one side or tilted slightly or sitting at a different height or even rotated into portrait mode, gives your body a *different frame* to look at. Your neck turns. Your eyes refocus at a different depth. Your shoulders adjust. Your spine notices it can do something other than hold the primary-monitor posture.

That is not a small benefit. Over a year of working days, the engineers I know who respect the second-monitor rule have fewer neck and shoulder complaints than the engineers who use two identically-positioned monitors as *one wider primary monitor.* The second monitor is not an extension of the first. It is a different surface, at a different orientation, used in a different posture. Treat it that way. Your future self, the one whose neck still works in a decade, will thank you.

A Soft Word, because You May Need One

If you have read this far in the chapter, I suspect you may be someone for whom work has started to take up more of life than you would like. The primary monitor, in this method, is where you do excellent work. The second monitor is a small, deliberate piece of your workspace that gives your attention — and your eyes, and your neck, and the part of you that still wants to read an article about something unrelated — a place to go without guilt during the working day.

That is not an indulgence. It is a structural piece of the system, as load-bearing as the lease file or the validation ritual. It exists because the best-designed system in the world will still grind you down if you live inside of it without a door to the outside. The second monitor is the door.

Put something on it right now that has nothing to do with the seven slots. A photo. A bookmark bar of things you like reading. A playlist. A game you play for ninety seconds at a time between rotations. Make it visibly, obviously, *yours*. The primary monitor is the dish by the door. The second monitor is the window over the sink — and every kitchen that has been a kind place to work in has one.

On to the two-session rhythm, which is the other non-negotiable, and which is the section where I am going to be, briefly, a little strict with you.

The Two-Session Rhythm

A note before this section gets practical. If you are reading it on a bad day — a Friday after a Thursday that broke you, or a Monday morning when the laptop felt heavier than usual — please know that what follows is not a rebuke. It is not a list of things you have been doing wrong. It is a description of a rhythm that, when honored, makes the next decade of your work sustainable, written by someone who has gotten the rhythm wrong many times and is offering you the cleaned-up version of what he learned. There is no test at the end. There is no catching up to do. Read what you can. The rest will be here.

You have read about the two-session day twice already. Once in Chapter 3, where it was a piece of arithmetic — *seven tickets per session, two sessions, fourteen tickets a day, the math just works.* Once in Chapter 6, where it was a piece of choreography — *9 to 12, lunch, 1 to 4, wind down by 5.* Both treatments were honest, and both were incomplete.

The reason the two-session rhythm earns a third treatment, here in the burnout chapter and not back in Chapter 6 with the rest of the daily mechanics, is that the rhythm is not actually a productivity device. It looks like a productivity device. It produces productivity numbers. But the thing it is *doing*, underneath the throughput, is keeping you intact for next Monday. And next Monday after that. And the Monday in March of 2032, by which time the agent on your screen will probably have a different name, possibly a different face, and almost certainly a different press release attached to it. The rhythm is what's between you and that distant Monday. Treat it as a piece of operational hygiene that happens to have throughput as a side effect, not the other way around.

There is a way of running this method that hits fourteen tickets on a Tuesday and produces a developer who, by the following October, can no longer remember why they wanted this job. There is also a way that hits twelve tickets on a Tuesday, calmly, with a forty-five-minute lunch outside, and produces a developer who is still doing this in eight years and looking forward to the work most mornings. The two ways look almost identical from the outside. The difference is the rhythm, applied with discipline and held lightly.

This section is what discipline-applied-and-held-lightly looks like in practice.

What the Morning Session Actually Feels Like

A morning session, done right, has three phases of its own — and the phases are not productivity phases. They are *attentional* phases, and the difference matters.

The first thirty to forty-five minutes are *loading*. You are getting back into the workspace after a night away from it. You are reading NOTES.md from the slots you kept warm overnight. You are loading new tickets into the slots you released yesterday afternoon. You are firing the first prompts and watching the first responses come back. You are, mentally, settling into the rotation. Throughput during this stretch is low. That is correct. Throughput during the loading phase is supposed to be low. If you are closing tickets before 9:30 you are either coming off a sprint into something genuinely simple, or you are not loading carefully enough and you will pay for it at 11:15 when a slot you didn't read carefully turns out to need a conversation you weren't ready to have.

The next hour and a half — roughly 9:45 to 11:15, give or take — is the *fast water*. The rotation hums. Slots are finishing. You are validating, releasing, loading. You are answering the question Slot 4 asked at 10:08 with a one-sentence answer, you are noticing Slot 2 has been silent for three minutes longer than it should be and probing it, you are watching Slot 6 finish a clean piece of work and feeling the small honest pleasure of promoting it. This is the part of the day the chapter on Execution was describing. This is also the part of the day a lot of engineers, on a good morning, mistake for the *whole* day. They feel it. They want more of it. They start arranging their afternoon to maximize it. This is exactly the wrong impulse, and the next paragraph will be about why, but first the third phase.

The last thirty to forty-five minutes — roughly 11:15 to noon — are *settling*. The good morning sessions taper deliberately. You stop loading new tickets after about

11:15. You let what's in flight finish or reach a natural pause. You note where each slot is so that, if you decide to keep one warm through lunch (you usually shouldn't, but occasionally), you have a clean handoff point. You start to disengage. By 11:50 your hands are off the keyboard for periods of fifteen, twenty seconds at a time and the rotation continues without you. By noon you are closing things.

A morning that is loading, fast water, and settling is a morning the body can do five days a week. A morning that is fast water, fast water, and *more* fast water is a morning that compounds against you. The difference looks small in any single day. After three weeks, the difference is the difference between sustainable and not.

The signal that the morning is going well is, unromantically, that you are slightly less impressed with yourself than you thought you'd be at noon. You have closed seven tickets, give or take. You are not exhausted. You have finished, in the sense that a morning can finish — there is a closed shape to it. The mug is empty. The notes are in NOTES.md. Lunch is the next thing.

The Lunch Hour as a Load-Bearing Beam

There is a building-trades phrase for the parts of a structure that hold the rest up: *load-bearing*. You don't notice them when they're working. You notice them, vividly and quickly, when they aren't. The lunch hour, in the two-session rhythm, is load-bearing in exactly that sense.

I want to be specific about what counts as a real lunch, because the phrase has been worn down by years of corporate misuse. A real lunch, for the purposes of this method, is a continuous block of time — at least forty-five minutes, ideally sixty — during which your eyes do not look at a screen, your hands do not type, and your attention is not on tickets. That is the entire requirement. The food itself is not the load-bearing part. The screen-lessness is.

Things that are real lunches: a sandwich at the kitchen table while you read three pages of an unrelated book. A walk to the corner and back with a coffee. Sitting in the backyard. Eating with a partner or a child or a friend without the laptop in line of sight. Going to the gym for forty-five minutes and eating something afterward. Putting on headphones with music and lying on the couch.

Things that are not real lunches: scrolling phone email at the table. Sandwich-over-keyboard. The famous *I'll just check one thing* that turns into nineteen minutes inside Slack. A "lunch meeting" that is, mechanically, a meeting with a sandwich. Watching a YouTube video about the framework you've been wrestling with all morning. Reading the engineering newsletter that just landed in your inbox. Reviewing a PR while you eat. The bar is high, and it is high on purpose.

The reason it is high on purpose is that the rotation phase you just left consumed a specific, depletable resource — the part of your attention that holds seven streams of work in the same head at once — and the only known way to refill it is to put it down completely for a stretch. Putting it down for nine minutes does not refill it. Putting it down while still glancing at the same screen it was just running on does not refill it. The afternoon session draws from this same well, and a half-empty well at 1 PM produces an afternoon that looks fine for the first forty minutes and then degrades into something the engineer at the desk will not be able to name but will feel for the rest of the week.

I have watched this go wrong in a hundred small ways in colleagues over the years, and the failure mode is almost always the same. The morning goes well. The fast water keeps moving past noon. *I'll just finish this slot, then go to lunch.* Then *I'll grab a sandwich and eat at the desk for a quick five.* Then 1:15 arrives and the engineer realizes they've been at it continuously since 9:00 with a snack break, and they've blown through their afternoon's reserve, and they will pay for it, and they don't yet know they will. They notice on Thursday, when a perfectly normal Thursday feels harder than it has any business feeling. By that point the receipt is overdue and bigger than the meal.

The simplest discipline I have found, after years of getting this wrong: *the desk is closed at noon.* The phrase is yours, said internally, with as much firmness as a phrase said internally can have. At noon you stand up, you turn the chair around, you eat, you move, and you do not return until 1 PM. If a slot is mid-thought at 11:58, that's fine — the slot will still be mid-thought at 1:00, and Claude will have either solved it itself or sat patiently with the ambiguity. The agent does not require your supervision through lunch. The agent is fine. Your nervous system is the constraint, and your nervous system needs the hour off.

What the Afternoon Session Actually Feels Like

The afternoon session is shorter than the morning session by design — three hours instead of three-and-a-bit — and it has a different attentional shape.

The first thirty minutes of the afternoon, roughly 1:00 to 1:30, are a *re-load*. They are the lunch-version of the morning's loading phase. You come back, you read what happened in the slots you left running, you check Communications for anything urgent, you load any new tickets into freed slots. The re-load is faster than the morning load — you don't have NOTES.md to read for slots that are continuing from before lunch — but it is real, and skipping it costs you. Engineers who treat 1:00 PM as *just resume* underrate how much state has shifted in their own heads during the lunch hour. (That shift is exactly what you wanted; honor it by re-loading.)

The middle of the afternoon, roughly 1:30 to 3:15, is the second fast water. It feels almost the same as the morning's fast water but it is *not* the same — it is drawing from a partially refilled well, and the partial refill is what determines how cleanly it runs. A good lunch produces an afternoon that hums. A skipped or polluted lunch produces an afternoon that *appears* to hum and then collapses around 2:50 when the engineer, mid-validation, suddenly cannot remember what they were validating and realizes they have been staring at the same diff for ninety seconds.

The last forty-five minutes of the afternoon, roughly 3:15 to 4:00, are *closing*. By 3:15 you are not loading new tickets. You are letting in-flight slots finish or reach a clean pause. You are validating final pieces. You are starting to write NOTES.md entries for the slots you'll keep warm overnight (usually one or two; never all seven). You are noting which tickets to load fresh tomorrow morning. By 3:45 the rotation has slowed to a walk. By 4:00 you are at the wind-down ritual itself — the short NOTES.md per active environment, the lease releases, the last glance across the screen.

The hard stop matters. If you have read this book carefully and adopted nothing else from it, adopt the hard stop. Four PM is when the desk closes the second time. Not 4:15 *because the slot is almost done*. Not 4:30 *because there's just one more thing*. Not 4:45 *because Slack pinged*. Four PM is when you stand up, the way you stood up at noon, and you go and live the rest of the day. The afternoon session has a closed shape to it the same way the morning session does. Closing it cleanly is what makes tomorrow morning possible.

The signal that the afternoon is going well is again unromantic: you have closed five or six more tickets, your eleven-or-twelve-ticket day is in the bag, you are slightly tired in the right way, and you have left enough of yourself that you can be a person for the next sixteen hours before the rotation starts again. If you have closed *more* than seven tickets in the afternoon, you probably ran the morning hot or skipped the lunch beam, and you will feel the receipt by Wednesday.

Friday Afternoon, Deliberately Light

Friday afternoon is a special case worth its own paragraph.

By Friday afternoon you have run roughly four full days of rotation. You are not depleted, if the week has gone well — but you are not at Monday's reserves either. The body knows this, the nervous system knows this, and treating Friday afternoon as Tuesday afternoon plus three days of fatigue is exactly the wrong move. Friday afternoons are designed for *lighter* work. Reading a paper that has been sitting in your tabs all week. Writing the week's `reflections.md` entry. Refining `CLAUDE.md` based on the corrections that surfaced during the week. Reviewing teammates' PRs without the pressure of also running your own rotation. Cleaning up branch state, archiving closed tickets, doing the quiet maintenance the rotation produces but doesn't pay for.

If you happen to close two more tickets on Friday afternoon, that's a bonus, not the goal. If you close zero tickets and instead spend the afternoon turning a week's worth of corrections into three durable rules, that is a *better* Friday afternoon than the fourteen-ticket version, because what you spent Friday on shows up in next Monday's tickets without you having to think about it. Refinement is its own session shape, and Friday afternoon is its native habitat. (Chapter 8 has more to say about why this works and when it doesn't.)

The trap to avoid here, which is real and which I have watched engineers fall into, is using Friday afternoon to *catch up*. As in: I had a slow week, my ticket count was light, I will hammer Friday afternoon to make the numbers look right. This always backfires. A hammered Friday afternoon eats into Friday evening's recovery, eats into the weekend's recovery, and shows up as a sluggish Monday morning that more than erases whatever it earned. If your week ran light, your week ran light. The system is calibrated for averages, not for individual-day heroics. Let Friday close the way the rest of the week did — with a clean shape — and trust that next week will average out.

What Going Wrong Looks Like

For honesty, here is a short catalogue of the specific ways engineers I've worked with have wrecked the two-session rhythm. Each one looks reasonable in the moment.

The skipped lunch. Discussed above; it is the single most common failure mode. The story you tell yourself is that you'll just keep going while the morning is hot. The story is wrong. The hotter the morning, the more important the lunch. The lunch is not what's *interrupting* the rotation; the lunch is what makes a sustained rotation possible.

The three-session day. The engineer who, energized by a good Tuesday, decides to add a third session from 7 PM to 9 PM. Just an hour or two, just to chip ahead. Almost always works once. Sometimes works twice. By the third or fourth time, Wednesday morning has a fog in it the engineer can't account for. By the second week, the original two-session rhythm has degraded — the morning starts later, the lunch gets shorter, the afternoon ends ragged — because the body is unconsciously protecting itself from a workday that has expanded past its budget. Three sessions is not two-plus-extra. Three sessions is two-minus-degraded, dressed up in additional hours that do not produce additional throughput, only additional fatigue.

The squeezed weekend. The engineer who *just checks Slack on Saturday morning, just reviews one PR on Sunday afternoon, just opens a slot on Sunday night to get a head start on Monday.* Each individual touch is small. The aggregate effect is that the rotation never actually went offline, and Monday morning's reserves are the reserves of an engineer who has been rotating for nine consecutive days, not five. The rhythm requires the off-period the way breathing requires the exhale. Skipping the exhale does not get you ahead; it asphyxiates the next inhale.

The phantom three-session week. Five real days of two sessions plus a hidden sixth day pieced together from squeezed weekend touches and stolen evening hours. This is the hardest pattern to spot in yourself because no individual day looks pathological. The signal is on Wednesday or Thursday of the *following* week, when you cannot quite explain why the rotation feels harder than it should. The engineer who can recognize this pattern early and add a real off-day mid-cycle to compensate generally recovers. The engineer who explains it away as *just a tough week* generally accumulates more.

The deadline override. The single hardest case. A real deadline, real stakes, the rest of the team genuinely needs the work shipped. The temptation to abandon the rhythm for the duration of the crunch is enormous, and sometimes it is also correct. (The next subsection.) The version that goes wrong is the deadline that becomes a *shape* — the engineer who runs deadline-mode for the duration of a release, then for the duration of the next release, then realizes six months later that the rhythm has been quietly missing for most of the year and the only thing keeping the tickets shipping is willpower, which is finite and which is starting to run out.

The on-call slide. Closely related. The engineer on a primary on-call rotation who, week after week, treats every page as license to extend the rotation past 4 PM, every alert as license to skip lunch, every weekend incident as license to stay engaged through the next morning. On-call is genuinely an exception (subsection below). The slide is what happens when the exception becomes the default and stays the default after the on-call rotation ends.

The pattern across all of these is that small overruns are not free, even when they feel free. The receipt comes due, but it comes due unevenly — usually a day or two later, usually accompanied by a plausible alternative explanation, and usually paid for in the form of a slightly worse week than you'd otherwise have had. This is also why the rhythm is hard to defend in the moment. The cost is delayed and diffused. The benefit of the overrun is immediate and concentrated. The system looks like it's rewarding the overrun. It isn't. It is loaning at a rate you cannot see written down, and the rate is high.

Exceptions Worth Honoring

There are exceptions worth honoring, because pretending there aren't is its own failure mode. Two in particular.

On-call. When you are the primary on-call engineer for a service, the rhythm has to bend. There is no version of "I am off the keyboard from noon to 1" that survives a 12:14 page. What you can preserve, even on-call, is *most* of the rhythm: a reduced rotation (four slots instead of seven, so that being yanked into an incident doesn't strand multiple in-flight tickets), a deliberate lunch as long as the pager allows it, a hard stop that becomes "soft 4 PM, hard 6 PM" depending on what happened that day, and an end-of-rotation recovery — the day or two after on-call ends should be a real off-method day, not a *catch up on what I missed* day. The cost of on-call is paid in recovery, and the recovery has to actually happen for the on-call rotation to be sustainable across a year.

Real deadlines. A real deadline — meaning one that comes around once or twice a quarter, has stakes the team has agreed are stakes, and has a defined endpoint — can justify a stretch of three-session days, weekend touches, or a 5 PM hard stop instead of 4 PM. The two conditions that make this honest rather than corrosive are (1) it has to actually be a real deadline, by which I mean rare, agreed-upon, time-bounded, and not a thing that recurs every other Friday because of an ambient cultural urgency the team has not bothered to push back on; and (2) the recovery period after the deadline has to be honored *as the deadline*. As in: the two days after the release ships are not for catching up on the regular work; they are for the rotation re-equilibrating. Take them as low-load days. Take them as off-method days if necessary. The deadline includes its recovery; the recovery is part of the engineering.

If you find yourself in deadline-mode month after month, the deadline isn't the problem. The deadline-mode-as-default is the problem, and the answer is a conversation with your manager that Chapter 12 has more to say about, not another three-session week.

Permission to Start Smaller

A note for the reader who is building toward two full sessions and is not yet there.

The rhythm in this section is the steady state. It is what running the method looks like once the slots, the layout, the prompts, and the rotation have all clicked. If you are in your first week, your first month, your first quarter — running two sessions of seven slots each is not the *starting* shape. It is the *target* shape.

A perfectly respectable starting shape: one session a day, four slots, of two hours, with a real break afterward. From there, three slots becoming five, two hours becoming three, one session becoming a morning-and-an-afternoon — each adjustment held for two weeks before the next. The book tells the steady-state story because the steady state is what you are aiming for. The path to the steady state takes weeks, not days, and pacing the ramp matters more than the speed of the ramp.

The reader who tries to run the full two-session, fourteen-ticket rhythm on their second day with the method has set themselves up, almost by definition, for a bad third day. The reader who runs four slots for three weeks, six slots for the next three, and lands on seven slots a couple of months in lands cleanly. The slow ramp is not a sign that you are slow. It is the rhythm being correctly fitted to your physiology. The fast ramp is a sign that you are skipping the fitting and will pay for it within the quarter.

If you can only run one session a day right now, run one session a day. If lunch is forty minutes instead of an hour, that is a forty-minute lunch and it is closer to the right answer than ninety minutes of skipped lunch over the same week. The rhythm is the goal; the discipline is the practice; the practice is what gets you to the goal. None of those words are *velocity*.

Why This Section Sits in the Burnout Chapter

A closing note before we move on to the weekly and monthly habits.

The two-session rhythm is the most operational thing in this chapter. It is the thing closest to the daily mechanics of Chapter 6 and the planning shape of Chapter 3. It would have been defensible to put it back there with the rest of the daily-rhythm material. The reason it sits here, in the chapter on AI burnout, is that the rhythm is what protects the developer running this method from a slow accumulation of decision-load that would, over a few months, deliver them quietly to the territory the rest of this chapter is about.

The chapters before this one taught you what to do during the rotation. This chapter is about what to do *around* the rotation so that the rotation can keep happening. The two-session rhythm is the around. The lunch is load-bearing. The hard stop is load-bearing. Friday afternoon is load-bearing. The weekend is load-bearing. The vacation, when it eventually arrives, is load-bearing.

You built a workspace that lets you run seven Claudes in parallel. The most sophisticated thing you can do with that workspace, beyond any individual ticket it produces, is run it sustainably. The rhythm in this section is what sustainability looks like in clock-time. Honor it for thirty days and it stops being a discipline and starts being a habit. Honor it for ninety days and it stops being a habit and starts being the way you work.

Honor it for a year, and you will be able to look back at the year and notice, faintly, how few weeks of it felt bad — and how many felt like the kind of working life you used to imagine you might one day have, before the AI-coding years made it briefly hard to imagine such a life at all.

That is the rhythm. The next section is about the rituals that sit a layer outside it: the weekly and monthly habits that make the rhythm itself sustainable across the longer arcs of a career.

Figure 11.2 is the rhythm drawn to scale — morning session, real lunch, afternoon session, wind-down — with the decision-load curve overlaid on top. The curve is the point of the picture. The gap at lunch is what lets the curve dip; the wind-down is what lets it fall rather than end in a crash. Skip either and the curve stays high into the evening, and then the next morning, and the one after that.

Figure 11.2 — The Two-Session Day

Morning block, real lunch, afternoon block, wind-down. The gap is load-bearing — do not skip it.

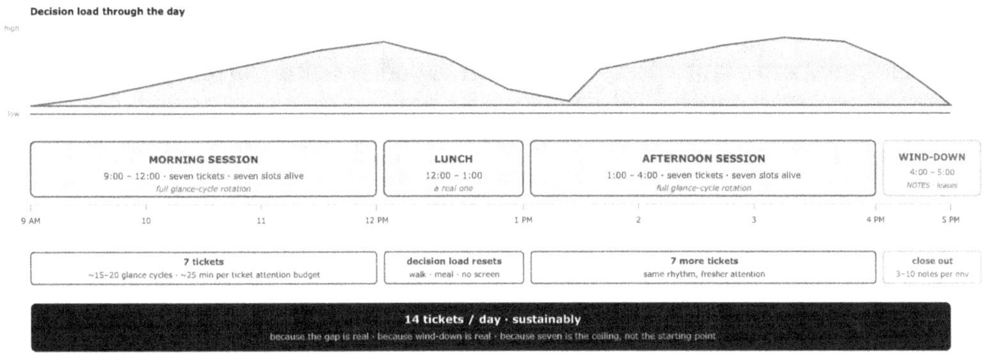

The lunch block is the most important rectangle on this diagram. Take it seriously -- a glance at Slack counts as work, and work means the load never resets.

Weekly and Monthly Habits for Sustainability

Why Weeks and Months Get Their Own Section

Up to this point in the chapter, the interventions have lived inside a single working day. The two-session rhythm. The glance cycle. The NOTES.md wind-down. The second monitor as escape hatch. Those are all *within-day* supports — daily maintenance for an attentional system that can be depleted and restored on a twenty-four-hour cycle.

But attention is not the only reservoir that matters, and twenty-four hours is not the only rhythm the body and mind run on. Underneath the daily budget there is a *weekly* budget, and underneath that a *monthly* one, and underneath that a seasonal one that I'll get to in the chapter on the long game. Each of those budgets has its own depletion curve, its own recovery mechanism, and its own failure mode when it runs dry. Engineers who have tuned their days beautifully and who still, somehow, end up exhausted after three months are almost always leaking at one of the longer timescales — at the weekly or the monthly level — without having a structure for noticing it.

This section is that structure. It is, honestly, the section of this chapter I am most worried you will skim, because weekly and monthly habits sound dutiful in a way that daily rhythms don't. Daily rhythms feel like engineering — they have mechanics, they are instrumentable, they feel like something a serious person does. Weekly reflection and monthly sabbaticals can feel, at first glance, like wellness-industry lifestyle accessories of the kind I promised in the Introduction I would not peddle. I know. I felt the same way for years, and I was wrong, and I want to make the case that these longer-cycle habits are not lifestyle. They are load-bearing infrastructure. They are the difference between an engineer who runs this method for six months and burns out quietly, and an engineer who runs it for a decade and is still sharp at the end.

I want to build the case carefully, in four movements.

First, I'll talk about what depletes on a weekly scale and what depletes on a monthly scale, and why these depletions don't show up in the daily instrumentation we've already built.

Second, I'll walk through the end-of-week ritual in detail. Thirty minutes, Friday afternoon, with specific prompts and a specific output. I'll tell you what I write, why I write it, and how the ritual compounds over a year.

Third, I'll make the case for what I call *off-method days* — days when you deliberately do not run the seven-slot structure. Why they matter, how often, and what protects you from the kind of guilt that normally prevents senior engineers from taking them.

Fourth, I'll make the most controversial argument in the chapter, which is the case for *AI-free sabbaticals* — days, occasionally whole weekends, where you sit down and code by hand, without any agent, on something you actually care about. This is the part of the chapter where I expect the most pushback, because we live in a moment where anyone suggesting occasional unplugged coding is easily dismissed as nostalgic or even reactionary. I want to preempt that framing, because I think the case is actually forward-looking: the skills you protect on those days are exactly the skills you need to run this method well for the next thirty years.

Take a breath. Settle in. This section is longer than the last one looked in the outline, and that is not a mistake. It is, in my view, the most practically important section in the chapter, because the people I have watched succeed at this work over long horizons are the people who treated their weeks and months as seriously as they treated their days.

What Depletes at Longer Timescales

Start with the mechanism. The within-day burnout we've already discussed — decision-load fatigue, the exhaustion of the checking-and-holding faculty — recovers on a sleep cycle. A full night's sleep, eight hours, does an enormous amount of restoration. Decision-load fatigue is, in that sense, the easy case: it has a built-in nightly reset, and most engineers who have been in the field long enough know roughly how to use that reset when they have to.

What does *not* reset overnight is a different set of quantities. I'll name the ones that, in my experience, run the longest cycles.

Judgment-shape drift. Over a week of running this method intensively, the shapes you recognize as *good code* and *good PRs* and *good design* start to subtly calibrate against what the agents are producing, because that is what you see most. If the agents you're running have a mild aesthetic preference — a particular kind of factoring, a particular idiom, a particular over-use of helper functions — and you see five hundred examples of that aesthetic in a week, your own taste begins to drift toward it, even if you started out not liking it. This is not vigilance failing; it is pattern-exposure doing the work it always does. It is how humans have always adapted to the dominant style around them. The problem is that the dominant style around you is now generated by a system whose aesthetic you did not choose and do not fully control, and over months, an engineer whose taste has been passively recalibrated by agent output no longer has a reliable compass for distinguishing what the agents are doing well from what they are doing badly.

Judgment-shape drift does not recover overnight. It recovers by *deliberate re-exposure* to human-authored code and to your own unaided work. That recovery happens, when it happens at all, at weekly or longer timescales. An engineer who never deliberately does it will, slowly and invisibly, lose the very judgment faculty on which the checking mode from the last section depends.

Deep-problem muscle atrophy. A working day inside this method is a day of many small checking acts and a handful of small briefing acts. It is emphatically not a day of *holding a hard problem in your head for six hours and emerging with a breakthrough.* That kind of deep solve, when it was the shape of your day, was doing something for you. It was exercising a specific cognitive capacity — sustained holding under novel conditions — that the method does not exercise at all on a typical ticket. A muscle you do not exercise atrophies. Engineers who run this method for months without ever doing a long solo solve notice, after a while, that when they do need to sit and think hard about something — a real design problem, a truly novel bug, a question with no agent-shaped answer — they are rustier than they used to be. That rustiness is not tragic; it is exactly what the exercise literature would predict, and it is exactly what you avoid by working that capacity every so often on purpose.

This, too, does not recover overnight. You don't get your deep-solve capacity back from a weekend of rest; you get it back from occasionally *doing* a deep solve.

Identity-and-meaning load. This one is harder to measure, and I want to be careful about how I name it, because it shades into territory where I am not competent to speak with medical authority. But it is real and it shows up, and I would be dishonest to leave it out.

Engineers who came to this profession because they loved the craft — the specific feeling of writing a function and watching it work, of designing a subsystem and seeing other people use it, of being the person who could make the computer do the thing — carry, underneath everything else they do at work, a relationship with their own identity as a *practitioner*. Not as an employee. As a craftsperson. That relationship is what got most of us through our first decade in the field when the pay was worse and the industry was saner and we were doing it for love.

A working week spent mostly in oversight of agents is not, from the craft-identity perspective, identical to a week spent doing the craft. It is a different relationship. It is a good relationship; the method is, I've argued at length, more sustainable than the alternatives. But it is different, and over long periods the difference accumulates into a quiet, subterranean question about whether you still *are* what you always took yourself to be. I have watched this in myself and in colleagues, and I am not going to pretend the question is silly. The question is serious. It is one of the things that burns engineers out most deeply, and it is one of the things most invisible to daily instrumentation.

The identity-and-meaning load recovers, when it recovers, on monthly or longer timescales. Not from rest. From doing some of the craft yourself, periodically, in a way that reminds the oldest part of your professional self that it is still in the chair.

Social-and-relational load. The last one is the one engineers talk about least, and it accounts for more of what destroys careers than all the others combined.

Decision-load fatigue and craft-identity drift are internal. They happen inside your head, in your working hours, against your own tools. The social load is different. It is what happens to the people who live with you and work with you while you are cognitively elsewhere for eight hours a day, five days a week, for years. It is what happens when your kids learn not to interrupt you when you have the "don't-interrupt face" on. It is what happens when your partner learns to start dinner conversations a certain way because they have found that you respond better when they don't lead with a question. It is what happens when your teammates learn that you are sharp on Tuesdays and unreliable on Thursdays, and they adjust around you quietly, and you never learn they did.

Social load accumulates in relationships on timescales that have nothing to do with your daily attention reservoir. It is not paid down by a good night's sleep. It is paid down, imperfectly, by *showing up* — fully present, camera on, attention unfragmented — for the people in your life at times that are predictable enough for them to count on. That showing-up has to be deliberately scheduled, because otherwise the method will eat it, and your day will refill, and no one in your household will ever know what you were like on Wednesday morning because your afternoon and your evening were the same distracted porridge you served them last week.

None of those four longer-cycle loads — judgment drift, deep-problem atrophy, identity-and-meaning load, social-and-relational load — shows up in the NOTES.md files. None of them gets better from a two-hour nap. All of them, if unattended, will eventually take the method down and you with it.

That is what the weekly and monthly habits in this section are for. They are not wellness decorations. They are the maintenance schedule for four specific reservoirs that have no daily refill and that together hold up the entire sustainability claim of this book.

The End-of-Week Ritual

Start with the simplest and most practically important habit.

Friday afternoon, in the last thirty to forty-five minutes of your working week, before you close the laptop, you run a specific ritual. It has a fixed shape, a fixed output, and a fixed place it lives on disk — because by now, eleven chapters in, you should recognize that pattern as the method applying to itself. The dish-by-the-door principle does not stop at tickets. It applies to your own reflection, and reflection that does not have a permanent home is reflection that does not actually happen.

I'll describe the ritual as I run it, and then I'll say why each piece is there.

Step 1: open `reflections.md`, which lives in your workspace alongside `CLAUDE.md`. It is a single markdown file, appended to at the end of every working week, with the date at the top of each new entry. Do not start a new file per week. You want a single continuous document because the cross-week patterns you will eventually find in it are the whole point of the exercise, and they are invisible if every week lives in its own file.

Step 2: answer four questions, in writing, by hand, in plain prose. Not bulleted. Prose. The bullets come later; the prose is the reflection. The questions are:

1. *What depleted me this week?*
2. *What energized me this week?*
3. *Where did I feel myself doing the work well?*
4. *Where did I feel myself going through the motions?*

You write two to four sentences on each. Not more. The ritual takes thirty minutes, not ninety, because if it takes ninety you will not do it, and a thirty-minute ritual done every week for a year is worth unspeakably more than a ninety-minute ritual done three times and abandoned.

I want to walk through each of the four questions individually, because the way they are phrased is not accidental.

What depleted me this week? — This is the one most engineers answer easily the first time they try this, because depletion is loud. You will write down the ticket that ate three days when it should have eaten one. The meeting that was a stakeholder debating a decision in front of you without including you. The reorg announcement. The agent that kept using the old API pattern even though you asked it not to. The Thursday where you did three sessions instead of two and paid for it Friday. Those will come to you quickly.

But the second and third time you run this ritual, you'll start to notice depletions that are quieter than any single incident. *"I think the validation zone setup isn't working for this kind of ticket."* Or *"I've been feeling low-grade anxiety about the model upgrade announcement and I haven't named it until now."* Or *"I had two really good one-on-ones this week but I'm tired from them in a way that surprised me."* Those quieter answers are the ones the ritual is mostly for. Loud depletions take care of themselves; quiet ones are what compound into next quarter's breakdown if you don't catch them.

What energized me this week? — This question is the one engineers initially think is a throwaway wellness question and that turns out, over months, to be the most practically important of the four. You will discover, by running this ritual for eight or ten weeks in a row, that there are *specific kinds of work* that leave you more energetic than you started, and other kinds that reliably drain you even when they're nominally small. The energizing kinds are not always the easy kinds. Sometimes they are the hardest kind — a week where you sat with a real design problem for two days and emerged with a clean answer, and felt, on Friday, better than you did on Monday. Sometimes they are small — a code review where you taught a junior engineer something and watched them light up. Sometimes they are very small — an hour spent refactoring one file carefully because you felt like it.

The data is not interesting in any single week. The data is interesting cumulatively. After two months, you will have a clear, unignorable portrait of what your specific brain and specific career currently love, and you can — this is the important part — *bias your next quarter toward more of those things.* Not all of it; you still have a job. But a little of it. An extra few hours a week. That bias, over a career, is the difference between an engineer who is alive in the chair at fifty and one who retired early because they couldn't face another Monday. The energizing-things data is the input to that bias, and most engineers never generate the data because they don't think to ask the question.

Where did I feel myself doing the work well? — This is a craft question. It is separate from *what went well this week*, which is a business question. *Doing the work well* is a subjective, internal, somatic sense — the moments when your hands and your judgment and your attention were all arranged in the right way, when the work flowed through you instead of past you. You know the feeling. It is rare enough that you can usually remember the specific moments.

Why this question: because craft identity, which I named above as one of the reservoirs the method can quietly deplete, is restored partly by *noticing* the moments when you are still doing the craft. Many engineers, running this method, start to forget that they are still doing it — the day's rhythm feels so managerial that the moments of real technical judgment get lost in the background. Asking the question weekly forces you to retrieve those moments from memory, look at them, and own them. Over a year, the habit of this retrieval is a substantial fraction of what keeps your craft-identity intact.

Where did I feel myself going through the motions? — The mirror image. Not where did I do badly — that's a different, more brittle question that invites self-flagellation and tends to produce generic answers like "my code reviews could have been sharper." Going through the motions is different. It is the specific sensation of *being present in name only* — the moments when you made a decision that did not feel like a decision, when you checked a PR and realized afterward you don't actually remember what was in it, when you rotated to a slot and noticed you had not processed the previous slot's output at all.

Those moments are real data about the edges of your sustainable practice. They are the moments just *beyond* the edge — the moments where the system started running you instead of the other way around, even if only briefly. You want that data. You want to know where the frontier is, because next week you want to pull back from it by five percent. The practice here is not self-criticism; it is edge detection. A good controller is not the one who never has a slot starting to drift; a good controller is the one who knows the feel of a slot starting to drift and clears the airspace before it becomes a real problem. This question is your drift warning.

Step 3: choose at most one adjustment for next week. One. Not a list. Not a plan. Not a manifesto. Write a single sentence describing what you're going to try differently next week, in response to what the reflections surfaced. If the reflections surfaced nothing that warrants change, write *"no adjustment this week"* and move on — the absence of a change is also a valid output.

The one-adjustment cap mirrors the weekly refinement cap from Chapter 8, and for the same reason: adjustments have a cost, the system compounds when adjustments are few and deliberate, and an engineer making five small changes a week on top of a system they are already stretched by is an engineer who will be thrashing by June.

Step 4: close the file, and close the laptop. Not metaphorically. Actually close the laptop. The ritual has to terminate at a specific observable moment, or it will slide into *"I'll just check one more thing"* and the weekend gets eaten.

That is the whole ritual. Four questions, two to four sentences each, one adjustment, thirty minutes, Friday afternoon. That's it. I have been running it, in some form, for most of the last decade, and the version of me that runs it is, by a considerable margin, more sustainable and more effective than the version of me that doesn't. The tools changed; the ritual did not.

Why Writing It Down Matters (and Why "Just Thinking about It" Does Not)

A practical aside before we move on to the monthly habits. Some readers, at this point, will be thinking *"I do a version of this every Friday already, in my head, in the shower, on the drive home."*

No, you don't. Not functionally. And I want to be gentle but direct about why.

What you do in the shower is reactive emotional processing. It is valuable — please continue to have long showers on Fridays — but it is not the same thing as structured reflection written in prose in a file that persists. The written version does three things the shower version cannot do, no matter how attentive you are.

First, writing forces specificity. The shower version of *"this week was kind of heavy"* becomes, on paper, *"Monday and Tuesday were fine. Wednesday I realized the API migration ticket was larger than I scoped and I spent the rest of the week rushing it. The rushing was what depleted me, not the work itself."* That second version is actionable. The first is not. Writing is what turns feelings into facts, and facts are what let you adjust.

Second, writing creates a record. In six months, you will be able to read back through the last twenty weekly entries and see patterns that were invisible at the time. *"Huh — three of my last four bad weeks involved the same subsystem. Maybe it is that subsystem, not my mood."* That kind of pattern recognition is flatly impossible without a record, because human memory is reconstructive and biased toward the recent. You cannot spot a quarterly pattern from quarterly memory; you need the weekly residue.

Third, writing terminates the reflection. When you have written the fourth paragraph and closed the file, the reflection is *done* for the week. When the reflection lives only in your head, it leaks — it recurs on Sunday night, it recurs at 2 AM, it recurs right before a difficult Monday meeting. A written reflection is a reflection that has a body and a location and a cover. An unwritten one is a reflection that follows you around.

This is not an argument for over-journaling. I am not suggesting you develop a practice that would look fine on the shelf at a bookshop's self-help section. I am saying the specific thirty-minute ritual of four questions on Friday afternoon does something mechanically different from thinking, and the difference is load-bearing. Do the writing.

Off-Method Days: The Necessary Release Valve

Move up a scale now. Beyond the weekly ritual, which is itself an intervention on a longer cycle than a single day, there are the *monthly* rhythms — and at the top of that list, the most practically important one, is the institution of what I've come to call **off-method days**.

An off-method day is exactly what the phrase suggests. It is a working day, at your normal start time, at your normal desk, that you spend deliberately *not* running the seven-slot system. You do not open seven terminals. You do not lease seven environments. You do not glance-cycle. You pick one ticket, and you work on it one ticket at a time, the way you used to, before. Or you pick no tickets, and you do something else entirely — catch up on a piece of research, read through a codebase you don't normally touch, fix a small annoying thing that nobody has been prioritizing. The constraint is only that you do not run the method.

The frequency I have settled on, after some years of experimentation, is roughly one off-method day every two to three weeks. Some people need them more often. Some need them less. The right cadence is the cadence that leaves you feeling like the method is a tool you choose to use rather than a treadmill you climbed onto and can't get off.

Let me say why these days exist, because the case is not obvious.

They prevent method dependency. An engineer who has run the method for three months straight develops a specific kind of fluency with it that is, paradoxically, a small risk. The fluency is good — it is what makes the method work — but it quietly trains a reflex: the first move of any morning is to lease environments. An off-method day breaks that reflex for a single day. It reminds you that the method is a chosen piece of infrastructure, not a law of nature. You are the engineer; the method serves you. An engineer who cannot work without their method has, in a small way, let the tool swallow the craftsperson, and the corrective is to spend a day, occasionally, demonstrating to yourself that you can still work without it.

They generate data the method cannot. A ticket done the old way — one agent, one screen, one problem held in your head — tells you things about the ticket that a seven-slot parallel day does not. It tells you how long the problem actually takes without the parallelization lifting the clock. It tells you what aspects of the problem the agents have been handling invisibly in your normal flow, because on the off-method day you will feel the absence of that handling and have to do it yourself. That felt-absence is information. It is how you learn which parts of the method are doing the most for you and which are decorative.

They restore the deep-problem muscle. This is the big one, and it ties back directly to *deep-problem atrophy* from the top of the section. A single off-method day every few weeks is often enough to keep the long-holding capacity from atrophying. The muscle does not need to be exercised daily; it needs to be exercised regularly, the way a recreational runner does not need to run every day but does need to run roughly weekly to stay in shape. A few off-method days a month is, for most engineers, the minimum effective dose. Fewer than that and the muscle slowly goes.

They restore the craftsperson's sense of control. I have named this one last on purpose, because it is the reason that most resonates with me personally, and I am not sure I can defend it rigorously. An off-method day, when the method has been running well, produces a specific, satisfying feeling — the feeling of being an engineer who has *chosen* their working arrangement, who can operate both ways, and who is therefore not hostage to either. That feeling is, I think, the clearest psychological counter to the industry's current tendency to tell engineers they are interchangeable cogs in a rapidly-reconfiguring machine. You are not interchangeable. You have tools. You choose among them. The off-method day is the weekly — or monthly — proof.

The discipline of taking them. You will not take these days unless you schedule them. That is not a character flaw; it is a structural fact. When every working day can produce fourteen tickets, it is very hard to voluntarily produce fewer on any given day, and the pressure of the next quarter's metrics will quietly erode every informal intention to "have a slower day sometime next week." So do not make it informal. Pick a day on the calendar — the first Wednesday of every month, say — and make it an off-method day in advance. Mark it in your calendar. If your team tracks capacity, note it as a low-capacity day. When the day arrives, honor the commitment the way you would any other commitment. The version of you who made the commitment knew things about sustainability that the version of you on that Wednesday morning may have temporarily forgotten.

AI-Free Sabbaticals: The Deliberate and Controversial Habit

Now we reach the part of the chapter I have been building toward, and the part I expect to be pushed back on.

An **AI-free sabbatical** is a working day — or occasionally a weekend — on which you code *without any agent at all.* No Claude Code. No Codex. No Copilot. No chat window. No autocomplete beyond what your editor ships with out of the box. Your hands, your editor, the language, the documentation, and whatever problem you chose to work on. Nothing else.

The frequency I suggest, and I will defend in a moment, is roughly one sabbatical day per month. Some engineers will want more. Some will start with fewer and increase over time. The right cadence is the one that keeps you feeling like a practitioner of the craft rather than a manager of agents who happens to have an engineering job title.

I want to anticipate the objections, because I've heard most of them and I want to address them in the order in which they're usually raised.

"Isn't this just performative Luddism?" No. Luddism was, famously and unfairly, caricatured as resistance to technology; actual Luddites were skilled workers protesting the specific use of technology to devalue their craft. You are not resisting AI tools. You use them every day — that is the whole book. What you are doing on a sabbatical day is maintaining, through deliberate practice, the specific skills on which your

effective use of those tools depends. That is not Luddism. It is the craft-level equivalent of a pilot maintaining hand-flying skills in the age of autopilot, precisely because when the autopilot fails — and it does fail — the hand-flying skills are what save the aircraft.

"Isn't this nostalgic?" Nostalgia is a longing for a past you cannot recover. That is not what the sabbatical is. The sabbatical is a *practice* in the present, conducted for present reasons. I am not suggesting you return permanently to the old way. I am suggesting you keep the old skill warm so that the new way, which depends on that old skill, does not silently hollow out the engineer using it.

"Won't this make me slow?" For the day, yes, slightly. That is the point. Slow is not a synonym for bad. Slow is a synonym for *fully cognitively engaged in a way the method deliberately does not require.* The slowness of a sabbatical day is not a tax; it is an exercise.

"Isn't this just signaling?" This one I'll take seriously, because there is a kind of engineer who announces their monthly unplugged day on social media and is transparently doing it to performatively differentiate themselves from their less disciplined colleagues. Don't be that engineer. The sabbatical, to do what it needs to do for you, should be quiet. Announce it to no one. Put it on no conference slides. Let it be a piece of your practice you maintain the way a musician maintains their scales — privately, consistently, because it keeps you sharp, not because it makes a good story.

With the objections out of the way, here is what the sabbatical day actually does, mechanically.

It maintains deep-solve capacity. The same capacity that off-method days exercise, sabbatical days exercise in a pure form. On an off-method day you may still have access to autocomplete, documentation search, reference material. On a sabbatical day those are dialed back to whatever your editor shipped with, and the whole act of solving becomes yours again in a way that exercises the long-holding muscle specifically. A monthly session of this is roughly what keeps an engineer's deep-solve capacity intact across a decade. I do not have rigorous data on that claim; I have my own career and the careers of several mentors, and in that sample it is a clean signal.

It maintains the specific skill of writing good prompts. This sounds paradoxical — the sabbatical is the day when you're not using agents, how does it improve your prompting? — but I'll stand by it. Your effectiveness at describing a problem to an agent is a function of how clearly you can state the problem yourself. Sabbatical days are the days when you re-encounter the problem without an intermediary, and the re-encounter refreshes your ability to *specify* the problem, which is the hard part of prompting. Engineers who never do this tend to develop, over months, a kind of mushy prompting — long prompts that describe the shape of what they want in the agent's vocabulary rather than the shape of the problem in their own. A sabbatical a month is a cheap way to keep the specification muscle sharp, and specification is the thing that makes every other day of your month effective.

It resets your aesthetic calibration. This is the direct intervention against *judgment-shape drift* from the top of the section. On a sabbatical day, you write code that reflects *your* taste rather than the agents' taste. You notice what you like, what you don't, what you reach for when no one has pre-populated a solution. Those noticings are calibration data for your judgment. Over a year, regular calibrations of this kind keep your taste yours instead of the model's — and the engineer whose taste is still their own is the engineer who can tell a mediocre agent output from an excellent one on a Tuesday, which is the skill the whole method rides on.

It restores craft identity at a deep level. This is the part I find hardest to write about precisely, because it is the most personal. A sabbatical day — especially a good one, one where the solve was real and the code was yours — produces a specific feeling at the end that the method does not produce. It is the feeling, I think, that I had more often earlier in my career, and that I have had less often in the last few years. I am not nostalgic for having it every day. I would not want to give up the method. But I would not want to give up *this feeling*, either, and the sabbatical is how I keep it accessible. For engineers whose relationship to the craft is a significant part of who they are, the sabbatical is what keeps that relationship fed. For engineers whose relationship is more transactional — who came to the profession primarily for pay and are indifferent to the craft — the sabbatical will be less important, and I have no quarrel with engineers who honestly describe themselves that way. The sabbatical is specifically an intervention for the craft-oriented, and if you have read to the end of this chapter, statistically, you are probably one of us.

How to actually do it. Pick a day. Pick a project — ideally something you care about that is not work. A side project. A tool for your own use. A small utility that scratches a personal itch. The work is more important than the subject, but the subject matters: you want to choose something that will hold your attention for the whole day, because the point of the exercise is sustained engagement rather than task completion. At the start of the day, close every agentic tool. Open your editor. Open the project. Work. If you get stuck, search the documentation, read the source, think. Resist the urge to "just ask Claude quickly." The urge will be strong. It is the withdrawal you should expect; it passes within about an hour. After that hour, most engineers report a particular quiet settling of attention that they remember from before, and a specific quality of problem-solving that they had forgotten was theirs. That settling is the signal that the exercise is working.

At the end of the day, write a single paragraph in `reflections.md` — what you did, how it felt, what you noticed. The record, again, is what turns experience into data.

The Team Dimension, Briefly

A short but important detour before we close the section. Everything I have described here is an *individual* practice. I want to flag, because it would be dishonest not to, that there is a team-level version of every habit in this section, and it matters.

If you have the authority to do so — as a lead, a manager, or simply as a respected senior engineer — consider whether the team as a whole has the equivalent of an end-of-week ritual, an off-method rhythm, and an AI-free practice. A team that runs this method every day of every week without any organizational pause will suffer, collectively, all the same reservoir drains I described above, and it will suffer them in ways that compound across the group rather than dampening. The team burnout I have seen looks like the individual version, except that when it lands, it takes whole functions down at once.

At the lightest end, *normalize* the individual practices. Talk about your off-method day in public when you take one; don't evangelize, but don't hide. Mention your reflections ritual when it's relevant. Give junior engineers implicit permission to take occasional sabbatical days, because they cannot take permission they don't know exists. At the middle end, build some of it into the team's operating rhythm — an end-of-week retro that asks the four questions at a team level, a monthly "light Friday" where the team agrees no one will be expected to ship on that afternoon. At the heaviest end, if you have budget authority, advocate for one unplugged engineering day per quarter on the calendar, and defend it against the metric-pressure that will inevitably try to eat it.

I am not going to walk the team argument further in this chapter because Chapter 12 is about exactly that, and I want to save the depth for the right place. But it is worth noting now that the individual habits and the team habits are continuous with each other, and that a team running this method well will quietly have most of these rhythms built in.

What Happens If You Don't Do Any of This

Let me be blunt for a paragraph, because I have been careful for most of the section and I think you can handle it.

If you take from this chapter only the daily rhythms — the two-session day, the glance cycle, the NOTES.md wind-down, the second monitor — and not the weekly and monthly ones, the method will work for you for somewhere between three and nine months. You will be productive. You will feel, for a while, that you have found a way through the current state of the industry. And then, over a period of weeks that you will probably not notice in real time, one of the four reservoirs at the top of this section will run dry, and you will not know it, and you will start to experience the specific drag of that reservoir being empty. You will reach for the daily interventions because the daily interventions are the ones this chapter made you familiar with, and they will not work, because the depletion is at a longer timescale than the daily tools can reach. Somewhere around month six or nine you will conclude that the method doesn't work after all, or that you are uniquely unsuited to it, or that you are just getting older.

I have watched this specific arc happen to careful, conscientious engineers, and I have experienced a version of it myself. It is the reason this section of the chapter is longer than the quick bullet list the outline originally suggested. The weekly and monthly habits are not an add-on. They are the part of the method that makes the daily part durable, and without them the daily part has a shelf life.

What Happens If You Do All of This

The flip side. If you install these habits — the Friday reflection, the every-few-weeks off-method day, the monthly AI-free sabbatical — and you run them for a year, a specific thing happens that is hard to describe in advance and easy to recognize afterward.

You become, and stay, *stable*. Not euphoric. Not in a permanent state of flow. Stable — which is a quieter, more valuable state than the productivity literature usually sells. You have a working relationship with the method that is sustainable at the timescale of your career, not just at the timescale of a sprint. You have a craft identity that the method has not hollowed out, because you have been feeding it deliberately. You have a set of relationships with the people in your life that have not been quietly shaved by five years of distracted presence, because the reflection ritual and the sabbatical rhythm and the off-method days collectively build in enough structural margin that your presence at home is not the last thing to get cut. You have, crucially, a taste that is still your own — a sense of what good code looks like that has not been silently recalibrated by the pattern-exposure of the last quarter's agent outputs.

That is what the chapter is about. That is what I mean by sustainability. Not the white-knuckled *hanging on*, the hope that the tools will stabilize and the layoffs will end and the pressure will subside. That hope is, on the current evidence, a poor plan. What I am describing is a different thing: a way of doing this work, at industrial intensity, that does not consume the engineer doing it. Not because the engineer is superhuman. Because the engineer has built — day by day, week by week, month by month — a structure that holds what cannot be held, rests what cannot be skipped, and exercises what cannot be allowed to atrophy. The daily habits are where the structure starts. The weekly and monthly habits are where the structure becomes a life you can actually live.

Before We Close

The next section — on the warning signs to watch for — is the shorter, sharper one. It is, in effect, the diagnostic for when the habits in this section have slipped, or have not been installed, or have not been enough for the specific load you are currently under. Think of it as the check-engine light. Before we get to it, take a minute. Put a note in your calendar for this coming Friday afternoon — thirty minutes, end of day, `reflections.md`. Put a note for the first week of next month — off-method day, one ticket, old style. Put a note three weeks out — sabbatical day, project of your choice, no agents.

Three notes. Three minutes of calendar work. The difference between intending to do these things and actually doing them is almost entirely in whether you open the calendar right now and create the entries. Do it before the chapter convinces you later, because later you will be tired and the entries will not happen. Do it now.

Then we can talk about what it looks like when the warning signs light up, and what to do when they do.

Warning Signs to Watch For

Why This Section Gets Its Own Heading

Here is the thing about burnout, and specifically about the decision-load kind we've been talking about for the whole chapter. You almost never see it coming while it's happening. You see it in hindsight, usually several weeks after the fact, usually after a weekend when someone you live with asked you a mild question and you responded as if they had set fire to your car.

The reason you don't see it in real time is that the early signs are *quiet*. They are not the dramatic movie-burnout moments. They are not a breakdown in the parking garage. They are small shifts in mood and body and attention that are individually easy to explain away — you slept badly, the week was stressful, the cat was weird on Tuesday — and collectively add up to a pattern that only becomes visible once it's well-established.

This section exists because a pattern you can name is a pattern you can act on, and a pattern you cannot name is a pattern that acts on you. I am going to walk through the four most common early-warning signs in enough detail that you will recognize them the moment they appear, and I am going to give you a measured response to each one. Not a panic response. Not a *quit everything and move to a yurt* response. A proportionate response — the kind you would give any other signal your body was sending you, like a sore shoulder or a headache that has gone on a day too long.

Two things before we start.

First, these four signs are not a rank-ordered severity list. Any of them, on its own, is worth taking seriously. You do not need to be experiencing all four to act. You do not need to be experiencing any of them *badly* to act — early is better than late, and the cost of stepping back for a day when you didn't strictly need to is almost nothing, while the cost of pushing through when you should have stepped back can be weeks.

Second, noticing these signs in yourself is not a failure. The method is not an endurance test you either pass or fail. It is a durable workflow, and durable workflows include a feedback loop between the work and the worker. The feedback loop is what this section is describing. If you notice something, the loop is working; that's the whole point.

Warning Sign One: Dread When You Open the Laptop

This one is the clearest early signal, and it is also the one most people talk themselves out of, because the explanation *I just don't feel like working today* is plausible, familiar, and usually true of individual mornings in any career.

What I mean by dread is more specific than that. I mean: you sit down at the desk, your coffee is fine, your home is fine, the ticket queue is not unusually ugly, and yet something in you flinches at the thought of opening the seven terminals. Not a big flinch. A small one, at the level of a sigh. A preference, felt in the body, for almost any other activity — even other work activities. *Maybe I'll do email first. Maybe I'll clean up this directory. Maybe I'll read the Slack archive.* You are generating perfectly respectable reasons to delay starting, and the reasons have the texture of real reasons, and they are, nevertheless, reasons to delay.

One morning of this is nothing. Two mornings in a row is worth noticing. Three mornings in a row is a signal. Five mornings in a row — a full week of small flinches at the top of every day — is the system telling you something clearly enough that you should listen.

What it usually means. Decision load has accumulated past a level your nervous system finds tolerable, and the prospect of another day of seven-stream rotation is hitting a part of you that knows, before conscious thought, how tired it is going to be by four in the afternoon. This is not laziness. It is pattern recognition by the part of your brain that keeps track of energy budgets, and it is almost always correct.

What to do. Take a day off the system. Not a day off work — unless you can swing that, in which case, do — but a day off *this particular way of working*. Go back to the pre-AI rhythm for a day. One ticket at a time. Linear, slow, no parallelism. Close the other six terminals. If your organization's norms allow it, pick a day that's naturally lighter — a Friday, a day without big meetings — and use it as a small decompression. You will probably finish fewer tickets that day. That is fine. You will almost always come back Monday with the dread gone or noticeably lighter. If you don't, take a longer break; we'll talk about calibrating the length of breaks near the end of this section.

Warning Sign Two: Snapping at People

The second sign is interpersonal, and it is usually noticed first by the people around you rather than by you. A colleague on Slack asks a reasonable question and you feel a flash of irritation before you've even read the second sentence. A teammate in a standup offers a small correction and you find yourself wanting to argue with it out of proportion to its importance. A partner at home asks when you're coming to dinner and your answer has an edge that you can hear, faintly, as it leaves your mouth, but you can't quite stop yourself from putting it there.

This is *not* you being a bad person. It is a specific kind of depletion that shows up in social behavior before it shows up in productivity metrics. The part of your brain that handles interpersonal patience is the same general part that handles all the small moment-to-moment judgments of the seven-slot rotation — noticing a stuck slot, deciding whether to intervene or wait, choosing which prompt to fire, switching contexts to a second Claude to review a first Claude's work. That well is not infinite, and when it runs low, the first thing that gets rationed is patience.

What it usually means. You are over-parallelizing for your current state. You may have been fine at seven slots three weeks ago. You may be fine at seven slots again three weeks from now. Today, you are not fine at seven, and the signal is leaking out sideways into your relationships because those relationships are drawing from the same drained well as the rotation.

What to do. Step down for a stretch. Run four or five slots instead of seven for a week. Nothing about the system requires you to run the full seven at all times. Seven is the ceiling, not the mandate. Dropping to four is a respectable working day and it gives the interpersonal reserves time to refill. Apologize, quickly and plainly, to anyone who caught the edge of the irritation. Most people will understand. Then watch, over the next few days, whether the flashes of impatience ease. If they do, you've found the right dosage for this week. If they don't, the problem is deeper than parallelism and it's worth looking further — sleep, outside-of-work stressors, anything that might be stacking on top of the rotation load.

Warning Sign Three: The Loss of Coding Pleasure

This one is harder to notice, because its absence is quieter than its presence. There was probably, at some point in your career, a particular small pleasure in writing code. The feeling when a function you just wrote compiles the first time. The odd satisfaction of a refactor that makes something click into place. The specific minor joy of a green test suite. Those feelings were never the main reason you did the job, but they were *a* reason, and they were part of what made the work something you could imagine doing for decades.

If you notice that those feelings are gone — that closing a ticket now feels like closing a ticket, nothing more, not even the small background hum of a job well done — that is a signal, and it is not one to wave away.

What is usually happening, in my experience and in what I have watched happen to others, is not that you have fallen out of love with the profession. It is that you have *outsourced too much of the actual crafting* to the agent, and you have thereby outsourced the craftsmanship pleasure along with it. The method's productivity gains come from letting Claude or Codex do the writing while you do the directing, but directing is a different pleasure than making. If every ticket of every day is ninety-percent-directing, the making-pleasure atrophies. You become a project manager who happens to use AI tools, and you miss the reason you chose this profession in the first place.

What to do. Put some coding back in your week. Not during your fourteen-ticket rotation — that's not what the rotation is for — but somewhere else. A side project. A small refactor on your own codebase at home. An hour on a Saturday where you turn off the agent entirely and write a thing by hand, start to finish, because you want to see if you still can and because you want to feel the old feelings. One of the monthly habits I mentioned in the previous section — a day of coding by hand — is specifically aimed at this. If the pleasure comes back when you're coding without an agent, the diagnosis is confirmed and the prescription is simple: make sure your life has some not-directing time in it. Not because directing is bad — it isn't; it's the work — but because a career of *only* directing is a narrower diet than most engineers can live on happily for thirty years.

Warning Sign Four: Physical Tension

The fourth sign is the one most engineers ignore the longest, and it is the one that compounds the nastiest if ignored. It shows up in the shoulders, in the neck, in the jaw, in the eyes. A tightness between the shoulder blades that didn't used to be there. A headache that starts at 3 PM most days and stops when you leave the desk. A soreness at the back of the neck after an hour of rotation. Eye strain that wasn't there a year ago. A persistent, low-grade clench in the jaw that your dentist will eventually mention.

Your body is keeping score. It is always keeping score. The reason this sign is so often ignored is that engineers, particularly engineers who sit and think for a living, have been trained to treat physical sensation as something to route around — to sit through, to push past, to medicate mildly and move on. That training is expensive. The signals your body sends you are real information, and ignoring them does not make the information go away; it just delays the reckoning.

What it usually means. Your layout ergonomics have drifted, or were never quite right, and the compounded micro-strains of a seven-slot workflow have finally accumulated to a level that your body can no longer absorb silently. Staring at seven small regions of the screen involves more micro-saccades — tiny eye movements — than staring at one large region. Rotating attention across those regions involves more small head turns. The sustained focus pose involves more shoulder-hold than you realize. None of these are large individually. All of them add up over a month.

What to do. Several things, in order of cheapness. First, audit your physical setup: chair height, monitor distance, monitor height, keyboard position, lighting. Any one of them being wrong by a small margin will compound over the seven-hour day. A cheap monitor arm that lets you adjust height is one of the highest-return purchases you can make for this work. Second, take the ergonomic breaks seriously. The two-session day already gives you a real lunch; use the lunch to actually leave the desk, not to eat a sandwich over the keyboard. Third, if the tension has reached the point where it is persistent rather than evening-only, see someone — a physiotherapist, a massage therapist, your doctor — sooner rather than later. Engineers are notoriously slow to get physical tension addressed and notoriously quick to regret the delay. Fourth, and this is the one I want to say gently but clearly: if your eyes are strained, your glasses prescription may be wrong. It is worth a check. I have watched two colleagues attribute a year of afternoon misery to burnout when the actual cause turned out to be an outdated prescription.

Bad Day, or Pattern

One thing worth saying plainly, because otherwise this section is going to make every reader diagnose themselves with something inside of fifteen minutes: there is a difference between a bad day and a pattern, and the response to each is different.

Any of the four signs can show up on a single day for reasons entirely outside the method. Bad sleep. A fight with a family member. A difficult piece of news. A cold coming on. An argument with a neighbor about a fence. A day of dread after one of those is not a signal about your work; it is a signal about life. Take the day gently, don't push through with bravado, and watch whether the sign persists into the next day.

The rough test is: if a warning sign shows up for **one day**, it is a data point and worth noticing but not acting on. If it shows up for **two or three days in a row**, it is worth a small intervention — the dosage drop, the day off the system, the ergonomic audit. If it shows up for **a week or more**, or if multiple signs show up together in the same week, it is a pattern and worth a larger intervention — a longer break, a conversation with someone who knows you well, possibly a conversation with a professional.

Do not wait for a dramatic moment to act. The movie version of burnout, where everything crescendos into a single bad afternoon, is not the version most engineers actually experience. The real version is quieter and slower, and the window in which a small intervention still works is larger than people think, but it does eventually close. Act in the window. It is cheap there.

The Calibrated Response Ladder

To make this concrete, here is the ladder of responses I actually use and recommend, from smallest to largest.

Rung one: a day at half speed. Four slots instead of seven. Lighter ticket choices. Leave at three. Costs you one afternoon of throughput, repays you with a softer next morning.

Rung two: a day off the system. Linear work for a full day. No rotation. One ticket at a time. The old pre-AI rhythm. Costs you a day of parallel throughput, repays you with a resetting of the rotation load.

Rung three: a full weekend off everything. No laptop Saturday or Sunday. No *quick checks* on Sunday evening. A clean forty-eight hours where the part of your brain that runs the rotation gets to be completely offline. This one sounds trivial until you realize how many engineers quietly don't do it.

Rung four: a week off. Time off, ideally away from the house, ideally without a laptop. Nothing dramatic. Just a normal vacation, taken at a moment when the early warning signs have shown up persistently enough that dropping rungs one through three didn't clear them.

Rung five: something bigger. A longer break, a change of team, a conversation with a doctor or therapist, a sabbatical if your organization offers one. Most readers will never need this rung. Some will. If you're on it, you're on it; the shame is in refusing to get on it, not in being there.

Climb the ladder from the bottom. Most signs get resolved on rung one or two. If they don't, go higher. The ladder is there precisely so that you don't have to figure out the right response from scratch each time — you pick the smallest rung that fits, you try it, and you move up only if you need to.

What Not to Do

Finally, a short list of responses I want you to avoid, because I have watched each one backfire in engineers I care about.

Do not blame the method. The method is not the problem. Ripping out the seven-slot structure in a moment of fatigue and trying to go back to working one agent at a time, all at once, for ideological reasons, will cost you the sustainability benefits the structure provides while solving none of the decision-load issue that's actually tiring you. The problem, when it shows up, is almost never that the structure is too much. It is that the dosage is too high for this week.

Do not blame yourself. Noticing a warning sign is not a character failing. It is the feedback loop working. The right response is proportionate action, not self-recrimination. If you catch yourself sliding into *I should be tougher than this* territory, recognize that as its own warning sign — it is the one that stops all the others from being acted on.

Do not make dramatic changes in a low mood. Quitting a tool, quitting a job, quitting the method, selling the monitor — none of these decisions are improved by being made on a Thursday afternoon when you feel like your head is full of wet cardboard. If you feel the urge to make a big change, write it down and promise yourself you'll revisit it on a Saturday when you are rested. Almost always, the rested version of you has a smaller and better answer.

Do not hide it. The engineers who come through this well are, without exception, the ones who said something out loud to someone they trust. A partner, a friend, a therapist, a colleague who has been through it. The method is an individual practice. Recovery from its failure modes is almost never an individual practice. You do not have to make a grand announcement. You do have to make a small, honest one, to at least one other person who will hold it with you.

You built the workspace to keep yourself intact for a long career. The warning signs are the instruments on the dashboard that tell you how the engine is doing. Read them calmly, act on them proportionately, and you will be at this longer and happier than the engineers who ignored them in the name of one more ticket this week.

Which brings us, naturally, to the last section — the long game.

The Long Game

The Frame the Industry Does Not Want You to Hold

I want to close this chapter by stepping back and asking a question that almost nobody in the current AI discourse is asking, because it is an unfashionable question, and because the economics of the moment do not reward asking it.

The question is this: *what is the work for?*

Not this Tuesday's ticket. Not this quarter's release. Not this year's performance review. The work — the long arc of it, the decades of it, the entire working life of a software engineer. What is that for, and what do we need to be true about a methodology if it is going to serve that arc rather than quietly mortgage it for a short-term gain?

I ask because almost every piece of productivity writing you'll read about AI tooling right now is calibrated to a very specific, very short time horizon. Get more tickets shipped this sprint. Ten-times your output this quarter. Out-produce the engineer next to you before the next round of cuts. The articles frame productivity as a race whose duration is always assumed to be *right now*, and whose finish line is always whichever window the company has decided to measure you in. That framing is not neutral. It is the native framing of a labor market under pressure, and a labor market under pressure prefers you to think in short windows because short-window thinking is what keeps you running.

What I'm going to argue in this section, and what I hope you already half-believe because you've made it this far, is that the short-window frame is the wrong frame to use this methodology inside of. This method is not optimized for a single quarter. It is not even particularly well-optimized for a single year. It is optimized for the thirty-year career that you, I hope, are actually living, and the gap between those two optimization targets is enormous, and almost everything that matters about your life as an engineer lives in that gap.

The Arithmetic of a Long Career

Let me put a number next to the claim, because I find these arguments land better when you can see the numbers.

A sustainable pace of fourteen tickets a day, across roughly two hundred working days a year — that figure accounts for vacation, holidays, sick days, the occasional week that gets swallowed by something personal — is about twenty-eight hundred tickets in a year. Not every ticket is equal. Some are features, some are small bugs, some are the infrastructural work that will pay dividends for a decade. But the shape of it, across a year, is about twenty-eight hundred moves made in the direction of the thing you're building.

Now run that forward ten years. You have made on the order of twenty-eight thousand moves. If you grant that the quality of those moves, over a decade, averages out to *careful* — not heroic, not sloppy, just the careful default that a well-designed method enforces — you have, across that decade, accumulated the kind of impact that no single quarter's heroics can reach. Not because you were fast in any given week, but because you were still there, still moving, still shipping in the tenth year the way you were in the first.

Contrast that with the alternative the discourse currently encourages. Twenty tickets a day for four months, followed by an eight-month trough of decreasing output and increasing resentment, followed by a career change or a job change or a long, quiet stretch on a team where nobody is sure whether you've checked out or you've simply run out. The published throughput number in that story is higher — *twenty* is a bigger number than *fourteen* — but the integral, the area under the curve across ten years, is smaller by a wide margin. Often much smaller. Sometimes negative, if the trough gets bad enough to cost you the career.

The math of sustainability is cruel to our instincts because our instincts were shaped for short horizons. A predator chasing prey, a farmer racing a storm, a trader reacting to a market move — all of those are domains where the optimal strategy is *go hard now and recover later*. Software is not that domain. Software is much closer to the work of a violin teacher or a family doctor or a long-married couple — it rewards the person who is still doing it thoughtfully in year twenty, and it penalizes the one who treated year two like a sprint and exited year four.

I have watched both versions of this play out across the careers of people I know, and I want to be honest with you: the sprint version usually loses, and it usually loses badly enough to be visible from the outside, and the person running it almost never sees it at the time.

The Research on Sustainable Careers

This is not only my pet theory. It lines up with a growing body of organizational-psychology literature on what researchers have, in the last decade or so, begun to call **sustainable careers**. Ans De Vos, Beatrice Van der Heijden, and Jos Akkermans, working in a network of Belgian and Dutch universities, laid out the canonical version of the argument in a 2020 conceptual paper in the *Journal of Vocational Behavior*: careers should be evaluated across three intersecting dimensions — *happiness*, *health*, and *productivity* — and a career which optimizes any one of those at the sustained expense of the other two is, by definition, not a sustainable career but an unsustainable one that happens to look impressive on LinkedIn for a while.

The research is unequivocal that the three dimensions compound. A healthy engineer does better work. An engineer doing meaningful work maintains their health. A productive engineer who is not running on empty tends to be a happier one, and happier engineers burn out less often, and so the loop reinforces itself in the direction of longevity. Break any one dimension hard enough and the other two follow it down. That is not a motivational poster. It is a finding that has been reproduced, in various forms, across dozens of studies since the mid-2010s.

Jeffrey Pfeffer at Stanford — whose 2018 book *Dying for a Paycheck* collected much of the harder evidence in one place — went further, and documented how workplace practices that optimize for short-term output while eroding workers' health are, in aggregate, killing people. Not metaphorically. Literally. His analysis of the epidemiological data put the number of premature American deaths attributable to workplace stress, each year, on the order of one hundred and twenty thousand, which would make work-related stress a leading cause of death in the United States, on a scale comparable to diabetes. That number is an estimate, and estimates can be argued with, but the shape of the finding — that how we work is not cosmetically relevant to how long we live, but structurally relevant to it — is at this point beyond serious dispute.

I bring those up not to scare you. I bring them up because our industry has spent the last several years pretending, not always intentionally, that this research does not apply to software. That because we work in offices and earn well and have excellent health insurance, the mechanisms Pfeffer documented in warehouses and call centers and truck cabs somehow stop at the edge of the tech campus. They do not. The mechanisms follow the stress, and the stress has followed us. The layoff waves I mentioned earlier in the chapter — the roughly quarter-million tech layoffs in 2023 tracked by layoffs.fyi, the continued cuts through 2024 and 2025, the ongoing framing of engineering work as something to be compressed by tooling rather than sustained by it — those are not distant financial-press stories. They are the proximate cause of a rise in exactly the health markers Pfeffer's framework would predict, inside our own profession.

The American Psychological Association's annual *Stress in America* reports, and Gallup's *State of the Global Workplace*, and the Microsoft/LinkedIn annual *Work Trend Index*, have all been converging in recent years on a picture that is consistent across methodologies: knowledge workers are more tired than they were a decade ago, more anxious about their jobs than they were a decade ago, and working in environments that are producing measurable health consequences earlier in their careers. Our profession is a salient subset of that population, not an exception to it.

What Thirty Years of Method Actually Looks Like

I want to bring the frame all the way down from statistics to daily life, because statistics are a poor teacher on their own.

Picture the engineer this method is for. Not you on Monday. You at fifty-five.

That engineer, thirty years from the afternoon you're reading this, is still sitting down at a workspace. The tools in the slots are not Claude Code and they are not Codex and they are probably not anything I could name if you asked me. But the shape on the monitor is familiar. Seven slots. Three zones. A lease file. A validation ritual. A two-session day whose second session ends in time to eat dinner with whoever the engineer is eating dinner with by then. The method has outlived several generations of tooling by not being specifically about any of them, and it has outlived them without having to be updated in any way that broke the continuity of the engineer's working life.

That engineer's body is still mostly intact. The wrists are not destroyed. The posture is recognizable. The eyes work. The sleep is okay most nights. The weekends contain activities that are not the job. The relationships with the people they share a house with have had room to persist across decades because the job never expanded to consume all of them. The engineer has hobbies, and the hobbies are not *a side hustle* and not *an investment thesis* and not *building in public*, but simply things the engineer does because they enjoy doing them. Some of those hobbies involve absolutely no screens at all. That engineer reads books.

The engineer is still curious. Thirty years in, they still find it interesting when a junior colleague shows them a new pattern. They still watch conference talks on Saturdays sometimes because the talk looked interesting, not because a performance review is coming. They still have opinions about the work, and the opinions are nuanced, and they can still generate new ones because the generative faculty — the part of them that makes judgments at all — has not been ground down by two decades of unceasing decision load.

The engineer is, importantly, not heroic. Nothing about that picture is heroic. The fiftieth year of a career is not where the heroic stories come from. The heroic stories come from the first five years and the last three months, and both of those make for thin lives. What the middle years of a long career produce, when you run them carefully, is the quiet and underrated thing: a person who has gotten better continuously for a long time, who is still good, and who is not yet done.

That person is what the method is for.

The Version That Is Not You, Whom You Know

I promised earlier in this chapter that I would ask you to think about people around you as well as yourself, and I want to honor that one more time before we close.

Somewhere in your circle — extended family, college friends, that cohort of people who all came up in the industry at the same time you did — is a person who is not going to make it to year thirty at the pace they are currently running. You do not have to have a crisp picture of who. A face probably arrived as you read the previous paragraphs. The person has been, for some years now, in the sprint version of the career. They post updates from conferences they did not need to attend. They ship in a blur of energy that has started to look, when you catch them at it, slightly haunted. Their relationship is showing small strains. They have said, in passing, things like *I'll rest when I finish this,* about a project that does not appear to have a finish. They are impressive. They are also running hot, and the running hot has started to cost them in ways that are visible to you, and that are probably not yet visible to them.

You cannot rescue that person. I want to say this plainly, because the instinct of good people is to try, and the attempt usually does nothing except generate resentment on both sides. Nobody has ever been talked out of the sprint version by an essay or a DM or a well-intentioned older colleague. The person has to arrive at the frame themselves, and the best thing we can do, as the colleagues and friends and family members around them, is to make sure the alternative frame exists in public, legibly, without moralizing, so that when the person eventually looks up, it is there waiting for them.

That is, in part, what this book is. It is not an argument that the short-horizon frame is evil. It is an argument that another frame is available, and that the people currently working inside that other frame are, by and large, still going to be doing good work in the year 2055, and that that is a quietly radical claim in an industry whose public discourse has difficulty imagining anything past next quarter.

If the person you are thinking about comes to you, eventually, some late evening, and says they are not okay — please be the person who has a frame ready for them. Not advice. Not a fix. A frame. The one where the work can be done for thirty years at a sustainable pace, inside a structure that protects the worker from being consumed by the work. That is a gift you can give them by having organized your own life in a way that demonstrates the frame is livable. You do not need to preach it. You need to live it, visibly, so that the alternative stops being theoretical.

The Quiet Goal, and the End of the Chapter

The goal of this method is not to make you the most productive engineer at your company this quarter. It is not to help you out-ship your colleagues. It is not to get you promoted, although those things may happen as side effects, because careful work tends to be rewarded in ways that rushed work does not.

The goal is to make you a useful, curious, happy developer for as long as you want to be one.

That sentence is unglamorous, and the industry will not put it on a landing page, and no thread about it will go viral. I am okay with that. The people in the profession who I most admire, across thirty-two years of looking, have all lived some version of it. They have been the quiet colleagues who were still there in year twenty-five, still doing good work, still having opinions, still able to mentor the juniors without condescension, still recognizably the person their family married. None of them got there by sprinting. All of them got there by having a method — sometimes an explicit one, more often an implicit one — that took the pressure off the individual day so that the accumulation of days could be kind.

That is what you have built, if you built what this book describes. A structure that takes the pressure off the individual day. A structure that makes your fourteen tickets feel boring in a good way. A structure that lets you close the lid at 4 PM without guilt, because the system will still be there in the morning, and so will the work, and so, critically, will *you*.

Buzzy knew two things well, drinking beer and smoking, and he knew them across enough years that when he picked up the hammer the hammer landed exactly where it needed to without any drama whatsoever. A long career is the same shape. Not heroism. Not velocity. Just the continuous, patient, well-rested repetition of a practice you have made sustainable, for as many years as you choose to practice it.

Pick up your hammer. Or your sword. Tomorrow morning, at nine, in a slot whose position you already know by heart, in front of an agent whose name will probably change twice before you retire. And then — this is the only thing I actually need you to promise me — stop at four. Eat dinner. Sleep through the night. Come back Tuesday.

That is the long game, and it is the whole game, and you are already playing it.

Chapter 12
Bringing Your Team Along

If you've followed the book this far, one of two things is true about you right now.

Either you've been reading along as a curious outsider, collecting ideas for a workspace you may or may not build, in which case welcome — this chapter will make more sense once you have actually sat down and done the work, and you can treat it as a gentle preview of a phase you may reach in a few months.

Or you've actually built the thing. You have seven environments on a laptop somewhere with Norse names you've stopped thinking of as clever. You have a CLAUDE.md file you've tuned over several weeks. You have a screen layout that your spouse or roommate has stopped asking questions about. You are closing tickets at a rate that, if you described it honestly at a company all-hands, would cause at least one person to insist you're lying and at least one other person to quietly open a spreadsheet and start doing math. In short: you have, mostly by yourself, become a team of one.

This is the chapter about what happens next.

I want to start by saying what this chapter is *not*, because the pitfalls in this particular territory are deep and the temptations are real. This is not a chapter about how to convince your teammates to adopt the method. It is not a chapter about running a workshop, giving a lunch-and-learn, starting an internal Slack channel called #ai-power-users, or putting together a slide deck for your skip-level. It is not a chapter about evangelism at all — and if you came to this part of the book hoping for a rhetorical toolkit for winning colleagues over to your new way of working, I'm going to disappoint you, on purpose, because I have watched that approach fail so many times that I now consider it a reliable *anti*-pattern. The single fastest way to kill adoption of a good idea inside a team of software engineers is to show up one Monday morning vibrating with conviction about it. Engineers have excellent pattern-matching for zealots, and the pattern-match, once triggered, is very hard to undo.

Instead, this chapter is about what actually happens — organically, over months, in a way that can feel disappointingly slow while it's happening and then surprisingly fast in retrospect — when one person on a team starts quietly outperforming the old baseline and doesn't make a big deal of it. It's about the arc of curiosity that forms around you whether you want it to or not, the specific questions your teammates will actually ask (which are almost never the questions you expect), the small generosities that make a real difference when someone finally does ask, the awkward moment when a manager notices and wants to roll the whole thing out across the team, and the quietly important question of what to do about teammates who, for perfectly legitimate reasons, will never adopt any of this and shouldn't be pressured to.

A note on tone for this chapter specifically, because it is the closing chapter of the book and I want to be honest about the kind of closing chapter I am trying not to write. I am not going to lecture. I am not going to give you a rousing call to action about changing the industry, evangelizing best practices, leading by example, or any of the other phrases that get printed on posters in open-plan offices by people who have never had to sit under the poster. Your job, as far as this book is concerned, is to do good work, to stay intact while doing it, and to be a decent coworker. That's it. If the method spreads from you, it spreads because it worked and people noticed. If it doesn't, you still have the benefit of it, which was the whole point.

Here's how the chapter is structured.

We'll start with the single most important piece of advice in it, which is *don't evangelize*, and I'll explain at length why the impulse to evangelize feels so right and produces such reliably bad outcomes. Then we'll walk through the natural curiosity arc — roughly what happens in weeks one through four, month two, month three, and beyond, as your throughput quietly accumulates and people start to notice. I'll give you some specific language for the moment when a teammate finally asks what you're doing differently, because that moment matters more than you'd think and most developers fumble it. We'll then cover the more complicated case, which is when a manager notices and decides the whole team should be doing this — because that introduces a completely different set of problems, most of them political, and you need a plan for them before the meeting starts. We'll talk about the teammates who won't adopt, ever, and how to be a good colleague to them without either pitying them or getting defensive about your own choices. We'll touch briefly on what it looks like when a team has, in fact, fully adopted the method, because that's the rarest and most interesting case and it comes with its own small set of coordination problems. And then we'll close the book.

One more thing before we begin, and then we'll get on with it. You have done, over the course of reading this book, something genuinely difficult. You have taken a shapeless, noisy, faintly humiliating situation — the industrial-scale shove of AI tools into our daily work, without a manual, without a method, and with a running commentary about how we'd all be replaced by next quarter — and you've built something humane and practical inside of it. That matters. It matters for you, and it probably matters, in smaller and quieter ways, for whoever is around you. You don't need to advertise it. The work itself is the advertisement. Let's talk about how to keep it from curdling into something worse.

Don't Evangelize

The Temptation, and Why It Feels So Righteous

Let me describe a moment you may have already had, or that you are about to have, and walk it carefully into the light.

You have been running the method for a few weeks. The seven slots hum along. Your daily ticket count has drifted up to something your pre-method self would not have believed, and the drift has been accompanied by a real and surprising improvement in how you feel at the end of the day. You are sleeping better. You closed the laptop at 4 PM last Thursday without guilt. Your partner noticed, gently, that you were in a better mood on Saturday. You have, for the first time in a couple of years of slow industry grinding, the quiet feeling of *being ahead of the weather* — of having found shelter inside something that everyone else is still standing out in.

And then, one morning, a colleague messages you in Slack. They are complaining about the AI tooling at the company. They say, with the particular flatness of someone who has been at it too long, that it is not helping them, that it is actually making their work worse, that they do not understand how the leadership expects them to hit their numbers when the tools are like this. They ask if it is the same for you.

Reader, I want you to feel the pull in this moment, because it is almost impossible to resist. You have, in your hands, a thing that would genuinely help this person. You know it would help. You have the proof in your own Friday ticket count. You care about them — they are a coworker, maybe a friend, someone whose pain is adjacent to your own recent pain and whose relief you would, if it were free, gladly hand them. Every instinct that makes you a decent colleague is lighting up at once, saying: *tell them, tell them, tell them what you figured out, save them the months you spent figuring it out alone.*

Do not tell them.

I mean this plainly, and I am going to spend the rest of the section making the case for why, but I want the rule to land first, before the argument dilutes it. When a tired colleague is complaining to you about their tooling, the correct response is to commiserate, to listen, to perhaps mention that you have found a few rhythms that are helping, and to *stop there,* even if they seem curious, even if the conversation has opened a door that you can see all the way through. Do not walk them through the method. Do not send them a link to this book. Do not offer to pair with them on a workspace setup over lunch. Do not, and this is the one I watch the most engineers fail at, *describe your own daily ticket count.* Nothing you do in this conversation is going to help the person more than leaving them the dignity of arriving at the method on their own terms, in their own time, if they ever do at all.

I know. I know how that sounds. It sounds like I am asking you to watch a friend drown while holding a life preserver behind your back. It is not that. The life preserver, in this metaphor, is attached to a very particular kind of rope, and throwing it hard at someone who did not ask for it is how you end up fishing a tangled mess out of the water while the person you were trying to help is angrier at you than they were at the ocean. The psychology of this is specific, it is robust across decades of research, and it deserves an honest look.

Why Evangelism Reliably Backfires

Social psychologists have been studying the shape of persuasion for a long time, and the most durable finding in the whole literature is one that almost every new convert to anything — a diet, a religion, a political framework, a software methodology — has to learn the hard way. **People do not change their minds because they were argued into it. They change their minds when the conditions for changing are right, and they arrive at the new position in a way that feels, to them, like their own discovery.**

Jack Brehm's theory of **psychological reactance**, first laid out in 1966 and replicated in many forms since, captures the mechanism with unusual precision. When someone perceives that their freedom to choose is being constrained — even by a well-meaning suggestion — they experience a measurable motivational state that pushes them *away* from whatever is being recommended, often regardless of whether the recommendation would have benefited them. The stronger the push, the stronger the reactance. A colleague you are pitching the method to is, whether you intend it or not, being invited to conclude that their current way of working is wrong. That invitation lands as a threat to autonomy, the reactance fires automatically, and the internal argument for sticking with the old method gets stronger than it was before you opened your mouth.

This is not a character flaw in your colleague. It is a default feature of the human motivational system, and it is there for a reason — it protects us from being talked into things by people with more rhetorical energy than we have resistance. You, in the moment of evangelism, are exactly the kind of person the reactance system evolved to protect against. You are vibrating. You are specific. You have an answer. You have, in the worst case, a little missionary gleam in your eye that you probably cannot see but that your colleague can see from across the room.

William Miller and Stephen Rollnick, working in the addiction-counseling literature from the 1980s onward, built an entire therapeutic practice — **motivational interviewing** — on the observation that direct persuasion of people who are not ready to change produces almost uniformly worse outcomes than patient, non-directive conversation that leaves the person's autonomy intact. The central move in motivational interviewing is not to advocate for the change; it is to ask questions that let the person articulate their own reasons, if they have any, for moving in that direction. The therapists who got this right had dramatically higher success rates than those who simply explained, clearly and correctly, why the behavior needed to change. The ones who explained were doing the right thing on paper and failing in practice. The ones who listened were doing something that looked, from outside, like doing nothing, and they were in fact doing the most effective version of the work.

Carnegie's *How to Win Friends and Influence People*, which you probably read in your twenties and possibly rolled your eyes at, makes essentially the same point in a less clinical register. His advice, reduced to a single line: *the only way to get somebody to do something is to make them want to do it.* You cannot make a person want to adopt a workflow by cornering them at standup. You can only create conditions in which wanting is possible, and then you can wait.

This is the general shape of the problem, and it is why evangelism, however well-intentioned, does not work. But there is a more specific version of the problem inside software engineering teams, and I want to name it explicitly because I have watched it hurt people.

Why Engineers in Particular Defend Against Zealots

Software engineering, as a culture, has a long and not-entirely-unjustified immune response to methodology zealots. We have all sat through some version of it. The colleague who came back from a conference convinced that the right way to do everything was hexagonal architecture, or event sourcing, or TDD, or actor models, or whatever the most recent crystallization of a perfectly fine idea had become after passing through enough conference talks to lose its nuance. That colleague often made a public nuisance of themselves for six weeks, produced exactly zero converts, introduced a faint background irritation into the team's culture, and eventually either quieted down or left.

Every engineer on your team has seen that movie. Some of them have been the zealot in a prior role. The movie trains a specific response, which is: *when a coworker starts preaching a methodology, raise shields, evaluate skeptically, do not engage emotionally, wait them out.* You may have the correct methodology this time. That does not matter. The social protocol is already in motion the instant your voice takes on the slightly-too-energetic tone that all methodology pitches take on. The team is now listening to you the way engineers have learned to listen to zealots. They are not hearing the content. They are hearing the posture.

The method in this book is particularly susceptible to this failure mode, for three reasons that I want to name clearly so you can see the trap.

First, it works. A methodology that did not work would be dismissed casually and the conversation would end there. A methodology that actually works — that is visibly producing a doubled or tripled ticket count for its practitioner — triggers a much stronger defensive reaction, because the alternative to adopting it is to accept that the practitioner has found something real that everyone else is missing. That is a harder thing to accept, and the default defense against it is to find a reason to discount the evidence. *They work longer hours. They have easier tickets. They cut corners. They are burning out and just don't know it yet.* The harder your evidence is to dismiss, the more creative the dismissals will become, and the more your evangelism will have created enemies of the evidence.

Second, it is opinionated enough to sound prescriptive. Seven slots. Norse names. Two-session day. The specifics — which are part of what makes the method usable — will sound, when you describe them out loud, like a set of requirements. Engineers bristle at requirements that are not load-bearing. Your teammate will fixate on *why seven* and *why Norse* and *why two sessions* and spend the conversation arguing against the specifics rather than engaging with the structure underneath. You will feel frustrated. They will feel lectured at. Nothing will have improved.

Third, it touches a tender place. The decision-load burnout we covered in Chapter 11 is not something people talk about publicly. If your coworker is carrying it, they are carrying it quietly, and they are probably not fully aware of the shape of what they're carrying. When you arrive with a methodology that *addresses a burnout they have not yet named*, you are — without meaning to — asking them to admit in the same conversation that they are tired in a specific way and that there is an answer that you already have and they do not. That is a lot of admission to stack on top of a cold Tuesday. Most people, reasonably, will reject the whole stack rather than accept any one piece of it.

These three forces compound. The methodology works; the specifics sound like rules; and the problem it solves is one your colleague has not given themselves permission to acknowledge. Evangelism, into that terrain, is almost optimally designed to fail.

What Actually Works: Let the Work Speak

Here is what I have seen work, across several teams, over a number of years. It is unglamorous. It will not feel, while you are doing it, like you are doing anything. That is the point.

Do your work well, and do not hide it. Ship your tickets. Let your PRs land in the queue at the pace the method produces. Do not volunteer the ticket count, but do not artificially slow down to avoid making anyone uncomfortable either. Your output is a legitimate artifact of the team's work and you are entitled to produce it at your pace. The team's standups, sprint reports, and PR dashboards will carry the signal without your having to say a word about it. A teammate noticing, at the three-week mark, that you have closed twice as many tickets as the sprint average is a much more persuasive event than you mentioning the same number in a Slack thread. The noticing happened *to them*; the mentioning would have been done *at them.*

Be generous when asked, and only when asked. The difference between evangelism and teaching is who initiated the conversation. If a colleague comes to you and asks *hey, what are you doing differently*, you are now in a completely different conversation than the one I spent the last pages warning you against. That conversation is welcome. That conversation can produce real learning. We'll spend the next section talking about exactly how to handle it — because the first question they ask is almost never the question they actually want the answer to, and there are ways to answer that open doors and ways that slam them — but the gating rule is simple. If they asked, you can share. If they did not ask, you cannot.

Build a quiet trail of breadcrumbs. This is the one place I want to offer a specific practice that has worked for me, because it is the middle path between zealotry and silence, and it is surprisingly effective.

A few years ago, on a team I was on, I started an internal Slack channel — call it `#how-im-working` or something similarly low-key — and used it as a kind of public engineering microblog. I did not invite anyone. I did not announce it. I just posted to it, once or twice a week, small factual notes about things I was trying. *Tried leaving the Claude terminal open overnight and checking it first thing. Didn't help; went back to starting it fresh. Prompt template for PR self-review seems to cut my review time in half. New rule in CLAUDE.md about always using v2 API paths; caught three drifts this week.* No opinions. No advocacy. No *you should try this.* Just a working journal of one engineer's experiments, visible to anyone who wanted to look.

What happened over three or four months was instructive. For the first few weeks, the channel had one member — me — and that was fine, because it was also serving a personal function of forcing me to articulate what I was trying, which made me think more clearly about it. Then, slowly, people joined. Not because I invited them. Because somebody mentioned in a DM that they'd seen the channel and found an idea useful, and somebody else saw that person was in the channel and joined out of curiosity, and eventually the member count was up to ten and some of those ten were quietly adopting pieces of the method — never the whole thing, never formally, but a prompt template here, a memory-rule there, a piece of layout reasoning that worked for them in their own setup.

Nobody evangelized. Nobody had to. The channel was a trail the curious could follow if they chose, and the people who chose were exactly the people for whom the method was going to work. The people for whom it wouldn't, or who weren't ready, scrolled past it or never joined, and nothing was lost because I had never asked them to engage in the first place. That asymmetry — the method becoming discoverable without being imposed — is the single most useful social pattern I have found for this kind of adoption, and it costs almost nothing to run.

Let the compounding do the persuading. A methodology that produces durable, visible results will, over a quarter or two, become conspicuous without anyone having to point at it. A manager will notice. A teammate on a parallel project will notice. A tech lead comparing sprint velocities will notice. The compound interest of steadily shipped tickets is the loudest possible advertisement, and it requires nothing from you except continuing to do the work. You may be tempted, in the months when no one has asked yet, to help the signal along — to say something in standup, to drop a hint in a retro, to leave a link in the team's shared docs folder. Resist. The signal is arriving at its own pace. Impatience on your part only raises the volume to the level where reactance fires; patience keeps the signal quiet enough to actually be heard.

A working engineer who has been in this industry long enough will tell you that the most persuasive colleague they ever had was the one who never tried to persuade them of anything — who simply worked visibly, answered questions honestly when asked, and left the final inference to be drawn by the person doing the looking. You have, in your hands, the opportunity to be that colleague. It is an unusual opportunity. Take it.

A Thought About the Teammates Who Are Struggling

Before we move on to the natural curiosity arc in the next section, I want to come back to the coworker from the beginning of this section — the one who DM'd you in Slack, tired, complaining about the tooling. I have not forgotten them, and I want to close this section by saying how to actually be a good colleague to that person, because *don't evangelize* is not the same as *don't care.*

The people around you who are struggling are, with high probability, some version of what Chapter 11 described. They are decision-load exhausted. They are economically anxious because the layoff news of the last three years has lodged in a permanent way in their nervous system. They are working inside a tooling situation they did not choose, with a productivity narrative imposed on them that is unforgiving, and they are holding it together with whatever materials they have at hand. What they need from you, in the moment, is not a methodology. It is a person who acknowledges the texture of the problem without making it their responsibility to solve it.

When that colleague DMs you, the warmest thing you can do — the warmest and also the most effective thing — is to meet them where they are. *Yeah, it's been rough. I hear you. I've been finding a few small rhythms that help but honestly the whole thing has been a lot for everyone.* That is not hiding the method. That is not being dishonest about your own experience. That is recognizing that their complaint is, first, a need for solidarity, and only second and much later, maybe, a potential opening for a practice conversation. If they later come back and ask specifically *what rhythms*, you are in the *when asked* category, and you can share generously. If they don't, they didn't need the information — they needed the solidarity, and you gave it to them, and that was the whole ask.

This is how you hold both things at once. You have a good method; your colleague is struggling; evangelizing at them would hurt both of you; and the answer is not to pretend you don't have the method but to let it come up, if it comes up, in the slow organic way that adoption actually happens. You can be generous without being pushy. You can be helpful without being prescriptive. You can, quietly, over long enough time, become the person on the team whom the struggling come to when they are finally ready to consider a different way, *because* you never insisted on one.

That is the whole lesson of this section, and it is the foundation for everything else this chapter will discuss. Do good work. Make it visible without advertising it. Be generous when asked. Be kind when someone is tired. Trust the compounding. Do not, under any circumstances, become the zealot.

The next section is about what to expect, week by week and month by month, as that approach plays out on a real team — the natural curiosity arc that forms around a quietly competent practitioner whether they invite it or not. It is slower than you'd like. It is more reliable than anything else available. And when it works, it works in a way that leaves the relationships intact, which is, in the end, the part that matters.

The Natural Curiosity Arc

Why the Arc Is a Thing at All

If the previous section was about why you should not evangelize, this section is about what happens instead. Because something does happen. Adoption of a genuinely useful practice in a team is not usually the result of someone standing up in a retro and making a speech. It is the result of a slow, largely invisible process in which one person's work starts to look different from the work around them, and the difference becomes the subject of a series of small, private curiosities that the people having them mostly do not mention out loud until they have turned them over in their own heads for weeks.

That slow, largely invisible process is what I am calling the **natural curiosity arc**. I want to walk through it in detail, because if you understand the arc — its shape, its tempo, the specific kinds of signals that emerge at each phase, and the specific mistakes that are easy to make at each phase — you will be in a much better position to let it do its work, without accidentally interrupting it with your own helpfulness.

A quick warning before we start. The timelines I am going to give below — *week one, week three, month two, month six* — are approximate. Some teams move faster.

Some move slower. A team under a merger will not run the arc at all for six months because everyone is too busy hiding. A team that just hired three people will run it in fits and starts because the new hires reset the social observation. What matters is not the specific week. What matters is the *order* in which things happen, because the order reveals the mechanism, and once you see the mechanism you can work with it rather than against it.

The arc has, in my experience, roughly seven phases. I'll walk each of them in sequence. And because this is a chapter about people and not about tooling, I want to spend more time than usual on the *psychology* of each phase — on what your teammates are actually thinking when they do or don't say something — because your behavior should be shaped by what's going on inside them, not by what you wish was going on inside them.

Phase One: Invisible Competence (Weeks One Through Two)

In the first two weeks after you have installed the method and are running it well, nobody on your team notices anything. Not anything. Zero signals.

This will feel strange. You have just done something genuinely ambitious — you have reshaped your entire working day, you are closing tickets at a rate that would have embarrassed the pre-method version of you, and you are almost certainly experiencing a kind of quiet elation that you want to share with someone. And nobody is saying anything. Standups come and go. Your PRs get reviewed and merged. Someone leaves a heart react on one of them. No one asks.

This phase is, psychologically, where the single most damaging temptation of the whole arc lives, and it is where most developers who try to bring the method to their team accidentally kill its spread. The temptation is to *tell someone*. To mention it in standup. To share the CLAUDE.md in the team channel. To post a tweet thread. To reply to a question about an unrelated topic in Slack with "by the way, I've been doing this thing" and dropping a link. Every one of those moves feels like generosity. Every one of them is, in fact, the evangelism failure mode from the previous section, and every one of them will slow the actual arc down by weeks or months.

Let me explain why nobody notices, because the explanation is the whole foundation of the arc.

Your teammates are not watching your metrics. They are watching the *rhythm* of their own work, and their peripheral awareness of you is a subconscious average of the last six or eight weeks of how often they had to wait on you, how often you got a PR up in time for their review block, how often you did or didn't make a meeting. That peripheral awareness updates slowly. Two weeks of improved throughput is, statistically, indistinguishable from a good two weeks against your older baseline. They haven't observed long enough to register a pattern. Their brains, like yours, are pattern-matchers, and the matcher needs more data.

There is a second reason, which I want to name plainly because it is psychologically real and not often discussed. Your teammates are busy, and their emotional bandwidth for noticing a colleague doing well is, right now, lower than it normally would be. We are in a period when a lot of engineers are carrying around a low-grade fear about their own performance and their own job security. In that state, a colleague who is visibly outperforming can register, initially and below conscious awareness, as a mild *threat* rather than a *curiosity*. The noticing, when it comes, is often delayed by a few weeks of the noticer's own emotional processing — during which they are, without realizing it, resolving a small internal question about whether they are comfortable being curious about you or whether they would prefer to find a reason to dismiss you.

This is not cynical. It is human. I have felt it myself about colleagues who got good at things before I did, and the right response to noticing it in yourself is not to be ashamed but to be patient. The curiosity will form. It will just take a beat.

So your job in phase one is to *work, and be quiet about it*. Specifically:

- Close your tickets.
- Maintain a normal, slightly unremarkable social presence. Don't go dark on Slack; that reads as aloof. Don't blast Slack; that reads as desperate. Behave like the version of you that existed four months ago, but ship more than they did.
- Do not mention the method. Not in standups, not in retros, not in 1:1s, not in the team channel, not over coffee, not in DMs. Not in any form. The only exception to this rule is if your manager directly asks you what you are doing differently, in a private setting, in which case you may answer honestly and briefly. No one else.

- Do not act humble in a way that calls attention to itself. “Aww, I just got lucky with the tickets” is not humility; it is a humble-brag, and it reads like one. If asked, say “it was a good week” in exactly the same tone you’d say “it was a rough week,” and move the conversation on.

That is phase one. Two weeks. Possibly three on a slow team.

Phase Two: The First Quiet Noticer (Weeks Three Through Five)

Eventually, and it is almost always one specific person on your team, someone will notice. This is the pivotal phase of the arc, and the specific shape of what happens — and of your response to it — matters a lot.

The first noticer is rarely the most senior person on the team. It is rarely the team lead. It is rarely the loudest. It is, in my observation, almost always a very specific archetype: the mid-career engineer who has been on the team for a year or two, is genuinely good at their job, is slightly unsatisfied with how their own practice is going, and has a capacity to pay quiet attention to their colleagues that most people lack. If you think about your team for a second, you can probably guess who it will be.

What the first noticer does, when they notice, is not what you expect. They do not come up to you and ask what you are doing. They observe, for another week or two, on their own. They look at your PR history. They look at your commit messages. They look at your CODEOWNERS responsibilities versus what you are shipping. They form a hypothesis, quietly, that *something* is going on with you — not necessarily the method, just *something* — and they wait for more data.

During this phase — which you will not be aware of, because it is happening inside someone else’s head — your behavior has to remain exactly what it was in phase one. Close your tickets. Be normal. Above all, *do not notice that someone is noticing you.* If you start behaving self-consciously in response to a silent observation, the observation will curdle. The noticer will start to feel that you are performing for them, and the curiosity will die. This is subtle and it is important. The arc works precisely because it is not being forced. Your job is to keep the forcing mechanism off.

The phase ends when the noticer finally asks. The question is almost never "how are you doing that." The question is, at its most common, a sideways one. Examples of the actual first-question phrasings I have heard over the years:

- *"Hey, quick question — what's your setup look like these days? I'm thinking about redoing mine."*
- *"How are you finding the new agent stuff? I've been trying it and I feel like I'm using it wrong."*
- *"You've been cranking through tickets. Anything you'd do differently if you were onboarding onto the team today?"*
- *"What does your CLAUDE.md look like? Just, like, generally."*

Notice the shape of these. They are not demands for a methodology. They are tentative probes, wrapped in plausible deniability. The noticer is giving you a face-saving way to talk about your practice without either of you having to acknowledge that they have been watching you for three weeks. Accept the face-saving. Meet them on the ground they are offering.

The worst possible response to any of the above is to say *"actually, I've been doing this whole method — here, let me show you everything."* That response turns a small, mutual act of curiosity into a lecture, and the noticer will back away, smile, and change the topic. The curiosity will shut.

The correct response is small. Answer the specific question they asked, with one or two sentences more than you strictly needed to. For example:

- *"My setup? Honestly, the biggest thing I changed is I keep seven terminals open now and let a different agent run in each one. It's made a weird amount of difference."*
- *"Yeah, the agent stuff is hit and miss. The thing that helped me most was writing out a CLAUDE.md with the rules of the codebase. Happy to show you mine if you want."*
- *"Onboarding — hmm. I'd tell onboarding-me to not try to prompt everything fresh each time. The persistent memory file does more than it looks like."*

Each of those answers invites a follow-up without demanding one. Each of them signals that you are willing to share more if asked, without demanding that the asker commit to learning the whole system. Each of them respects the other person's own pace.

And then — this is the part that takes discipline — you stop. You let them decide whether to ask the follow-up. You do not volunteer the next piece. You do not send them a link to this book. You do not offer to pair-program. You give them the small answer to the small question, and you wait.

Most of the time, in my experience, phase two ends with a short, pleasant exchange and no follow-up for another week or two. The noticer has taken the tiny piece you gave them and is now running *that* against their own practice, and your patience is doing real work during that interval.

Phase Three: The Second and Third Noticers (Months Two Through Three)

Here is where things get genuinely interesting.

Once the first noticer has seen your answer, something subtle happens on the team. The noticer, in their own work, starts to experiment with whatever small piece of your answer they took home — maybe the CLAUDE.md, maybe a dedicated terminal, maybe nothing more concrete than a slightly more deliberate approach to their agent interactions. And because they are doing it, quietly, in their own space, their own output starts to shift. Not dramatically, but enough.

Now there are two people on the team whose rhythm is slightly different from the baseline. Not coordinating. Not collaborating on this. Each of you is running your own version. But the *team's* aggregate throughput has nudged up, and the social observation field of the team has gained a second anomaly.

This is the phase where the second and third noticers appear. They tend to come in a small cluster — two or three people within a couple of weeks of each other — because the statistical threshold for noticing one person is higher than the threshold for noticing two, and the second noticer will usually trigger a third. Unlike the first noticer, these later ones are often more willing to ask directly, because the first noticer has (unintentionally) normalized the conversation.

Their questions are also shaped differently. The first noticer's questions were oblique and face-saving. The second-wave questions are more specific. Examples:

- *"I saw Jamie mention you've been running multiple agents. How do you actually keep them from stepping on each other?"*
- *"Can I see what your CLAUDE.md file looks like?"*
- *"How do you decide which agent to use for what?"*
- *"Do you have a setup I could copy?"*

At this phase, the generosity imperative kicks in. The next section in this chapter is about how to be generous when someone asks. I will not preempt it, except to note one thing here that belongs specifically to the *arc* rather than to the response: the cluster of noticers will, by themselves, start to form a small, informal cohort of practitioners on your team. They will not call themselves that. They will not announce anything. They will simply start to ask each other questions as well as asking you, and they will start to share small discoveries among themselves.

The correct thing for you to do as this cohort forms is *join it as a peer*, not preside over it as a founder. You are not the expert. You are one of the people in it who happens to have been doing it longer. Participate in their small discoveries. Be genuinely interested in what *they* are finding out. Answer questions, ask questions, treat their half-formed variations of the method as interesting data about the method rather than as misunderstandings to be corrected.

This part is psychologically demanding, because there will be moments when a teammate is doing the method wrong in a way that you are certain will cost them, and your instinct will be to correct them. Sometimes you should — quietly, peer to peer, once. Most of the time, you should let them discover their own correction, because the lesson is sticky when it is theirs and loose when it is yours. There is a deep finding from pedagogy research here, sometimes associated with the term *productive struggle*, sometimes discussed in the literature on *generation effect* in learning, that supports this: people learn more durably when they have to figure something out themselves, even imperfectly, than when they are told the correct answer upfront. Your role as the more-experienced practitioner is to provide the safety net, not to fill the gap.

One practical habit that has served me well during this phase: when a teammate asks me a question about my setup, I try to also ask them a question about *theirs*. Not rhetorically. Honestly. Their setup has things in it that mine doesn't, and some of those things are good. A real exchange of practice, in which they are teaching you something too, does more to cement the cohort than any amount of your teaching alone would. It also keeps your ego in the right place. You are a practitioner, not a guru. Keep it that way.

Phase Four: The Plateau of Partial Adoption (Months Three Through Six)

Here is where most books about team adoption would jump to the happy ending. I am not going to, because the actual arc has a boring middle, and the boring middle is the part where most adoption efforts quietly die.

Somewhere around month three to month four, the cohort will have stabilized. A handful of your teammates — let's say three or four — have adopted some meaningful subset of the method. Two or three others have heard about it, tried a piece, and drifted away. The majority of the team, which is still the plurality, is doing their own thing.

Aggregate team throughput is up, but not uniformly, and the delta is diffuse enough that management has not yet articulated it to themselves. The cohort's members are happier than they were three months ago and are saying so quietly to each other. The non-adopters are, variously: (a) unaware, (b) aware and indifferent, (c) aware and skeptical, (d) aware and secretly curious but not ready to ask.

This is the plateau. It can last a long time. Months. I have seen it last a year. It is, in its way, the most important phase of the arc, because what happens during it determines whether the method gets embedded into the team or whether it remains a curiosity that a handful of people do and the others view as a quirk.

Your behavior during the plateau is critical and counterintuitive.

Don't push. The plateau is not a problem to be solved. It is a natural phase. The non-adopters who are aware and curious but not ready to ask are, right now, in the same private phase the first noticer was in during phase two. They are watching. They are not ready. Pushing them will flip their private state from "watching with interest" to "watching with resistance" in a single conversation, and you will have lost them for the rest of the arc.

Maintain the quality of your own practice. The plateau is also, for you personally, the phase when the novelty of the method has worn off and the temptation to skip the refinement cadence or the weekly ritual is highest. Don't. The quality of your own practice is, in a very real sense, what sustains the cohort. If you slip, the cohort's implicit model of "we are doing a real thing" degrades, because their model was partly anchored on you doing it convincingly. This is a responsibility you signed up for when the first noticer asked, and it is a responsibility worth taking seriously.

Continue to be a peer to the cohort. Do not become its leader. Do not host weekly practice meetings. Do not propose a team channel. If the cohort wants to coordinate, they will coordinate themselves — and if they don't, they don't want to. Your role remains peer-level. Any gravitational pull you exert toward being the leader will make the method *yours* in the team's eyes rather than *theirs*, and that transition is fatal to adoption. The moment a practice stops being "what good practitioners in our team do" and starts being "Blake's thing," it is dead.

Do the work you would be doing if the cohort didn't exist. This is a subtle one. Some engineers, during the plateau, start to over-prioritize mentoring the cohort at the expense of their own tickets. Don't. Your own output is the advertisement. An engineer who has gotten so invested in spreading the method that their own throughput has sagged is, ironically, making the strongest possible negative case for the method. Close your tickets. Be the practitioner the cohort is modeling themselves on.

The plateau ends when one of two things happens: a natural event triggers the next phase (a team reorg, a shift in tooling, a high-profile project that the method helps deliver), or the slow accumulation of the cohort's visible improvement finally reaches the threshold where management formally takes an interest. I'll describe both transitions in the next two phases.

Phase Five: The Management Noticing (Months Six Through Nine)

Eventually, a manager will notice. This is a different event than a peer noticing, and it has to be handled very differently, because the social dynamics around a manager asking about a practice are not the same as the social dynamics around a peer asking.

The most common way this surfaces is through a one-on-one. Your manager, who has been watching the team's aggregate numbers without articulating a theory, will say something like: *"I've noticed you and a few others have been shipping at a different pace lately. What's going on?"*

Or, sometimes more bluntly: *"Can you walk me through your workflow? I want to understand what's making the difference."*

The temptation at this moment — and this is the single largest judgment call in the whole arc — is to tell the manager everything. To pull up CLAUDE.md, to walk them through the seven-slot layout, to show them the whole apparatus, to propose a team-wide rollout. Every instinct you have will be pointing that direction, because you have finally been asked by someone with the authority to make the thing real.

I want you to not do that. Not in the first meeting. Here is why.

A manager asking about a practice has two kinds of consequences that a peer asking does not. The first is that whatever you tell them is, implicitly, a recommendation for team-wide adoption, whether or not you meant it that way. Managers are authorization-machines by profession; they will hear "this is what's working for me" as "this is what the team should do next quarter," and a fraction of them will act on that interpretation before the week is out. If the team is not ready — and at month six, it usually is not, because the plateau has not fully played out — the forced rollout will fail, and the failure will take the whole method down with it. Not just for the new adopters. For you too.

The second consequence is more subtle. A manager who has been told about a practice by an individual contributor has a small, usually unconscious obligation to credit or attribute the practice. Some managers do this gracefully; some do it by announcing it at a team meeting with your name attached in a way that turns you into the team's "AI person"; some do it by quietly proposing the practice to their own manager as something *they* discovered. None of these outcomes are what you want. You want the practice to belong to the team, not to you.

The correct response to the first manager-asking is calibrated, specific, and deliberately understated. Roughly:

- Acknowledge that you have been running a particular working setup that has been working well for you.
- Mention that a few of your teammates have been trying related things on their own.
- Offer to share what you've been doing with anyone on the team who is interested, on a one-to-one basis.
- Gently decline, for now, any suggestion of a formal rollout or team meeting.

Specific language that has worked for me:

"Yeah, I've been running a different workspace setup for a while. I'm happy to walk anyone through it who wants — I've done that for a few folks already. I'd probably rather not formalize it as a team thing yet, because I think it works best when people find it on their own timeline. Let me know if you want me to share it with you personally, though; I'd be happy to."

Notice the shape. You've acknowledged the reality. You've signaled generosity. You've named, without explaining at length, that a forced rollout is likely to be counterproductive. You've offered the manager their own one-to-one walkthrough, which flatters them appropriately and gives them a route that does not require them to impose it on others. You've kept your own practice peer-level in how you framed it.

The manager will usually accept this. Some won't. The ones who don't accept it, and who press for a formal rollout, are the subject of another section of this chapter — *rolling out to a team when asked by a manager* — and I won't preempt it here, except to say that even when the rollout is going to happen anyway, the right move at the first conversation is to slow it down by a beat, not to accelerate it. The arc needs time, and time is the one thing a manager with a fresh enthusiasm cannot imagine needing.

Phase Six: The Team Inflection (Months Nine Through Eighteen)

This is the phase most engineers who read books like this one never see, because it happens slowly, in the background, and only on teams where the earlier phases were handled well. I want to describe it anyway, because knowing it exists is motivating during the slower months.

Somewhere between nine and eighteen months after you started the method, if the earlier phases went well — if you did not evangelize, if the cohort formed organically, if the plateau was survived patiently, if the manager was handled deftly — a quiet tipping point occurs. More than half of the team is running some meaningful version of the method. The holdouts are either people who for legitimate reasons will never adopt it (covered in a later section) or new hires who have not yet been exposed.

At this point, the method stops being *your* practice and becomes *the team's* practice. You will stop being the person teammates ask about it, because there are now several people who can answer. You will stop being the reference implementation, because several people have refined things that you haven't. You will become, in a pleasant and slightly disorienting way, one member of a group that does this work well together.

A specific thing you should do at this phase: *notice the moment when a new hire is being onboarded and someone other than you is the person teaching them the method.* That moment is the arc completing itself. The practice has escaped your hands. It now lives in the team. You are, paradoxically, freed from it — you can stop worrying about sustaining it, because it is being sustained by others.

There is a small grief that sometimes comes with this moment, for the original practitioner. The practice was yours. You spent quiet months refining it and tending it and worrying about whether it would take. Now it is no longer specially yours. The engineer who learned it last week has it, and their version is slightly different, and everyone around you refers to "how we work" rather than "how Blake works." I want to name the grief because pretending it doesn't exist is dishonest, and because the grief is actually evidence that the arc went well. If you still owned the practice, the arc had failed.

Let the grief pass. Enjoy the team you are now on. This is the part of a career where the work gets good in a way that earlier work was not, because you are not doing it alone and you are not carrying the whole rhythm yourself.

Phase Seven: The New Baseline (Eighteen Months and Beyond)

Finally, after the inflection, there is the new baseline. This is what the team looks like once the method is simply how they work.

A few things are different now, and they are worth naming because they are what you have been building toward without necessarily realizing it.

The team ships at a pace that is structurally higher than the pre-method baseline, but without the individual heroics that used to power occasional bursts. The pace is *sustainable*, which is a word I have used a lot in this book and which, in this context, means *it does not destroy the people producing it*. Nights are quiet. Weekends are quiet. People are still intact at the end of quarters. The team can absorb a high-stakes project without the usual crunch, because the method's slack is built in.

Onboarding is faster. New hires ramp to full productivity in weeks rather than months, because the method gives them a structure they can inhabit rather than an ambiguous set of cultural expectations they have to guess at. The CLAUDE.md tradition has become a team artifact; new hires read it on day one and are productive by day three in a way that would not have been possible before.

Management's relationship to the team is calmer. Metrics are predictable. The need for emergency interventions is lower. Managers who have watched the team for a year can, if they are thoughtful, articulate why — and those managers become important allies in defending the team's rhythm against the kind of external pressures (quarterly goal increases, unrealistic roadmap asks) that previously would have broken it.

And — this is the part I find it hardest to write about without sounding sentimental, and I am going to risk it anyway — there is a *tone* on a team that has gone through the full arc. A specific kind of ambient calm. A tendency to joke about the work rather than about the struggle. A willingness to be curious about new tools without panicking about them. A way of carrying oneself into a Monday morning that says, plainly and without performance, *we know how we work, we are good at it, and we are not at the mercy of whatever the industry does this week.* That tone is the real payoff. More than the throughput numbers. More than the calmer managers. More than the faster onboarding. It is a team that has remembered how to *be a team* in a period of the industry when a lot of teams have forgotten.

The Arc in Summary

Seven phases, eighteen or so months, mostly invisible while they are happening and clear in retrospect. Let me restate the whole thing compactly, because the full-length version is easy to lose the shape of.

1. **Invisible competence.** Weeks one through two. Nobody notices. Work quietly.
2. **The first quiet noticer.** Weeks three through five. One person observes. Answer their small question with a small answer.
3. **The second and third noticers.** Months two through three. A small cohort forms. Be a peer, not a guru.
4. **The plateau of partial adoption.** Months three through six. Boring. Essential. Do not push. Do the work.
5. **Management noticing.** Months six through nine. The first time authority asks. Go slow. Decline the team rollout, for now.
6. **The team inflection.** Months nine through eighteen. The practice escapes your hands. Let it.
7. **The new baseline.** Eighteen months and beyond. The team now works this way. You are one of them.

The whole arc is, in the most literal sense, an exercise in patience. Each phase has a specific mistake that is tempting and ruinous. Each phase rewards restraint that feels, in the moment, like passivity. Each phase's reward is invisible until the next phase begins.

The Psychology Underneath the Arc, One More Pass

I want to close this section with the most general observation about the arc, because I think it is what makes all the specific advice work.

The arc is not really about a method. It is about how a small, fragile, genuinely useful idea travels through a group of adult humans who are tired, suspicious, competitive, and kind, all at once. The travel of the idea looks, on the surface, like a communication problem — *how do I tell my team about this thing* — but it is not. It is a trust problem. It is a patience problem. It is, at its root, a question of whether you can hold a good thing lightly enough that the people around you get to come to it themselves.

Most of the literature on organizational change — from Kotter's *Leading Change* through the more recent writing on psychological safety and Amy Edmondson's work — comes down, when you boil it off, to a handful of overlapping insights about why top-down change fails and bottom-up change works. The arc in this section is one specific instance of that general finding, applied to the specific case of a developer adopting a practice and hoping that their colleagues will want to adopt it too. It has the same shape the general literature describes: the change has to feel safe enough to adopt, interesting enough to notice, and authored enough by the adopter that they own it rather than perform it. The method you have built, because it lives on disk and on your screen and nowhere near anyone else's authority, is almost uniquely well suited to this kind of travel. It is *exportable without being imposable.* That is a precious property. Treat it carefully.

And finally — because this is the closing part of a chapter on teams, in the closing part of a book about burnout, in the closing part of a year when a lot of teams have had a hard time — I want to say plainly that the real reward of the arc is not productivity metrics. It is the slow reconstitution of a working environment that supports the people inside it. The method is not, and never was, just a way to close more tickets. It is a way to be a developer, in a hard industry, at a hard time, without losing the part of yourself that made you want to be a developer in the first place. When you bring that to a team, carefully, patiently, over many months, and they come along with you — not because you sold it to them, but because they watched it work and asked — what you end up building together is not an AI-productivity team. It is something rarer. It is a team of craftspeople, supporting each other, in a period when support is scarce and craft is rumored to be obsolete.

That, more than any ticket count, is the thing worth being patient for.

Before We Move On

The next section — *when someone asks, be generous* — is the practical companion to this one. This section was about the *arc*, which is mostly about your discipline and patience as the practice travels. The next section is about the *exchange*, which is what happens in the actual conversation when a teammate finally does ask. They are meant to be read in sequence, and the one after it — on what to do when a manager asks for a team rollout — is the political companion to both.

Before you read the next section, do one small thing. Think of one specific teammate you believe, from the description above, might be your first quiet noticer. Not to target them. Not to orchestrate anything. Just to hold them in mind, so that when they come to you with their first small sideways question — the *"hey, quick question about your setup"* — you recognize the moment, and you give them the small answer rather than the large one. The arc will take care of the rest.

Figure 12.1 draws the arc from first noticing to local variant, with stage-by-stage markers for what each stretch actually looks like. The banner across the top is the list of things that *don't* accelerate the curve — evangelism, lunch-and-learns without invitation, quoting the book at people. Those are the flattening moves. The goal is to do none of them and trust the shape.

Figure 12.1 — The Natural Curiosity Arc

How teammates actually come to the method — slowly, sideways, when they decide to. Never by being sold to.

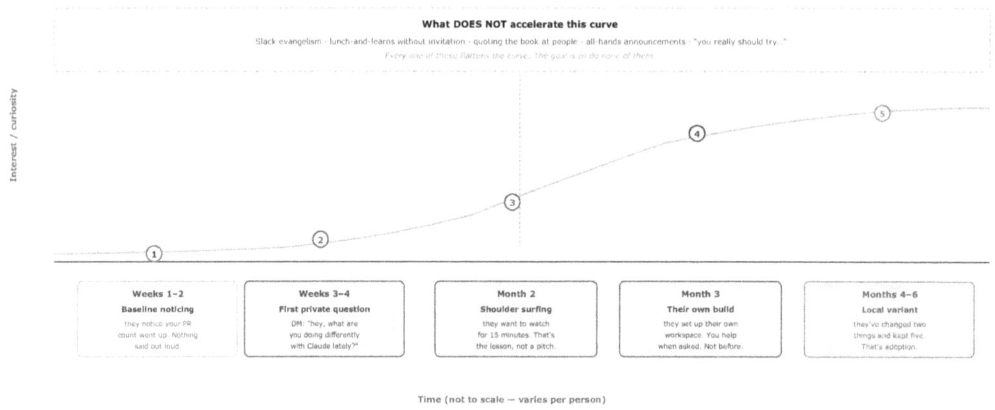

When Someone Asks, Be Generous

We spent the last two sections preparing the ground. First, a long argument for why you should not evangelize. Then, a walk through the curiosity arc — the specific pattern by which someone, eventually, is going to wander over to your desk and ask you a question.

This section is about that moment.

I want to take it slowly, because most developers I know fumble this moment, and they fumble it in ways that close off the conversation that was about to happen. The stakes are larger than they look. The way you respond the first time a teammate asks you what you're doing differently will, in my experience, determine whether that teammate ever asks you a second question. It will also, quietly, determine what the method looks like in their hands a month from now, if it gets into their hands at all.

So let's be careful about this. Let's not wing it. Let's think through, before it happens, what kind of person you want to be in that moment, and what that person actually does.

The Impulse You Are About to Feel

Imagine the moment concretely.

You are sitting at your desk, mid-rotation, maybe halfway through a Wednesday afternoon. tyr is ten minutes into a ticket. freya just surfaced a PR you're about to review. Your NOTES.md for odin is open in a scratch window because you were reminding yourself of where you left it yesterday. A teammate — let's call her Priya, because I need a name and I've known several Priyas who were excellent — rolls her chair over and says, "Hey, what are you doing over here? I've been watching you close like four tickets a day for a month."

Something is about to happen to your nervous system, and I want you to know about it in advance so you can notice it when it arrives.

You are going to feel *recognized*. Possibly for the first time in a long time. You have been quietly doing excellent work for weeks or months, no one has said much about it, and now someone — a colleague you respect, someone whose attention you value — has noticed and is standing at your desk asking you about it.

The impulse this triggers, for almost everyone, is to overshare. Dramatically. You will feel a warm burst of energy that wants to explain everything at once. The slots. The zones. The Norse environments. The lease file. The draft PRs. The glance cycle. The two-session day. The dish by the door. The history of how you figured it all out. The specific wrong turns you made in week two that taught you why the environments have to be named. The cognitive-economics argument from Chapter 11. The principles from Chapter 10. The way the CLAUDE.md file has evolved over the last six weeks. The entire contents of this book, delivered in seventeen minutes, in front of a whiteboard you grabbed on the way to the conference room.

Do not do this.

I am asking you, gently but firmly, not to do this. Not because the information is wrong — it isn't — but because the volume is wrong. Priya did not ask you for the book. She asked you one question, and she asked it during a working afternoon, standing at your desk, with work of her own that she needs to get back to. If you answer her question by giving her the book, you have not been generous. You have been overwhelming, which is a different thing, and often a worse one.

The warm burst of energy you are about to feel is real and it is not a bad feeling — it is the relief of having something genuinely good to share, after weeks of quiet. But the way to honor that feeling is not to let it flood the moment. It is to *match the moment* — to give Priya exactly what she asked for, at the pace she asked for it, with the warmth of someone who has found something lovely and wants her to find it too.

The Frame I Want You to Carry Into the Conversation

Here is the frame I want you to hold in your head before you open your mouth.

This is a *"hey, look what I found"* moment. It is not a *"let me teach you my system"* moment. Those two framings produce very different conversations, and the first one produces the conversation you actually want.

The difference is not merely tonal, though it is also tonal. The difference is about whose discovery the method ends up being. In the *"look what I found"* frame, you are sharing something exciting that came into your life, and Priya gets to decide whether any of it is exciting to her, and to pick up the parts that are. In the *"let me teach you my system"* frame, the method is yours, and Priya's job is to receive it from you in the shape you hand it to her.

The first frame makes Priya a fellow explorer. The second frame makes her a student. Engineers do not like being students of their peers, and they shouldn't — it is a subtly demeaning role, and their resistance to being put in it is a healthy instinct, not a character flaw. If you accidentally put Priya in the student role, she will quietly decide within ninety seconds that she is not interested after all, and the conversation will end with her saying something polite and rolling her chair back to her own desk, and you will wonder, later, why it didn't go better.

So before you answer, sit with the frame for a beat. *Hey, look what I found.* Not *let me show you how I work.* Not *I've figured something out.* Not *I've been meaning to write this up.* Just — look what I found. I am excited about it. Some of it might be useful to you. You tell me which parts.

If you can hold that frame — if you can come into the conversation from the posture of a person who found something cool in the woods, not from the posture of a person who runs a training program — most of the rest of the advice in this section will land naturally. If you cannot hold that frame, no amount of specific advice will save you from the student/teacher dynamic you'll accidentally trip into.

The Three Generosities

With the frame in place, here is what to actually do when Priya asks. I'll walk through three specific gestures, in order, because the order matters. Each one is small. Each one is load-bearing.

First: show her your layout.

Not a diagram. Not a description. The actual layout. On the actual monitor. Right now.

Turn your screen slightly so she can see. Point at Slot 1 and say something like — I don't want to put words in your mouth, but something in the shape of — "I keep seven of these going at once. Each one is a ticket. These three regions" (gesture at

Zone 1, Zone 2, Zone 3) "have fixed jobs — validation up here, comms up there, Jira down here. The rest is working slots." That's it. Five sentences. Thirty seconds.

Notice what you did not do. You did not explain why seven. You did not explain the staggering rule. You did not explain the stack geometry. You did not explain the Norse names. You did not explain the lease file. You showed her the *shape* of the thing, briefly, and now you get to see what she does with the shape.

If Priya asks a follow-up — *why seven? what are the names about? how do you not lose your mind?* — you answer *that* question, specifically, and then you stop. If she does not ask a follow-up, the shape alone was enough for today, and you have given her something she can now look at and think about on her own time.

The reason to start with the layout rather than with the philosophy is simple. The layout is *visible*. It is the part of the method that is easiest to share across the gap between two people who don't yet share the conceptual vocabulary for it. The philosophy will come later if she wants it. The layout, on the screen, now, is the entry point.

Second: show her your `CLAUDE.md`.

After the layout conversation settles — and only if the conversation is still going, not because you feel obligated to produce all three generosities in the same sitting — pull up your CLAUDE.md file and say something like, "This is where I keep the rules I've figured out. It lives in this directory, and every Claude I launch reads it automatically. Anything that has bitten me twice goes in here."

Scroll through it slowly. Let her read a few lines. Don't narrate. Don't explain every rule. Let her eye land where it lands.

This is a more intimate share than the layout. The CLAUDE.md is the artifact of all the small mistakes you have made over months of use. It has your voice in it — the voice of someone who got burned once by something specific and wrote down the lesson. That voice is what makes the file credible. Do not apologize for it being messy, and do not clean it up before you show her. Show her the real one. If the file has a rule that says *"do not ever again use the phrase 'it should just work' without actually running it first — you said that about ABC-1204 and we shipped a bug"*, leave it in.

That rule, in Priya's eyes, will do more to convince her the method is real than any polished description ever could.

The file also gives her a concrete thing to copy. Not the file itself — her rules will be different — but the *format*. She now knows what a CLAUDE.md looks like when it has been lived in. That visual model is worth more than any explanation of what permanent memory is.

Third: walk her through her own Step 1.

This is the generous part, and it is the part most developers skip, because it costs the most.

Sometime in the next day or two — not necessarily in the same conversation, not necessarily even in the same week — offer to help Priya set up her own workspace. Not yours in miniature. Hers. She will want some things different, and that's fine. What you are offering is an hour of pair-setup, during which she installs Claude Code (or Codex, or whatever she prefers), creates her own workspace directory, drafts her own first CLAUDE.md with three or four rules she already knows about how she likes to work, and launches her first agent.

That's Step 1. Not the whole method. Setup.

You offer Step 1 because Step 1 is the hardest moment in adoption — the moment where most engineers, on their own, stall out because the first thirty minutes of configuration are annoying and underdocumented. If you sit with Priya through those thirty minutes, you dramatically increase the odds that she will have a real workspace in her hands tomorrow. If she has a real workspace tomorrow, she will use it for something small this week. If she uses it for something small this week, she will probably still be using it next month. That is the arc. You can shorten the first gap by about an hour of your time, and that hour is one of the most generous hours you will spend on your career as a teammate.

And — this matters — *you help her set up her own Step 1.* Not yours. You do not talk her into seven environments on day one. You do not talk her into the Norse names if she wants English ones. You do not talk her into the two-session day if her schedule doesn't permit it. Her Step 1 is whatever gets her from nothing to a working first agent with a CLAUDE.md that reflects *her* preferences. Execution comes later, naturally, once Setup is real.

Don't Try to Convert Her to Everything at Once

This brings me to the last piece of advice for this section, which is partly a restatement of the *don't evangelize* principle and partly something slightly different.

When Priya has Step 1 running, resist — with everything you have — the urge to also hand her all the rest of it. The glance cycle. The two-session day. The validation ritual. The NOTES.md wind-down. The work-on-a-ticket protocol. The second monitor as escape hatch. The refinement cadence. All of those things are real, all of them matter, and none of them matter *this week* for someone who has just gotten their first agent running on their first ticket.

The method has a natural order of adoption, and the order is not the order of the chapters in this book. Setup comes first. Then a person works for a week or two with the equivalent of maybe two slots, because that's all their nervous system has bandwidth for on top of the novelty. Then they naturally start to want a third slot and a fourth, and they notice that they're losing track of which ticket is where, and at that point they will come ask you about the lease file, which is now a real problem they have, not an abstract solution to an imagined one. Then validation will start to feel missing, and they will come ask you about how you decide a PR is ready, and you'll tell them about the six-step ritual, and it will land because they are ready for it.

Each piece of the method becomes obvious to adopt the moment the problem it solves becomes real. Handed too early, before the problem is real, every piece looks like an arbitrary rule with no apparent reason to exist. Priya will nod politely, forget it by Friday, and never implement it, because she didn't need it yet.

So when she asks, "what about the rest of it?" you answer the specific question, and then you let her go back to working. She will come back when she has the next specific question. That is the rhythm. That is the whole way this spreads.

Why Generosity, Specifically

I want to close the section on the word *generous*, because I chose it deliberately and I want to be clear about what I mean by it.

Generous does not mean *give everything all at once.* It also does not mean *give it away because it isn't worth much.* The method took you months to build. It cost you real effort. You are not obligated to hand it out like free samples at a grocery store.

Generous means something more specific. It means: *when someone asks, you answer honestly, at the depth they asked for, with real detail, and without making them work harder than they should have to in order to receive the answer.* It means you show them the actual file, not a sanitized version. It means you spend the hour to get their Step 1 working, not because you owe it to them but because the method is genuinely better than what they had before, and sharing the good thing is the kind of coworker you want to be. It means you do not hoard. It also means you do not overwhelm. Both hoarding and overwhelming are failures of generosity, not successes of it.

The best model for the posture I am describing is not the tech-talk speaker, and it is not the mentor, and it is not the evangelist. It is the friend who, when you mention you are thinking about taking up cooking, says *"oh, here — this is the one knife you actually need, and here's how to hold it, and here's the first dish to try, and let me know how it goes."* That friend is generous. That friend is not trying to turn you into them. That friend is genuinely happy you are becoming interested in something they love, and they want your entry into it to be easier than their own was.

Be that friend.

Hey, look what I found. Not *let me train you in the system I have perfected.* A found thing, handed over in the exact shape and size the other person asked to see it, with the quiet understanding that they may take all of it, part of it, or none of it, and that any of those three outcomes is fine.

That is what makes this method spread, over time, without anyone ever evangelizing. It is also — and this is not a small point — what makes your workplace better to be in. A team of engineers who treat their hard-earned competencies as found things they are glad to share is a rare and precious team. Most teams are not that. You do not have to wait for your team to become that before you start behaving that way. You can just start. Today. With Priya, if Priya is real, or with the person who is about to roll their chair over to your desk, who may not be named Priya, but who is probably coming sometime in the next few weeks.

Be ready for them. Not with a slide deck. With the layout on your screen, the CLAUDE.md in your editor, and the quiet willingness to spend an hour helping them get their own first agent running. That is the whole answer. That is generosity, correctly sized.

On to the harder case — the one where a manager notices.

Rolling Out to a Team (When Asked by a Manager)

Sooner or later, if your throughput stays consistent and your team is paying any attention at all, a manager — yours, or theirs, or a manager two levels up who only sees you in dashboards — is going to ask you to roll the method out to the rest of the team.

The previous sections of this chapter were about the version of this conversation you actually want — the one where a teammate rolls their chair over and asks. This section is about the version of the conversation that is harder, riskier, and far more common than the curiosity arc would suggest. It happens in conference rooms, sometimes by Zoom, sometimes in a one-on-one in a glass-walled room with a small bowl of mints on the table. It usually starts with a sentence that has been workshopped — *We've been seeing some really good numbers from your work lately, and I want to talk about how we get the rest of the team there* — and it usually ends with you having agreed, in vague but warm terms, to *take point on rolling this out across the engineering org*.

I want you to not say yes to that ending, at least not in those words, and I want to walk through how to not say yes in a way that does not damage your relationship with the manager, your standing on the team, or the durability of the method itself.

Three things make this conversation harder than the peer-curiosity version. First, the power asymmetry — you cannot just decline the way you might decline a lunch invitation. Second, the metric pressure — the manager has been looking at your output, and the implicit framing of the conversation is that your numbers are the standard the rest of the team should reach. Third, the timeline assumption — managers tend to imagine adoption as a thing that takes weeks, when in fact it takes a quarter to two quarters per developer, and the gap between those two timelines is where most rollouts go wrong.

The good news is that the conversation is not adversarial. The manager is, almost always, trying to do something good — make the team more effective, give people access to the same leverage you have, justify a budget line. They are not trying to harm anyone. They are also not, by default, going to know how to roll this out well, because most managers have not done this before with this kind of methodology. Your job is to help them do it well, which is different from agreeing to do it for them, and different again from refusing to be involved.

This section walks through the shape of that whole arc — from the conversation you have not yet had, through the pilot you may help design, to the moment you hand the practice off to the team and step back so it can live without you.

What to Think About Before the Meeting

The manager has not asked you yet. They will. The week before they do, you will start to feel the conversation coming — a comment in a one-on-one, a slide in a quarterly review, an offhand mention of a colleague's recent retraining. Use that lead time. The single most useful thing you can do for the conversation that is coming is to walk into it with a few positions already worked out, so you are not improvising under pressure in front of a person whose support you need.

Three things to figure out before you walk in.

First, the line between what you will and won't take on. You can take on: a small, time-bounded volunteer pilot with two or three interested colleagues; an hour-by-hour budget for pair-setup with anyone who asks; a willingness to share your `CLAUDE.md`, your prompts, your layout files. You probably should not take on: a mandate to bring the whole team to your throughput; a teaching role that displaces your engineering work; a target date by which everyone will have adopted; a metric tied to other people's adoption. Know where your line is *before* the conversation, so that when the manager proposes something on the wrong side of it, you can name the alternative without it sounding like you're pushing back on the goal.

Second, the shape of the success the manager actually wants. Sometimes "I want to roll this out" means *I want the team's output to go up.* Sometimes it means *I have a mandate from above to demonstrate AI-tooling adoption and I need a story to tell.* Sometimes it means *I genuinely think this would help these people and I don't know how to get it to them.* Those are three different goals, and they call for three different responses. You will not always know which one is operating, and the manager themselves may not know — but you can tell a lot from how the request is framed, who else is in the room, and whether the manager has already decided what the rollout looks like. Listen carefully in the first few minutes. The framing of the goal will tell you what kind of conversation you are actually in.

Third, what you would refuse to measure. This is the most important pre-thinking, because it is the thing managers reach for first when a rollout starts and the thing that most reliably wrecks the team if you let it through. Tickets-per-developer-per-week is a poison metric. PR-size is a poison metric. AI-prompt-count is a poison metric. Self-reported satisfaction over a small number of weeks is a noise metric. You will be asked, with the best of intentions, to help define metrics for the pilot. Have an answer ready that says yes to the legitimate measurement question — *did this make the people in the pilot more effective at their actual work?* — and no to the surveillance metrics, framed in language that helps the manager see why the no is in their interest as well as yours.

If you have those three things worked out before the meeting, the meeting goes dramatically better. If you have not, you will agree to something in the room that you will quietly regret on the drive home, and undoing that agreement will take three subsequent conversations none of which will be as friendly as the first.

A Note for the Reader Who Is Already a Week Too Late

A word for the reader who is reading this section with the awkward feeling that the meeting has already happened, the broad strokes have already been agreed to, and the agreement was warmer and more aggressive than the section above would have advised. You are not the first engineer to have that feeling, and the situation is more recoverable than it currently looks.

The repair move is a short, specific email — to the manager, in your own words — that says you have been thinking more about how to make the rollout actually work and you would like to propose a slightly different shape than what you sketched in the room. Reframe what you said as a starting offer rather than a commitment. *Here's what I'd refine, having sat with it for a few days* is a perfectly normal sentence in any working relationship. Most managers will accept the revision. The ones who will not have already told you, in the original meeting, more than you knew at the time.

The discipline is to repair within a week, before the original framing has crystallized into a plan that is harder to undo. After that, the work is the volunteer pilot below. The conversation is repairable. The shame the engineer is most likely to feel — *I should have known to push back in the room* — is itself a sign the original instincts were good; the room just moved faster than the instincts could surface. That is a normal failure mode of meetings. Repair it the way you'd repair any other handoff that didn't go quite right: clearly, quickly, and once.

The First Conversation

Walk into the meeting with the goal of not closing the conversation. The manager will probably want to close it — to leave with a plan, a name, a date, a deliverable. That is a reasonable instinct from a person whose job is producing plans. Your job, in the first conversation, is to widen the frame enough that the easy plan is replaced with a slightly harder but much better one.

Here is roughly the arc I have found works.

Open by *thanking them for noticing*. This is not flattery and it is not strategic. The manager has paid attention to your work, has formed a hypothesis about why your output is what it is, and is bringing it to you with a constructive ask. That is a real act of management, and acknowledging it costs nothing. *I appreciate you bringing this to me — I've been trying to be careful about not pushing it on the team unprompted, so I want to be thoughtful about how we do this.* That sentence puts you and the manager on the same side of the table, which is where you want to be for the rest of the conversation.

Then *describe what the method actually is*, briefly, in language that helps them see why a fast rollout is the wrong tool. *It's a workflow that took me about three months to build and another two to settle into. It's not a tool I installed; it's a set of habits I built up around a couple of tools. The reason I'm cautious about rolling it out is that the habits don't transfer in a one-day training — they have to grow in each person, around their own work, and the gain only shows up after the habits are settled.* You are not arguing against the rollout. You are reframing the timeline, gently, from *weeks* to *quarters per person*, in a sentence the manager can repeat in their own next conversation upstream.

Then *propose the volunteer pilot.* This is the substance of the meeting and the thing you want the manager to leave with. *I think the right shape is a small voluntary pilot — two or three people who are already curious — for one quarter. I'll pair-set-up with each of them, share my* `CLAUDE.md` *and my prompts, and we'll meet weekly for an hour. At the end of the quarter we can look at how they're doing and decide what comes next.* The pilot is the offer. It is concrete, it has a timeline the manager can put in a slide, it commits you to real work, and it does not commit you to outcomes you can't deliver.

Then *name what you will not do*. Quietly, as a matter of working clarity. *I want to be careful about a few things. I don't want to set a target ticket count for the pilot people, because the gain shows up in different shapes for different engineers and a ticket target will skew their behavior in the first month. I also don't want to compare their numbers to mine in any forum, because my numbers are mine for reasons that have to do with me, and that comparison would be unfair to them. And I'd want to be cautious about extending the pilot to anyone who hasn't volunteered, because forced adoption of a workflow is the most reliable way to break it.* Three sentences. Each one names a thing you will not do, and each one comes with a brief, specific reason that helps the manager see why you are protecting the work, not protecting yourself. Most managers will hear those three sentences and update their model. The manager who cannot accept any of them is the manager you have to negotiate harder with, and we'll come to that.

Close by *offering to write up the pilot proposal.* This is a small generosity that pays off twice. It puts the proposal in your words, with your protections built in, and it gives the manager a document they can use to defend the slow-rollout shape in their own conversations. *Let me write up what we just discussed and send it to you by Friday — a one-pager on the pilot, who's in it, what we're trying to learn, and what the next checkpoint looks like.* That document, honestly drafted, is one of the most valuable artifacts in the whole rollout. It is what the rest of the conversation runs on for the next three months.

If the meeting goes well, you leave with a quarterly pilot of two or three volunteers, a written proposal you'll deliver in a few days, and a manager who has updated their timeline expectations from weeks to quarters. If the meeting goes less well — if the manager pushes for a faster rollout, larger numbers, a target date — the next two subsections cover both branches.

The Volunteer Pilot, in Practice

Assume the manager said yes to the pilot shape. Now you have to actually run it, and the running of it is where the rollout either lays down healthy roots or quietly poisons the team.

The pilot has four jobs, in roughly this order. *Help two or three colleagues build their own working version of the method. Learn what works about the method when it leaves your specific brain and lives in someone else's working life. Generate the artifacts the manager needs to defend the slow rollout to their own management. Establish the social precedent that adoption is voluntary and growth is gradual, not mandated and rapid.*

Pick the volunteers carefully. The right volunteers are colleagues who have *already* shown a flicker of interest, ideally before the pilot was a thing. The wrong volunteers are people the manager picked to round out a number, people who said yes because they thought it would help their performance review, or people who are quietly resistant but felt obligated. A pilot run with the wrong people is worse than no pilot at all, because it produces a record of failed adoption that anti-adoption arguments will use later.

When you sit down with each volunteer, do not start with your method. Start with their work. Ask what their week looks like, what kind of tickets fill their queue, where they currently feel the most friction. Their answers will shape what version of the method makes sense for them, which will not be your version. A backend engineer working on long-running ticketed refactors gets a different first slot than a frontend engineer doing many small pieces of UI work. Your method is not the method they will end up with; it is the seed they will grow their own method from. Your job in the pilot is to help that growth happen, not to clone yourself across two or three colleagues.

Be more conservative in scope than you would be for yourself. In the pilot, two slots is plenty for the first month. Three slots in the second month if it's working. Five by the end of the quarter is success — not seven. The volunteer who reaches seven slots at the end of a quarter has either burned out trying or got there honestly, and either way, that volunteer is the exception, not the standard. Your version of the method took months to settle. Theirs will too.

Meet weekly, in a real cadence, for an hour. Not as a class — as a working session where they bring a problem they hit and you walk through how the method addresses it (or doesn't). The weekly meeting is the load-bearing beam of the pilot in the same way the lunch hour is the load-bearing beam of the two-session day. Skipping it is what makes the pilot drift back into informal mentoring, which is harder to defend to management later.

Document, as you go, in a private channel — yours, not the volunteer's. Not their throughput. Not their pace of adoption. The *texture* of what's working: which prompts traveled cleanly, which ones needed rewording, which parts of the method clicked first, which parts each volunteer hit a wall on, what surprised you. That texture is the artifact you'll bring to the second conversation with the manager, and it will be the most valuable thing in the rollout because it will be the only document anywhere in your organization that takes adoption seriously as a process rather than as a yes/no event.

What to Measure, and What to Refuse to Measure

Managers ask for metrics. This is correct, and you are not going to argue them out of it. The question is which metrics, and the answers matter.

The *legitimate* questions a metrics conversation can answer are roughly these: *Did the volunteers experience a meaningful change in how they work? Are they sustainably more effective at the actual things they were trying to do? Would they recommend the pilot to a peer?* You can answer those, honestly, with three pieces of data: a structured one-hour interview with each volunteer at the end of the pilot, a quiet chat with each volunteer's manager about what they observed, and a self-assessment from the volunteer about what changed. None of those is a number. All three are real evidence.

The *poison* questions are the ones that turn the pilot into a surveillance exercise: *How many tickets did each volunteer close per week? How does their throughput compare to non-pilot teammates? How does it compare to your throughput? What's their AI-prompt count? What's their PR-size distribution?* Each of these looks like a metric. None of them is. They are pseudo-metrics that will distort behavior in the pilot, distort comparisons across the team, and pre-bake the wrong story for the rollout that follows.

I want to be specific about why each of those is poison, because the manager will need a reason if you are going to push back on any of them.

Tickets per week is poison because it incentivizes ticket-splitting, ticket-shopping, and the closing of trivial work to pad the count, which are exactly the behaviors the method is supposed to make unnecessary. The volunteer who is honestly slowing down to learn the method will look bad on this metric for the first six weeks, even if they are on track to triple their effective output by month four. You will lose them on the chart before they have a chance to land.

Throughput compared to peers is poison because it sets up an adopters-versus-non-adopters dynamic the entire prior section of this chapter argued against. A team in which half the engineers are on a pilot that is being scored against the other half is a team that is going to have a bad quarter regardless of how the pilot goes.

Throughput compared to you is the worst of these. You have run the method for years. The volunteers have run it for weeks. Comparing their pilot output to your steady state is a category error that will, predictably, make every volunteer feel like they have failed even when the pilot is succeeding. Refuse this comparison plainly, every time it comes up.

AI-prompt counts and *PR-size distributions* are poison because they are easy to game and uncorrelated with what you actually care about. A high prompt count with low ticket-completion is worse than a low prompt count with high completion; a high prompt count with high completion is fine; the prompt count itself tells you nothing without the rest of the picture, and including it on a dashboard guarantees a colleague will start optimizing for it within two weeks.

The frame I have found works for declining poison metrics is: *I want to give you the cleanest possible signal about whether this is helping these engineers, and the cleanest signal is going to be qualitative — interviews, manager observations, self-assessment — because the quantitative metrics we'd reach for here all distort behavior in ways that would mask the actual answer.* That sentence respects the manager's job. It also commits you to producing real qualitative evidence, which is harder than producing a chart but more durable.

If a manager insists on quantitative metrics anyway, the least-harmful one I have found is *retention of the practice three months after the pilot ends.* If a volunteer is still using the method three months after the pilot's formal end, the method took. If they aren't, it didn't. That metric does not pressure the volunteer during the pilot, does not compare them to peers, and answers the only quantitative question that matters: did this stick.

Cohabitation During the Pilot

A pilot creates a small but real social discontinuity on the team — three of your colleagues are spending an hour a week with you and trying something the rest of the team isn't. Manage that discontinuity carefully, because the rest of the team is watching, even if they pretend not to be.

A few rules that have served me well.

Do not let the pilot become a social club. A regular weekly hour with the volunteers is fine. Three lunches a week with the volunteers, while the non-pilot members of the team eat alone, is not. Whatever the casual social texture of your team was before the pilot, keep it the same during the pilot. The method is a workflow, not a clique.

Do not let the volunteers' tickets be flagged as special. If your team uses labels, milestones, or any kind of board categorization, the volunteers' work goes through exactly the same path as everyone else's. The minute a "pilot" label appears on a ticket, the work has been singled out, and the framing for the rest of the team becomes *us-and-them*. Avoid the label.

Do let the volunteers struggle in public. If a volunteer hits a wall in week three and posts in the team channel asking how everyone else handles a particular kind of code review, let them. Don't quietly DM them the method's answer. Let the team see them working through a real engineering problem the way anyone on the team would. The visibility of normal struggle is what prevents the pilot from looking like a magic-user club to the rest of the team.

Do let the non-pilot teammates ignore the pilot. This sounds obvious and is not. The non-pilot teammates may be perfectly aware that a pilot is happening and may have decided, for reasons of their own, not to be in it. Honor that decision exactly as the previous section of this chapter described — no pity, no defensiveness, no proselytizing. If the pilot is going well, the non-pilot teammates may eventually become curious; if it's not going well, they shouldn't have to share its failure. Either way, the right posture toward them during the pilot is the same posture as before the pilot. They are colleagues. The method is your thing. The pilot is the manager's thing. None of it is theirs to deal with unless they choose otherwise.

The Second Conversation

Three months in, you and the manager will have a second conversation. The shape of this conversation depends on what the pilot showed.

The good case: the volunteers have built their own working versions of the method, are reporting that it has helped, and the qualitative interviews land cleanly. The right move here is *to slow down*, not to speed up. The manager will be tempted to expand the pilot dramatically — five more volunteers, then ten, then a mandate. Argue for the same shape, slightly larger. *Two or three more volunteers next quarter. Same cadence. Same protections.* The pilot worked because of how it was run. Doubling the size doubles the strain on you, halves the per-volunteer attention, and starts to look like a deployment rather than a growth — and deployments are where the failures begin.

The mixed case: one volunteer thrived, one stalled, one went sideways. This is the most common outcome and the one most worth discussing honestly. The honest discussion is about *what shape of person and what shape of work* the method fit, and what didn't. The volunteer who stalled may have hit a real life-circumstance constraint (echoing the next section's discussion of legitimate non-adoption). The volunteer who went sideways may have run too hot too fast and burned through the goodwill the method needs to settle. Both are useful information. Neither is a verdict on the method. Bring this honestly to the manager and let it shape the next quarter's offer.

The hard case: the pilot didn't take. The volunteers tried, struggled, and decided it isn't for them. This sometimes happens, and the right response is to absorb it honestly. The method is genuinely not for everyone. A pilot that ends with three volunteers walking away politely is data — it tells you something specific about the fit between this method and these people in this context. Tell the manager that, plainly, and propose ending the formal rollout. Some managers will accept this gracefully. Some will not, and that takes us to the next subsection.

When Management Mandates Anyway

The version of this story I most wish I did not have to write is the one where the manager — yours, theirs, two levels up — decides, despite the conversations above, to mandate the method across the team or across the org. This sometimes happens. Pretending it doesn't is its own failure mode.

If a mandate comes down, three things matter.

Refuse to be the messenger. A mandate that is announced over the engineer who designed the slow-rollout protections is a mandate that has betrayed those protections. Do not present the mandate to the team as if it were yours. If you are asked to, decline politely and clearly. *I think this is the manager's decision to communicate, not mine. I'm happy to support people through the change if they want help, but I shouldn't be the face of a mandate that I argued against.* That sentence may have a cost. It is worth its cost.

Refuse to be the enforcer. You will be tempted, after a mandate, to start tracking who is and isn't adopting, to flag the slow ones in your one-on-ones, to nudge the resistant. Do not. The previous section of this chapter applies in full — adoption pressure from peers is corrosive, and the corrosiveness does not stop being corrosive because management requested it.

Be more useful to the resisters than to the mandators. A mandate creates, immediately, a quiet group of teammates who are now being asked to adopt a method against their better judgment. Some of them will need help. Some of them will need cover. Some of them will need someone to tell them, plainly, that the method is real and worth a try and that it is also not the only way to be a competent engineer. You can be that person. You will not be popular with management for being that person. You will be the right kind of useful for the team.

The question worth asking yourself, if a mandate sits on the team for a long stretch and is making things worse, is whether your continued presence is what's keeping the manager comfortable enforcing it. Sometimes the most generous thing an adopter can do, in a poorly run rollout, is stop being available. Take the role somewhere else. The team will recover faster without the implicit benchmark of your throughput hanging over it. This is a hard call and not always the right one — but it is a real option, and worth keeping on the table.

Protecting Non-Adopting Teammates From the Numbers

Whether or not a mandate happens, there is a quieter, more constant version of pressure that shows up the moment the method becomes legible to management: the comparison of your numbers to your teammates' numbers, often in performance contexts you are not in the room for.

You have more agency here than you might think.

In every one-on-one with the manager, decline to compare. *I want to be careful about putting my output up as a benchmark for the team. The method works for me for reasons that are mine, and the comparison would be unfair to people whose work has different shapes than mine has.* Say it the first time. Say it the second time. By the third time, most managers will have updated their model, and you will have established that the comparison is not yours to authorize.

If you discover that a non-adopting teammate is being held against your numbers in a review you are not in, decline to provide ammunition. If a manager asks you to characterize the gap between you and them, refuse the framing. *I'm not in a position to compare our work that way. We're doing different things at different paces, and the comparison would say more about our different roles than about either of our performance.* The manager may push. Push back. Your output is yours; their output is theirs; the team's output is the team's. None of those three things is the same kind of measurement, and conflating them is what poisons rollouts.

If the manager seems to be benchmarking you against the team in your reviews — a related but distinct danger — that is worth a separate conversation. *I'd like to understand what success looks like for me in your eyes that doesn't depend on my output staying inside this specific method, because the method is not guaranteed to be the right shape for me forever, and I'd like to know what would still count as good work if I changed how I worked.* That sentence relocates the conversation from your throughput to your judgment, where it belongs.

The Handoff and Stepping Back

There is one last move in a healthy rollout, and it is the move most adopters miss because it requires giving up something that has come to feel like part of their identity on the team.

If the pilot has worked and the method has begun to live in two or three of your colleagues' practices, *step back*. Not all at once. Not dramatically. But over the next quarter, deliberately reduce the amount of pair-setup you offer, the frequency of the weekly pilot meetings, the visibility of yourself as the method's author on the team. Let the colleagues who adopted it talk about it in their own words, in their own voices, with their own emphases. Their version of the method will diverge from yours, and that divergence is *correct* — it is the method finding its own shape in different practitioners.

The mistake I have watched adopters make at this stage is treating the method as a thing they own and need to keep authoring. They sit in on every retrospective where the method comes up. They correct misstatements about the method in writing. They appoint themselves the gardener of the team's CLAUDE.md files. This produces, quickly, a team in which the method is *yours-being-tolerated-by-everyone* rather than *the team's-now-with-its-own-evolution*. The first of those is a fragile situation that ends when you change jobs. The second is a durable practice that will outlive your tenure on the team.

Step back, and let the practice become the team's. If two of the colleagues who adopted it eventually develop opinions about the method that disagree with yours, that is a sign of health, not heresy. The method gets stronger by passing through more brains. Your specific version was always one shape of it; their versions will be others. None of them is wrong.

A Closing on the Manager-Rollout Case

I want to end this section by naming what success actually looks like, because the conventional answer is wrong.

Success in a manager-driven rollout is not *full adoption across the team in two quarters*. That outcome, if it happens, is almost always a mandate dressed up as adoption, and the team paying for it will pay quietly for years.

Success is something more modest and more durable. *A small number of colleagues, who chose the method because it fit them, are running their own working versions of it a year after the pilot ended. The non-adopters on the team are doing well in their own workflows and have not been treated badly by management because of the comparison. The team's overall health is the same or better than it was before the rollout. You, the original adopter, are still on the team and still using the method, and you are not exhausted from a year of carrying everyone else's adoption.*

That outcome is achievable. It is also rare, because the path to it is not the path managers default to, and it requires a long, quiet, occasionally awkward sequence of negotiations of the kind we just walked through.

If you can hold the shape of those negotiations — if you can say no to the wrong rollout in a way that doesn't damage the relationship with the manager who asked, if you can run the volunteer pilot honestly without it becoming a club, if you can refuse the poison metrics, if you can step back when the practice has taken root in someone else — you will have given your team something better than a mandated workflow.

You will have given them an example of how to introduce a new way of working into a team without breaking the team in the process.

That example, in my experience, ends up mattering more than the method itself. Methods come and go. The way teams treat each other when something new shows up — that, you keep.

Which brings us to the harder case, and the one I most worried about getting right: the colleagues who, having watched all of this, choose not to adopt at all.

Teammates Who Won't Adopt

A Note Before We Start

This is the section of the chapter I most worried about getting wrong, so I want to tell you plainly what I am trying to do and what I am trying not to do.

What I am trying to do is help you be a good colleague to the people on your team who, for their own perfectly legitimate reasons, are never going to adopt this method. That group will exist on any team you are on, probably for the rest of your career. It will include people you respect, people you are learning from, people whose code you think is better than yours, and people whose judgment you trust more than your own. It will not be a collection of dinosaurs.

What I am trying not to do is write the version of this section that would appear in a worse book — the version that frames non-adopters as a problem to be managed, a group to be patiently brought around, or (worst of all) a cautionary tale about who gets left behind. That framing is wrong on the merits, it is corrosive to your working relationships, and it is the exact kind of thinking that the entire chapter has been pushing back against.

If you have been reading carefully, you already know the principle at work here. The method is yours. You built it because it fit you. Other people have built other things — sometimes very different things — that fit them. A team is healthier when it contains a real range of workflows than when it contains one enforced workflow, and the durability of your own practice does not depend on anybody else adopting it.

With that on the table, let's look at the actual shape of non-adoption, because it is not one thing.

The Five Legitimate Reasons People Don't Adopt

I have watched enough teammates choose not to adopt, or adopt and then walk away, that I can now sort the reasons into roughly five buckets. They are not the same, and the distinction matters, because how you carry yourself around a non-adopter depends on why they are a non-adopter. Most importantly — and I want to say this before any of the buckets — *none of these reasons is a problem with the person.* All five are reasonable responses to a real situation.

First: they are already happy and productive in their current workflow. This is the biggest bucket, and it is the one most often dismissed by adopters. There are engineers on your team who were shipping excellent work before any of this arrived, who are still shipping excellent work, and who have correctly calculated that the two-to-three-month transition tax to rebuild their workflow around a new method is a bad trade given what they already have. They are right. The method gives you leverage you did not previously have; it does not automatically improve a practice that was already working at a pace the person was happy with. A senior engineer shipping thoughtful, well-tested, well-reviewed code at four tickets a day is not someone who needs to become a fourteen-ticket-a-day engineer. They are someone who is already good at their job.

Second: they have principled objections to the tools themselves. There are engineers who have thought carefully about the training data, the labor implications, the environmental cost, the accuracy ceilings, or the intellectual-property questions around AI coding agents, and who have concluded that they do not want to use them. These are not crank positions. Some of them are positions you may privately share. The fact that you have chosen to use the tools does not oblige you to argue your colleague out of their choice, and arguing is almost guaranteed to make the working relationship worse. You do your work. They do theirs. The shared codebase does not require you to have the same views on the ethics of the tooling.

Third: they are at a career stage where the investment doesn't pay back. Someone two years from retirement, or someone winding down to part-time, or someone planning a career transition that will take them out of day-to-day engineering, is making a perfectly rational choice when they decline to spend a quarter rebuilding their workflow. The method's payoff curve is shaped for people who will be in this kind of work for years after the adoption cost clears. If that is not the shape of the person's next few years, the math simply doesn't work, and they are right to notice.

Fourth: their life circumstances don't support the attention pattern. This one is the quietest and also, I think, the one most often mishandled. The seven-slot rotation asks something specific of your nervous system — sustained parallel attention across multiple streams, hour after hour, with frequent context switches. Not every working adult has a life that affords that kind of attention budget. A colleague who is the primary caregiver for young children, a colleague with a chronic illness that already eats a daily allotment of attention, a colleague whose brain simply does not split attention in the shape the rotation requires — any of those people may be doing excellent work with a single-stream workflow specifically *because* that is the workflow their life and body allow. Asking them to take on the method's attention load would be asking them to spend reserves they do not have. The method would make them worse engineers, not better ones. Honor that without needing it explained.

Fifth: they have tried it and it didn't suit them. A small number of people will actually try the method, work through the setup, live with it for a few weeks, and conclude it isn't for them. This one is sometimes the hardest for the adopter to accept, because you cannot wave away the experience — they know what they're passing on. But the honest truth is that not every workflow fits every practitioner, and a person who tried it, found it didn't fit, and went back to their previous way of working has done something admirable. They have checked for themselves instead of taking anyone else's word for it, and they have made the call their own experience supported. Your only job, if that is the story, is to respect the call and move on.

There may be other buckets. Those are the five I have seen recur. What matters is that every one of them is a reasonable position held by a reasonable person, and your working life goes better when you treat them that way from the start.

How to Carry Yourself Day to Day

Concretely, here is what being a good colleague to a non-adopter looks like in the small moments of the working week. None of these are dramatic. All of them compound.

Don't bring it up. If a teammate has not asked you about your workflow, do not tell them about your workflow. Don't mention it in standups. Don't reference it in retros. Don't drop a *well, with my seven slots I was able to...* into a hallway conversation. If they are curious, the curiosity will find you — we walked through the natural curiosity arc earlier in this chapter, and that arc does its work whether you help it along or not. What it does not survive is the adopter pushing the topic into rooms where it wasn't invited.

Don't turn tickets into a competition. The ticket board is a shared tool for coordinating the team's work. It is not a leaderboard. A team where one person's throughput becomes the implicit benchmark for everyone else's is a team where the non-adopters feel surveilled and the adopter starts getting treated with suspicion. Close your tickets, let the counts fall where they fall, and if your ticket count is conspicuously higher than everyone else's, consider whether there's work you could voluntarily share — documentation, reviews, mentoring, the unglamorous backlog — instead of just stacking more wins on your pile.

Don't complain about their pace to others. This is the failure mode I have seen break more working relationships than any other, and it almost always happens in private — a slightly rolled eye in a one-on-one, a sentence beginning *I just wish X would...*, a comparison between their week and yours in a conversation with their manager. Those small gestures are poison. They will, eventually, get back to the person, and they will, permanently, change how that person thinks of you. You are not their manager. You are not their pace-setter. If you have genuine feedback for them about their work, give it to them directly and kindly, in the language of their work, not in the language of your workflow.

Don't volunteer their work onto your plate. There is a specific trap that hits high-throughput engineers after a few months of the method: your capacity is higher, so you start absorbing tickets other people would normally have picked up. Two months in, you are doing forty percent of the team's output, your non-adopter colleagues are doing proportionally less, and everyone involved is subtly unhappy — you resent the load, they feel sidelined, and the team's overall health is worse than it was before your throughput went up. Pick your tickets at roughly the pace you would have without the method, and let your *extra* capacity buy you margin — time for review, time for mentoring, time for the harder problems that nobody had room for before — rather than letting it quietly absorb the work your teammates needed to do.

Do the unglamorous team work. The best use of the throughput gain, in my experience, is to be the person who finally has time to pick up the things no one ever had time for. The flaky test nobody wanted to investigate. The documentation that's been three versions out of date. The onboarding guide that broke last year. The quiet review of a junior teammate's PR that you now have a spare half-hour for. These are the things that make a team healthier and, not incidentally, the things that most reliably demonstrate that the method is about capacity rather than competition.

The Two Traps: Pity and Defensiveness

There are two failure modes in how adopters hold themselves in their heads — not in what they say, but in what they think — toward non-adopters, and I want to name both because both are corrosive and both are easy to fall into.

The pity trap. This is the quiet, private thought that goes: *poor X, they just don't know what they're missing, they'd be so much better off if they'd try it.* That thought is almost always wrong and it is always patronizing. It treats the non-adopter as someone who has failed to discover a truth you have discovered, rather than as a peer who has made a considered choice. It also has a nasty habit of leaking out, in tone of voice, in how you respond to their questions, in the small condescensions that are impossible to hide from a colleague who has known you for years. If you notice the thought, notice it, and correct it. Pity is not the same as respect, and colleagues can tell the difference.

The defensiveness trap. This is the opposite failure mode, which shows up when a non-adopter gently questions something about your workflow — a wasted prompt, a code smell, a pattern that didn't quite land — and you hear the question as an attack on the whole method. It isn't. It is almost always a specific, limited, correctable observation, and your job is to hear it as the specific observation it is. If you find yourself getting defensive about the method in response to ordinary engineering feedback, that is a signal that the method has become identity-adjacent for you, which is exactly the place you should not let it get to. The method is a workflow. It is not who you are. Feedback about the workflow is feedback about the workflow, and usually useful.

Noticing either trap in yourself is worth doing regularly, because both grow slowly and neither announces itself. A good check: if you imagine being reviewed by a thoughtful non-adopter on your team, do you feel a small flinch? If yes, examine what you're flinching from. It is probably one of these two.

The Asymmetric Cases

A particular kind of awkwardness shows up when the non-adopter is your senior, your reviewer, or a person with more authority on the codebase than you have. You may be the throughput leader on the team, but they are still, for perfectly good reasons, the person whose sign-off the PR needs.

Two things to hold in mind here.

First, their authority on the codebase is not something the method gives you or takes away. They have it because they have earned it over time, through judgment you have benefited from whether you realized it or not. Your higher ticket count does not translate into higher standing on the codebase, and it should not. The work still has to meet the standard they have been holding, and the standard is not lower because you produced the work faster.

Second, they may — entirely fairly — want more, not less, explanation from you about how a PR got made. If the PR was largely written by Claude under your direction, that's something they may want to know, both so they can calibrate their review and so they can develop their own view of what the team's code is going to look like going forward. Tell them. Not defensively, not evangelically — just as a matter of honest work reporting. *This one was mostly me directing Claude through the refactor; I reviewed it shape-level before sending it up.* That kind of plain statement respects their role as reviewer and respects their right to have opinions about how the work was produced.

If a senior teammate asks you to change something about your workflow — to run tests locally before pushing, to include more context in PR descriptions, to slow down on a specific area of the codebase — take the request seriously, even if the method made the current pattern easy for you. They are not obligated to rearrange the team's review norms to fit your throughput.

When Management Pressures Them

A harder case, which I want to flag briefly here even though it mostly belongs in the manager-rollout section, is what to do when management starts pressuring a non-adopter to adopt — sometimes using your throughput numbers as the lever.

This will sometimes happen. It is almost always bad. And you have more agency in these moments than you might think.

If a manager uses your numbers in a conversation with a non-adopter, say something. Not publicly, not in a dramatic way — a quiet note, in your next one-on-one, that you do not want your output to be held up as a standard for teammates whose circumstances are different, and that the method works because it suits *you*, not because it is universal. Managers, generally, will respect that. If they don't, that tells you something about the manager, which is also useful information.

If a non-adopter confides in you that they are being pressured, listen. Don't counter-pressure in the other direction. Don't promise to fix it. Do, quietly, decline to participate in framings of your own work that make their work look worse by comparison. You are not their advocate, and you should not try to be — but you can refuse to be weaponized against them, and that refusal is worth more than it sounds.

The Long View on Non-Adopters

I want to end this section on a quieter note, because the whole topic rewards quietness.

The first year of any new methodology in an industry is the year of loud adopters. The tenth year is the year of settled practice, in which some people use the thing, some people don't, and almost nobody thinks about it as a topic of debate anymore. We are currently, give or take, in year two. Most of the non-adopters on your team are not making a permanent declaration about their relationship to AI tooling. They are making a decision about this quarter, or this year, under the current conditions, with the current versions of the tools, inside the current shape of their life. Some of those conditions will change. Some of those people will, three years from now, quietly build a version of this method for themselves because a problem in front of them finally made it the right tool.

You do not need to be waiting for that. You do not need to be working toward it. You do not need to believe it will happen to any specific person. What you need to do is *not* be the person whose behavior, in year two, made them less willing to ever look at this work when the conditions in their life changed.

Be a good colleague. Do your work. Share generously when asked. Let the rest of it be.

The Team That Has Fully Adopted

The Rarest and Most Instructive Case

Most of this chapter has been about the quiet arc of one careful practitioner, working alone at first, letting the method speak through shipped tickets, fielding the occasional curious question, and watching a small number of teammates pick up pieces over time. That is the common path, and it is the path I would bet on for most readers of this book.

But there is a rarer, more interesting case, and it deserves its own section, because I have seen it happen twice now and I have heard about it happening a handful of other times, and the dynamics inside a team that has *fully* adopted the method are different enough from a team with one or two practitioners that the lessons of the earlier sections stop quite applying. A team in which every engineer is running seven slots, every engineer has their own tuned `CLAUDE.md`, every engineer is shipping at the pace the method produces, and every engineer is closing the lid at 4 PM without guilt — that team is in a different regime. It is also, for reasons I want to walk through carefully, a much better place to work than almost any other regime currently available to a software engineer, and it is worth understanding what it looks like on the inside so that you can recognize it if you ever find yourself building one.

I want to start with the arrival at that state, because the arrival is almost always a surprise. No team I have seen get there was aiming for it. In both of the teams I watched reach full adoption, it happened the same way: one person was working the method quietly, a couple of people got curious, a Slack channel that started as one engineer's working journal grew slowly to five, then eight, then twelve members; over a quarter or two, without any formal push, a critical mass of the team was running some version of the practice. At some point a new member joined the team and discovered, with what I remember as genuine amazement, that *this is just how the team works*. The method had become the local culture. Nobody had taken a vote.

That path — organic, gradual, unannounced — is the only path I have ever seen actually work at the team scale. The other path, in which a manager tries to mandate adoption after seeing one engineer's numbers, almost always fails in the specific ways the manager-rollout section described. The difference between a team that got there by compounding curiosity and a team that got there by mandate is the difference between a healthy adult and someone who has been force-fed a meal. The calories may be the same. The digestion is not.

The Social Psychology of Collective Adoption

When a team fully adopts a practice, something happens that social psychologists have been studying since the 1930s, and that is worth naming because it will shape your experience of being on such a team in ways that will otherwise feel mysterious.

Descriptive norms — the perception of what people around you actually do — are, in the research literature going back to Muzafer Sherif's autokinetic experiments in 1935 and continuing through Robert Cialdini's work on social influence in the late twentieth century, the single most powerful predictor of individual behavior in group contexts. This finding has been replicated so many times, in so many domains, that it is no longer controversial. When a new engineer joins a team and observes that *everyone here closes fourteen tickets a day and stops at four*, they will, almost without conscious deliberation, begin adjusting their own behavior toward that norm. The adjustment happens much faster than any training program could produce. It happens in weeks, not quarters. It happens because humans are, at a very deep level, calibrated to match the behavior of the groups they belong to, and the calibration mechanism is mostly not available to introspection.

This is why the organic-adoption path works where the mandate path fails. A mandate creates an *injunctive norm* — a rule about what should be done — and injunctive norms, the research consistently finds, tend to be weaker behavioral drivers than descriptive ones, especially when you're trying to get people to *take up* a new positive practice rather than to suppress an undesirable one. Telling people what they *should* do produces compliance at best and reactance at worst. Showing people what their peers *are* doing produces alignment, quickly and durably, without anyone having to be told to align.

The team that has fully adopted the method has, somewhere along the way, flipped from a descriptive norm of *everyone is exhausted and underwater* to a descriptive norm of *everyone ships, sustainably, at the pace the method produces.* Once that flip has happened, new arrivals stop being a challenge; they become self-calibrating. You do not have to onboard them into the method. The method is the water they are swimming in, and they will learn to swim in it faster than they could ever be taught to.

Irving Janis, whose mid-twentieth-century work on group dynamics is still required reading in serious organizational-behavior curricula, warned about the failure modes of cohesive teams — the conformity pressures that can produce bad collective decisions when the group's norms go uncorrected. That warning is real, and I'll come back to it at the end of this section, because a fully-adopted team has some specific ways of going wrong that every member of such a team should keep an eye on. But for now I want to stay with the upside, because the upside is genuine and is the main reason to aim for this regime if you have the chance.

What the Team Actually Looks Like From the Inside

Let me describe a fully-adopted team from the inside, because I want you to be able to picture it.

Every member of the team has their own workspace. The workspaces are not identical — people have preferences, and some engineers end up running five slots where others run seven, and the Norse names sometimes get swapped for other mythologies that the individual engineer found more congenial — but the *shape* is the same across everyone. Seven environments, a lease file, a CLAUDE.md or

AGENTS.md, a two-session day. The individual variations are just that: individual variations, inside a shared structure.

Stand-ups take fifteen minutes, not thirty. They go fast because everyone has a clear picture of what they shipped yesterday, because the NOTES.md wind-down forced them to articulate it at 4 PM the night before, and because nobody is pretending to have finished work they had not finished. You notice, if you come from a different team, that the stand-ups contain almost no hedging language. People say what they did and what they are doing next, without the small linguistic shields — *I was kind of looking at, I started to maybe* — that drained the time out of your old team's stand-ups.

PR review is *fast*. This is the single most visible difference, and it is the thing that new team members comment on most often. When everyone is producing PRs at the method's pace, reviewer capacity becomes the binding constraint, and teams that figure this out early enough build explicit rotating review responsibilities into their week. On a healthy fully-adopted team, every PR gets reviewed within two hours of being promoted out of draft, because everyone on the team has committed to spending a portion of their attentional budget on reviewing rather than writing. Review is not an afterthought; it is first-class work, equally compensated in the team's informal accounting with writing, and the reviewers understand they are doing the load-bearing work that keeps the whole pipeline moving.

The ticket count per engineer stops being interesting. On a team where the *median* is fourteen tickets a day, nobody is impressed by fourteen tickets a day. It is the baseline, and attention moves to the shape of what is being shipped rather than the count. The team culture shifts toward caring about the *quality* of tickets — whether they are well-scoped, whether they address the right problem, whether the solutions are durable. That shift is itself an enormous improvement over the ordinary software team, which spends a surprising amount of its cultural energy on arguing about velocity metrics. Velocity has, on a fully-adopted team, been solved; the conversation can move up a level.

Layoff anxiety does not vanish — no team culture can fully inoculate against the broader industry news, and the layoffs.fyi counters that crossed a quarter of a million tech positions eliminated in 2023 and continued through 2025 are a constant background pressure for every engineer paying attention — but it attenuates. When your team is shipping conspicuously more than peer teams, when your output is visible in every direction the executive eye cares to look, the conversation about your team's headcount becomes very different from the conversation about underperforming teams. That protection is not guaranteed, and it is not a reason on its own to adopt the method, but it is real, and I have watched it materially affect how teams felt through bad quarters at companies undergoing restructuring.

The dinner-at-home rate goes up. This is, honestly, the finding I care most about. On the two fully-adopted teams I have watched, the percentage of engineers regularly eating dinner with their families rose measurably over the first six months of adoption. No one measured it formally. Everyone noticed it informally. A team in which nine people stop at 4 PM, reliably, five days a week, is a team in which nine families, nine marriages or partnerships, nine sets of children or pets or friendships, get a version of the engineer home at a civilized hour. That external-to-the-workplace benefit is not a perk. It is, I think, the *point*, and it is the quiet consequence of the method that I would want every executive reading this book to understand is part of the bundle being adopted.

The New Problems, Honestly

I do not want to paint a rosier picture than is warranted, so let me name the problems that appear when a team has fully adopted the method, because they are real and they caught me by surprise the first time I saw them.

Product cannot keep up. A team shipping at fourteen tickets per engineer per day, across seven engineers, is producing roughly ninety-eight tickets a day of completed work. No product organization in the world is generating well-scoped ninety-eight tickets a day of work for a seven-person team. Within two or three weeks, the team drains its backlog. Within a month, the team is inventing its own tickets — infrastructural work, tech-debt paydowns, tooling improvements, things the product team never would have prioritized — because there is simply no formal work to do. This sounds like a nice problem, and it mostly is, but the team has to develop the discipline of self-directing its excess capacity in productive directions rather than drifting into gold-plating or re-architecture theater. I have watched teams handle this well and teams handle this badly, and the difference is whether the team has a shared frame for what *quality of investment* looks like when they get to pick their own tickets.

The company's planning processes are calibrated to old throughput. If leadership expects a team to ship a feature in a quarter, and the team actually ships it in three weeks, the quarterly planning meeting is now slightly broken in ways that can become politically interesting. The team may find themselves in the awkward position of needing to either slow down visibly to match planning expectations, or accept that they are going to ship three times the planned scope per cycle and force the planning process to adapt. Either option has costs. The team members, individually, have a strong self-interest in the second option; the organization, institutionally, will often prefer the first; and navigating that tension takes a diplomatic hand that not every team has.

Reviewer fatigue is a real thing. I mentioned above that review becomes first-class work. The other edge of that sword is that sustained review, day after day, week after week, can produce its own specific exhaustion — not the decision-load exhaustion from Chapter 11 exactly, but an adjacent one, the fatigue of a reviewer who has evaluated three or four hundred PRs in a month and whose taste is beginning to calibrate toward whatever pattern the reviewees' agents favor. Rotating the review responsibility, so that no single engineer is doing review for more than a few hours a day or a few days a week, is load-bearing. Teams that do not rotate discover that the reviewers quietly burn out in a specific way that the method was otherwise supposed to prevent.

New-hire onboarding has to be reinvented. Traditional onboarding assumes a junior pace — a new engineer ramping up over six to twelve weeks to a normal output level. A fully-adopted team compresses that curve dramatically, because the infrastructure the new engineer is stepping into is so structured that a capable adult can be productive in days rather than weeks. But the *reasoning* behind the structure takes longer to internalize than the mechanics do, and a new hire who reaches method-pace without understanding *why* the structure exists can develop habits that drift from the principles over time. Teams that handle onboarding well build in a dedicated conceptual pass — a week of mentorship specifically about the principles of Chapter 10 — before letting the new engineer run at full parallelism.

Janis's warning, updated. The cohesion risk I flagged earlier is worth taking seriously. A team in which everyone has adopted the same method, uses the same conventions, and shares the same framing can, over time, lose the ability to critique the framing from the inside. The method becomes assumed, and assumed things stop being examined, and examinations that should happen simply don't. The protection against this is *deliberate external exposure* — rotating engineers through other teams periodically, reading broadly outside the team's practice, inviting outside review of the team's conventions once a year or so. None of that is automatic. A healthy fully-adopted team builds it in on purpose.

A Gentle Word to the Reader Who Is Not There Yet

Most readers of this book are not going to be on a fully-adopted team any time soon, and I want to close this section by speaking directly to that reader, because I think the vision of the fully-adopted team can land in two very different ways and I want to nudge it toward the one that serves you.

It can land as a dream that makes the present feel worse — as a picture of a place you are not, surrounded by colleagues who are not ready, watching the hours of your life compress under a different team's pressure while someone else is hypothetically eating dinner with their family in a team you are not on. That framing is corrosive, and I would rather you not hold it.

It can also land as a direction of travel. A thing that is possible, that is being done somewhere in the industry right now, that can be pointed at when someone asks you what you are aiming toward. You are allowed to have a direction of travel. You are allowed to work in the way the method teaches even while the rest of your team has not caught up, and to take the individual benefits of doing so — the decision-load relief, the sustainable pace, the quiet evenings — while you wait for whatever organic adoption is or is not going to happen around you. The benefits of being a practitioner do not require the team to cooperate. They arrive on day one, for you, as a function of how you personally structure your work.

And here is the genuinely encouraging thing, and the thing that I most want to leave you with. Every team I have seen fully adopt the method started with one practitioner. Not a coalition. Not a mandate. One person, working quietly, shipping reliably, posting occasionally in a microblog channel, being generous when asked, and waiting. If you are that person right now, you are the beginning of a story that has, elsewhere, ended in nine families eating dinner together on a Wednesday. The story does not always end that way. Often it ends with you alone, shipping well, going home at a reasonable hour, nobody else on your team adopting anything. That outcome is also fine. You still got the benefits. The method is complete at the scale of one.

But sometimes — sometimes — the curiosity takes. The Slack channel grows. A teammate picks up a piece of the practice. Then another. Then, eighteen months later, a new hire joins and finds themselves bewildered by how much the team ships and how calm everyone is about it, and they begin to calibrate, without anyone telling them to, to a new norm.

If that happens, you will be the person it happened around, and you will not have evangelized once. You will have simply done the work, visibly and well, for long enough that the work became the argument.

That is the quiet radicalism of this method. It does not require anyone to be converted. It only requires one person to be faithful to it, for long enough, with patience, and it builds the rest of the conditions around itself when the conditions are ready.

You can be that person. You may already be.

Closing Thought

We have, across many pages, covered a lot of ground.

We began with a split cedar shingle and an old carpenter named Buzzy, who knew exactly two things well and whose single swing of a hammer was the closing argument for every page that followed. We talked about why AI feels different even when the fundamentals aren't — about context loss and decision load and the Before Times, when all any of this meant was whether to take lunch at noon or at one. We set up a workspace, named seven environments after Norse gods for reasons that stopped feeling arbitrary roughly one week into the using of them, and arranged seven slots across a single monitor in a pattern whose geometry you now know by muscle memory. We taught Claude a lease protocol. We sat through a six-step validation ritual. We glanced across the seven slots every few minutes like an air traffic controller and let the agents fly. We watched Monday morning turn into the single most comfortable moment of the working week, because the system held the state we no longer had to hold. We proved the method was portable by dropping it onto Codex and watching the forms survive the swap. We talked about burnout honestly, and about the long game, and about the engineer you hope to still be in the year 2055. And in this last chapter, we walked through the slow, quiet, many-months arc by which all of this spreads — if it spreads — to the people around you, without any of the evangelism that would have poisoned it.

If you have done the work, and not merely read about it, you are a different engineer than you were when you opened the book. Not louder. Not faster in the breathless public way the current discourse rewards. Something more interesting than either: quieter, steadier, better-rested, still curious, and doing more genuine work than you were doing before while carrying less of the weight in your body at the end of the day. That is not a small thing. In this particular year of this particular industry, it is something close to a radical thing, even though it does not feel radical from the inside. Most radical things don't.

Musashi closes the *Book of Five Rings* with a chapter he called *The Book of the Void*, in which he argues that the highest expression of a mastered practice is the moment when all of its techniques have become invisible — when the swordsman stops thinking about stance and distance and timing because the stance and the distance and the timing are simply how the swordsman now moves. That is, I suspect, the shape this method takes in a body over years. The seven slots stop feeling like seven slots and start feeling like the shape of a working day. The lease file stops feeling like bookkeeping and starts feeling like the weather you work inside of. The validation ritual stops feeling like six steps and starts feeling like the single motion of finishing a thing. None of that will happen this week, or next week, or this quarter. But it will happen, if you keep at it patiently, and when it happens you will understand what Buzzy had with the hammer. He was not deliberating about whether to blunt the nail. He reached for a nail and blunted it, in a single motion that had long since stopped being a motion he thought about, because that was simply how Buzzy picked up nails. The technique had disappeared into the practitioner, which is, in every tradition I know of, the definition of mastery.

The tools will keep changing. You already know this. Claude Code will get better in ways you are glad for and worse in ways you are annoyed by. Codex will update and sometimes surprise you. Some agent you cannot currently name will launch in the middle of 2027 and be briefly everywhere, and you will try it for a week, notice what it does well and what it doesn't, and either keep it in one of your seven slots or you won't. None of that will disturb the structure underneath, because the structure was never really made of any particular tool. It was made of a workspace, a file, a rhythm, a rule, a glance, a dish by the door. Those do not update. Those outlast the churn.

I want to say two last things before I stop, and they are the sentences that have been waiting at the end of this book since the Introduction.

Take care of yourself. You are the instrument on which all of this runs, and if you break, the method does not save you — the method was built to keep you from breaking in the first place, which is a different kind of usefulness and not a substitute for your own ordinary human care. Sleep. Eat meals that are actually meals. Close the laptop at four. Spend a weekend without opening it. Read a book about something that has nothing to do with software. Remember that the job is a part of a life, not the shape of one. The engineer at fifty-five from Chapter 11 — the one who is still sitting at a workspace, still curious, still intact — does not arrive at fifty-five by accident. They arrive there because of ten thousand small kindnesses extended to their own body and mind across the years, the first of which is the one you give yourself tonight, when you close this book and, if it is late, go to bed.

Take care of each other. The teammate rolling their chair over to your desk with a sideways question is a person, first, before they are a potential adopter or a cohort member or anything the arc in the previous sections would describe them as. Meet them as a person. The hour you spend helping them set up their Step 1 is the kind of hour the industry does not know how to count, will not put on any performance metric, and will not reward at a promotion cycle, and it is also, more often than not, the hour that mattered most in their year. Be the colleague willing to spend it. Be the friend who says *here, this is the knife you actually need, and here's how to hold it.* Be the quiet steady person in the slot next to theirs whose practice, a year from now, they will recognize as the thing they unknowingly modeled their own on. None of that requires you to teach, or preach, or lead anything. It requires only that you be present, and patient, and generous when asked. Which, if you made it this far in this book, I suspect you already are.

That is, I think, the whole book. A method that holds up under pressure. A career that holds up under decades. A team that holds up under the industry. And a person — you — holding up inside all of it, still recognizably yourself, still curious on a Saturday, still doing good work in the thirtieth year, still able to stop at four.

When I began this book, I promised you a specific thing. I promised that I could show you how to use these tools without being consumed by them. I promised that the method was learnable, that the learning was sustainable, and that the sustainability was the point. I have tried, across every chapter that followed, to keep that promise honestly — to give you the working details rather than the inspirational gloss, to show you the file formats and the keyboard habits and the awkward early mistakes alongside the philosophy, and to trust that you are the kind of reader who wants the real thing rather than a polished version of it. If any of it has worked, it worked because you were willing to meet it seriously, and the credit for that belongs to you, not to me. Books do not change readers. Readers change themselves, sometimes with a book in hand. Thank you for letting this one be the one in yours for a little while.

Pick up your hammer. Or your sword. You know by now that they are closer to the same thing than they looked at the start, and you know by now how to hold cither one without splitting the wood. Tomorrow morning, at nine, in a slot whose position you already know by heart, with an agent whose name will probably change twice before you retire, begin.

And then — because this is the only thing I have actually asked of you, and because the whole rest of the book was designed so that you could honor this one small ask without it costing you anything — stop at four. Eat dinner. Sleep through the night. Come back Tuesday.

I'll see you there.

— Blake Tullysmith Santa Rosa, California

Appendices

The chapters are the book. The appendices are the workshop behind the chapters — the prompt library you'll keep reaching for, the install instructions you'll consult once and then ignore, the troubleshooting recipes for the moments when something has gone visibly wrong.

Read the chapters in order, more or less. Read the appendices the way you'd consult a hardware-store reference: when you need a thing, when you need it now, and not a moment before.

A short index, in case you arrived here looking for something specific:

- **Appendix A — Prompt Templates.** Roughly thirty patterns across eight sections — opening, ticket work, validation, summarization, memory, scripts, research, and meta-prompts. The single most-used appendix in this book.
- **Appendix B — Installing Claude Code.** Per-platform install steps. Read this before Chapter 4 if `claude --version` doesn't return a number on your machine yet.
- **Appendix C — Dark Mode and Terminal Setup.** Eye-strain physiology, dark-mode defaults, terminal colors, font choices. Quietly the most-research-grounded appendix in the book.
- **Appendix D — Parallel Instructions for Ubuntu, Bash, WSL2, Fish.** Everything in Chapter 4 that's macOS-specific, translated for the rest of the world.
- **Appendix E — Workspace File Reference.** Annotated examples of every file the workspace contains, top to bottom.
- **Appendix F — Troubleshooting Common Failure Modes.** When something has clearly gone wrong, start here.
- **Appendix G — Glossary.** Every term defined in the book, alphabetized. Flip back when a word in a chapter has slipped its mooring.
- **Appendix H — Layout Variants.** Variations on the seven-slot layout for the readers whose hardware or workflow doesn't quite match the canonical setup.

If you have a specific failure on your screen right now, jump to Appendix F. If you forgot what *holding* meant, Appendix G. If you're trying to install something, B (or D for non-macOS). Otherwise, the chapters are still where the work is.

Appendix A: Prompt Templates

This appendix is a working reference. It collects the prompt patterns referenced across the book, along with a broader set I have accumulated over a few years of daily use, and presents them in a form you can copy, adapt, and keep in a single markdown file on your second monitor — parked at a protruding edge, as described in Chapter 11, so that any template is a single click away.

A note on how to read this appendix before you dive in.

The templates below are *patterns*, not incantations. Every one of them has worked for me, reliably, across many tickets. None of them will work exactly the same way in your codebase the first time you use them, because your codebase is not my codebase and the specific shape of the problems you want the agent to solve is different. Treat each template as a starting shape. Run it once. Watch what the agent does with it. Edit the template in response to what you saw, and then run the edited version. Two or three iterations in, the template will have calibrated to your specific use, at which point it becomes genuinely yours.

The appendix is organized in eight sections, starting from the smallest and most utilitarian prompts and ending with the most ambitious. By the time you reach Section 8, you will have seen how to prompt your way into prompts that write prompts — which sounds recursive and mildly ridiculous, and which turns out to be one of the most useful patterns in the entire toolkit.

One more thing. I have kept all of these in plain language. You will not find elaborate XML tag structures, triple-backtick-wrapped JSON schemas, or role-play preambles longer than the actual request. Those patterns exist and they have their place in certain production pipelines, but for interactive coding work inside this methodology, plain English beats structured-prompt-engineering almost every time. Short sentences. Clear asks. No ceremony.

Section 1: Opening and Orientation

These are the prompts you send in the first thirty seconds of a session. Their job is to establish context, confirm the agent is reading what you think it is reading, and set the protocol for the rest of the conversation.

The Multi-Step Conversation Opener

Used at the start of any setup or multi-phase task. Tells the agent that you will feed work in chunks and that it should wait for your signal before proceeding to the next chunk.

> **Prompt to Claude Code:**
>
> `I'm going to walk you through a multi-step process. After each step, complete the work and ask me for the next step. Do not run ahead or assume what the next step is. Confirm when you're ready for step one.`

The Orientation Check

Used at the start of any session on an unfamiliar environment. Confirms the agent knows where it is and what it has access to.

> **Prompt to Claude Code:**
>
> `Before we start, tell me: 1. Your current working directory. 2. The top of the CLAUDE.md file you can see from here. 3. The name of the environment you're leased to, if the lease file indicates one. 4. The branch you're currently on in the repo. Answer in four short lines. Don't elaborate.`

The "Where Were We" Prompt

Used at the start of a session that is resuming work from a previous day. Builds on the NOTES.md wind-down ritual from Chapter 6.

> **Prompt to Claude Code:**
>
> `Read NOTES.md in this environment. Summarize in three lines: 1. What was done last session. 2. What's in progress and where it got interrupted. 3. What you'd recommend picking up first. Then stop. Don't begin work yet.`

Section 2: The Work-on-a-Ticket Protocol

This is the spine of an executing day. The prompts below implement the protocol described in Chapter 6: the agent leases an environment, reads the ticket, summarizes back, creates a branch, makes an empty commit, pushes, opens a draft PR, and only then begins actual work.

These templates assume the canonical lease protocol from Chapter 4 (Step 7) and the work-on-a-ticket protocol from Chapter 4 (Step 8) are already in your permanent memory. The prompts are the short triggers that invoke them, not re-transcriptions. If you have not yet recorded those protocols in `CLAUDE.md` or `AGENTS.md`, send the Chapter 4 Step 7 and Step 8 prompts first; everything below depends on them being in memory.

The Lease Claim

> **Prompt to Claude Code:**
>
> `Claim an environment for a new ticket, following the lease protocol recorded in permanent memory. Tell me which environment you took once you have copied in the repos and read the repo memory files.`

The Work-on-Ticket Bootstrap

> **Prompt to Claude Code:**
>
> `Work on {TICKET-ID}, following the work-on-a-ticket protocol in permanent memory. If you don't already hold a lease on an environment, claim one first. Before writing any code: fetch the ticket, summarize it back to me in five lines or fewer (problem, acceptance criteria, affected area if you can infer it, anything ambiguous, a one-sentence plan), and wait for me to confirm.`

The Branch and Draft-PR Opener

> **Prompt to Claude Code:**
>
> `Good. Now: 1. Create a branch named {TICKET-ID}-{short-slug} off the default branch. 2. Make an empty commit with the message "{TICKET-ID}: begin work". 3. Push the branch. 4. Open a DRAFT pull request against the default branch. Title it with the ticket ID and the short slug. In the PR body, paste the five-line summary from the previous step and add a checklist of the acceptance criteria. 5. Report back the PR URL. Then begin work.`

The Release-a-Lease Prompt

Used at the end of a session or when a ticket is completed.

> **Prompt to Claude Code:**
>
> ``` Release the lease on this environment, following the release rule in the lease protocol recorded in permanent memory.
>
> Steps: 1. Write a 3–10 line wind-down note to NOTES.md in this environment, following the template at the top of the file. 2. Check out the default branch and use git stash to clear local state. 3. Clear this environment's row from ENVIRONMENT_LEASES.md. 4. Print a one-line confirmation: "Released {env}". ```

Section 3: Validation Prompts

These are the prompts you fire during the six-step validation ritual. Each one isolates a single kind of scrutiny so that the agent's attention — and yours — lands cleanly on it.

The Self-Critique Prompt

Used as step one of the validation ritual.

> **Prompt to Claude Code:**
>
> ``` Before I look at what you've done, critique your own work.
>
> Pretend you are the most skeptical senior engineer on the team. You have not seen this code before. You don't know the author. You have no emotional investment in the solution.
>
> Walk the diff top to bottom and tell me: 1. The three things you'd most likely push back on in code review. 2. Any change you made that you cannot justify against a specific acceptance criterion. 3. Any test that is present but wouldn't fail if you broke the behavior it's supposed to test. 4. Any edge case you thought about and decided not to handle — and why.
>
> Be specific. Cite file paths and line numbers. Do not modify any files during this review. ```

The Test-Running Prompt

> **Prompt to Claude Code:**
>
> ``` Run the test suite. Specifically: 1. Run the tests for the files you changed. 2. Run any integration tests that exercise the affected paths. 3. Run the full suite last.
>
> Report the results in three lines per bucket. If anything fails, stop and show me the failure before attempting a fix. ```

The Acceptance-Criteria Check

> **Prompt to Claude Code:**
>
> ``` Walk each acceptance criterion in the PR body one at a time. For each criterion, tell me: 1. The criterion, verbatim. 2. Where in the diff it's satisfied (file path and the specific change). 3. Whether a test exists that would fail if that criterion were broken. If not, note it.
>
> Format as a numbered list. No prose between items. ```

The Multi-Perspective PR Review

One of my most-used templates. Produces a review that catches things a single-lens review misses.

> **Prompt to Claude Code:**
>
> ``` Review this PR from the following four perspectives, in order. Write a separate section for each, labeled.
>
> 1. **Cybersecurity.** Auth, authz, input validation, injection surfaces, secret handling, dependency risk.
> 2. **Implementation.** Correctness, edge cases, error handling, test coverage, readability.
> 3. **Operational.** Observability, rollback safety, performance under load, resource usage, configuration exposure.
> 4. **Product alignment.** Does this actually address the ticket's acceptance criteria? Is there scope creep? Is there scope creep I should argue for keeping because it's load-bearing?
>
> Be concise. No more than six bullets per section. ```

The Panel-of-Experts Review

This is the most powerful review prompt I use, and it is worth reading twice. The trick is to stage a diverse set of reviewers with different priorities, and to require convergence before the review is considered complete.

> **Prompt to Claude Code:**
>
> ``` Assemble a review panel for this PR. The panel consists of:
>
> - Three principal engineers (one distributed-systems specialist, one language-and-typing specialist, one API-design specialist)
> - One data scientist
> - Two cloud architects (one AWS-leaning, one multi-cloud)
> - Three lawyers (one privacy, one IP, one regulatory)
> - Four risk managers (operational risk, financial risk, security risk, reputational risk)
> - One UAT/QA expert
> - Several TypeScript experts (as many as needed for the diff)
>
> Run the panel in iterations. In each iteration: 1. Each member produces their own review independently. Do not let earlier members' comments contaminate later members. 2. Compile all reviews. 3. Identify disagreements between members. Resolve them by having the relevant members debate briefly in the next iteration. 4. Identify items that multiple members flagged. Those become consensus issues.
>
> Continue iterating until every member signs off with no further feedback. Report the final consensus issues, the resolved disagreements, and the sign-off confirmation. Do not modify any files during this review. ```

The Diagnostic Prompt

Used when a test starts failing and you want to slow the agent down before it speedruns through speculative fixes.

Prompt to Claude Code:

``` A test is failing. Diagnose what went wrong. Do not fix anything yet.

Specifically: 1. State the failing assertion, verbatim. 2. Trace the production-code path that feeds into that assertion. 3. Identify the specific place where the actual behavior diverges from the expected behavior. 4. Propose three possible root causes, ranked by likelihood. 5. For each, name the single change that would confirm or rule it out.

Stop after step 5. Do not write code. I'll choose which hypothesis you chase. ```
```

Section 4: Summarization and Status

These are the prompts that compress state into something you can scan in ten seconds. They matter more than they sound, because the method's throughput depends on your ability to rotate across seven slots without losing track of what's in each one.

The End-of-Session Summary

Used before the NOTES.md wind-down ritual.

> **Prompt to Claude Code:**
>
> ``` Summarize where you are on this ticket in three blocks, each no longer than three lines.
>
> 1. What's done.
> 2. What's in progress, and the exact state of the in-progress work.
> 3. What I'd want to pick up first tomorrow.
>
> Then write those three blocks to NOTES.md in this environment, appending to the dated entry for today. ```

The Slot-Glance Prompt

Used in the glance cycle when you want a fast status read from a slot without opening its full transcript.

> **Prompt to Claude Code:**
>
> `One line. What are you doing right now, and is there anything that needs me.`

The PR-Description Auto-Draft

Used after work is complete and before promoting the PR out of draft.

Prompt to Claude Code:

``` Rewrite the PR description. The description should contain:

1. A one-paragraph summary of what this PR does.
2. A bulleted list of the specific changes, grouped by file area.
3. The original acceptance criteria as a checklist, with boxes checked where the criterion is now satisfied.
4. A "Notes for reviewer" section with anything I'd want a teammate to know before reading the diff — including any deliberate non-changes where I considered changing something and chose not to.

Keep the tone matter-of-fact. Do not use exclamation marks or marketing language. ```
```

Section 5: Refinement and Memory

Prompts that act on permanent memory — CLAUDE.md, AGENTS.md, the `.claude/` config directory. These are the templates you use during the refinement phase from Chapter 8.

The Add-a-Rule Prompt

> **Prompt to Claude Code:**
>
> ``` Please add the following rule to permanent memory:
>
> "{the rule, in one sentence}"
>
> Place it in the most appropriate existing section of CLAUDE.md. If no good section exists, create a new one with a reasonable heading. Match the tone of the neighboring rules. Report the diff before committing it. ```

The Prune-Memory Prompt

Used at cold starts or during occasional refinement.

> **Prompt to Claude Code:**
>
> ``` Read the current CLAUDE.md. For each rule in it, classify it as:
>
> - Load-bearing (I would bet money this still matters)
> - Uncertain (was useful once, unclear if it still is)
> - Probably dead (specific to a situation that no longer applies)
>
> Present the classification as a three-column table. Do not delete anything. I'll decide which of the Uncertain and Probably-dead items to remove. ```

The Memory-Conflict Check

Used after a refactor or a major rule update to catch contradictions.

Prompt to Claude Code:

```
Read CLAUDE.md end to end. Identify any pair of rules that could
conflict with each other, even in edge cases. For each pair, de-
scribe the conflict in one sentence and propose a reconciliation.
```

Section 6: Scripts and Tools

Prompts that produce executable artifacts — shell scripts, one-off tools, small utilities that don't deserve a ticket of their own. This is the kind of ad-hoc work that lives on the second monitor rather than in a slot, from Chapter 11.

The Quick-Script Prompt

> **Prompt to Claude Code:**
>
> ``` Write a shell script that does the following:
>
> {one-sentence description of the task}
>
> Constraints: - POSIX sh-compatible (no bash-isms) unless I tell you otherwise. - Exits non-zero on any error. - Prints what it's about to do before doing it. - Does not require sudo. - Is short enough to paste into a terminal and read in under a minute.
>
> Give me the script and a one-line invocation. No explanation. ```

The One-Off SQL Prompt

> **Prompt to Claude Code:**
>
> ``` I need a one-off query to answer the following question:
>
> {the question, plainly}
>
> The relevant tables are {list}. Before you write the query, tell me which tables and which join keys you'll use, and why. Then write the query. Then tell me what a sanity-check on the result would look like — what count or range you'd expect the answer to fall within if nothing has gone wrong. ```

The Throwaway-Tool Prompt

> **Prompt to Claude Code:**
>
> ``` I want a small command-line tool that does {one thing}. Requirements:
>
> - Single file. No build system. No dependencies beyond the standard library of the language.
> - Language: {pick one — Python, Go, whatever fits}.
> - Usable from a pipe. Reads stdin if no arguments are given.
> - Prints usage on -h or –help.
> - Has three test inputs at the bottom of the file in a comment, so I can sanity-check it without writing a test harness.
>
> Write it. Then run it against the three test inputs and show me the output. ```

Section 7: Research and Reading

Not every agent task is code. Some of the most valuable uses are research — reading documentation, understanding a new library, or summarizing a long PR that someone else wrote. These prompts keep the agent from free-associating.

The Library-Orientation Prompt

> **Prompt to Claude Code:**
>
> ``` I'm about to use {library name} for the first time in this project. In under 200 words:
>
> 1. What problem does it solve, specifically?
> 2. What's the smallest useful example of using it?
> 3. What are the two or three most common footguns?
> 4. What does it *not* do, that I might wrongly expect it to do?
>
> Cite the specific documentation page or source file for each claim. If you can't cite, say you can't. ```

The Someone-Else's-PR Summary

> **Prompt to Claude Code:**
>
> ``` Read PR #{NUMBER} on {REPO}. I am not the author.
>
> Produce a summary in the following shape:
>
> 1. One paragraph of plain English describing what the PR does.
> 2. The three most important files changed, with a one-line description of each change.
> 3. Anything in the PR that looks load-bearing for downstream work I'd be reviewing — i.e., changes that my own ticket depends on.
> 4. Two specific questions I should ask the author before I approve.
>
> Do not leave a review on the PR. Just report to me. ```

The Long-Document Reader

> **Prompt to Claude Code:**
>
> ``` Read {the document — path or URL}. Produce:
>
> 1. A three-sentence summary of what it's actually saying (not what it claims to be saying).
> 2. The single most load-bearing claim it makes.
> 3. Any claim it makes that contradicts something we have in CLAUDE.md.
> 4. What I should do differently, if anything, based on it.
>
> Be direct. No hedging. ```

Section 8: Meta-Prompts — Prompts That Produce Prompts

This is the section that changes what is possible with this toolkit, and it is the one most developers have never been told exists. The pattern is simple: you ask the agent to write a prompt you can use later. The result is a prompt calibrated to a specific situation you don't yet fully understand, generated by a system that is often better than you are at articulating the shape of the request.

The Prompt-Refinement Prompt

Used when one of your existing templates is not producing what you want.

> **Prompt to Claude Code:**
>
> ``` Here is a prompt template I have been using:
>
> ”“” {paste the template} “”“
>
> Here is what it produces that I don't like:
>
> {paste or describe the bad output}
>
> Here is what I want it to produce instead:
>
> {describe the desired output in plain English}
>
> Rewrite the template so that it reliably produces the desired output. Then explain, in three bullets, what you changed and why. ```

The Prompt-Factory Prompt

Used when you have a recurring task and no template yet.

> **Prompt to Claude Code:**
>
> ``` I keep finding myself doing the following kind of task:
>
> {describe the task pattern in plain English, including three examples of specific instances}
>
> Produce a reusable prompt template I can keep in my templates file. The template should:
>
> 1. Use {placeholders} for the parts that change between uses.
> 2. Be short enough to fit on a sticky note.
> 3. Produce output in a predictable shape that I can scan in under ten seconds.
>
> After you give me the template, run it once against one of the example instances so I can see what it produces. ```

The Summarizer-Factory

Used when you want a compressed view of something repeatedly — a log file, a daily digest, a recurring email.

> **Prompt to Claude Code:**
>
> ``` Write a prompt that, when given {the input type}, produces a summary in the following shape:
>
> {describe the summary shape — length, tone, sections, what to include and what to leave out}
>
> The prompt should be robust to variations in the input — missing fields, different lengths, occasional noise. It should fail gracefully and tell me what it couldn't find rather than guessing.
>
> Give me the prompt and one example input/output pair. ```

The Chain-Builder

Used when a task is too big for a single prompt and you want the agent to help you design the decomposition.

> **Prompt to Claude Code:**
>
> ``` I want to accomplish the following, but I think it's too big for a single prompt:
>
> {describe the goal}
>
> Design a chain of three to five prompts that would, in sequence, accomplish this. For each prompt in the chain:
>
> 1. State its job in one sentence.
> 2. Write the actual prompt text.
> 3. Describe what its output should look like, so I know whether the chain broke at that step.
> 4. Describe what the next prompt in the chain will need from its output.
>
> Do not run the chain. Just design it. I'll run it myself and report back what worked. ```

The Meta-Reviewer

The most recursive of the bunch. Takes the output of one prompt and asks whether the output actually answered the question.

Prompt to Claude Code:

``` I just asked the following prompt:

"“” {paste the original prompt} “”“

And got the following answer:

"“” {paste the answer} “”“

Evaluate whether the answer actually addresses the prompt. Be strict. Specifically:

1. Which parts of the prompt were fully addressed?
2. Which parts were partially addressed?
3. Which parts were not addressed at all?
4. Are there any places where the answer confidently states something that cannot be verified from the prompt's stated context?
5. What is the single most important follow-up prompt I should send to close the gaps?

Do not rewrite the answer. Just evaluate it. ```
```

A Closing Note on How to Use This Appendix

The templates above will not work the first time. Some of them will work, charmingly, on the second or third try. A few will need fundamental rewriting before they fit your codebase. That is expected. That is, in fact, the *point*.

A template is a hypothesis about a prompt that will work repeatedly. You confirm or refute the hypothesis by running the prompt, watching the output, and editing the template. Two weeks in, you will have a templates file that looks nothing like this appendix and that works better for you than anything I could have written. Your templates file is a product of your attention. This appendix is a starting point.

Keep the file on your second monitor. Keep it in plain markdown. Edit it freely. When a template stops working, don't archive it — rewrite it. When a template surprises you by working in a new domain, add a line under it noting what the new use was. The file grows with you, and the growth is the practice.

If you find a template that works so well you want to share it, post it in whatever microblog channel you are keeping in your workplace, without comment, and let the next person find it the way you found the ones above. That is how good practice travels inside a team — not through documentation bureaucracy, but through small acts of *here, look what I found.* This whole book has been an argument that those small acts, accumulated across years, are the structure a career is actually made of.

Pick up the templates. Edit them. Put them to work. And when one of them becomes invisible to you — when you reach for it without thinking, the way Buzzy reached for a nail — you'll know the practice has taken.

Appendix B: Installing Claude Code

This appendix is the definitive per-platform installation reference for Claude Code and its immediate prerequisites. Chapter 4 assumes you already have Claude Code installed, and defers the platform-specific install steps here, so that the main text could stay focused on the *method* rather than the mechanics. This is the mechanics.

A word before we start, because I want to set expectations honestly.

Installation instructions are the single most time-sensitive kind of technical writing. The recommended install path for a piece of software can change two or three times a year, sometimes without warning, and a printed book is an especially poor place to memorialize a specific command that may be stale by the second printing. I have tried to write this appendix in a way that ages gracefully — I lean on the official installation documentation for the canonical truth, and I give you the *shape* of the install rather than a command-by-command transcript where I can — but if you read something here that contradicts what Anthropic's official docs say, trust the docs. They are the source of record. This appendix is a guide.

The appendix is organized around four platforms — macOS, Windows, Linux (generalized), and WSL2 specifically — followed by a set of cross-cutting concerns: verifying the install, configuring `PATH`, upgrading later, sandboxing with Node version managers, and a troubleshooting section that covers the failure modes I have personally walked people through over the last two years. A separate concluding section covers the supporting command-line tools (`git`, `gh`, `jira`) at a reference level, complementary to the teaching treatment in Chapter 4.

Let's begin.

Prerequisites, on Every Platform

Before you install Claude Code itself, you need three things on your system. These are universal; the platform-specific sections below assume you have them.

Node.js, version 18 or newer (and in practice, Node 22 LTS or 24 LTS). Claude Code's npm install path requires Node 18 as a minimum floor at the time of writing, but Node 18 itself reached end-of-life in 2025. In practice, install Node 22 LTS or Node 24 LTS — newer is better, and Claude Code's own minimum tends to creep upward over time. If you have Node already installed and it is version 18+, you are fine for now. If you do not, the platform sections below will walk you through getting it. If your Node is older than 18 — a surprising number of engineers are still on Node 16 because they installed it three years ago and forgot — upgrading is part of the prerequisite work, not an optional improvement. (Note that Anthropic also publishes a native installer that does not require Node at all; see the macOS, Windows, and Linux sections below for that path.)

A terminal you are comfortable in. Claude Code is a command-line tool. You will spend most of your working day inside a terminal with a Claude session running in it. Pick a terminal you enjoy looking at; set it to a dark theme (see Appendix C); pick a font size you can read at 10 PM after a long day. The specific terminal emulator does not particularly matter — iTerm2, Ghostty, Warp, Alacritty, Windows Terminal, the default on whatever platform you use — but the terminal's *comfort* matters more than its feature list.

An Anthropic account. Claude Code authenticates against your Anthropic account via a browser-based login flow that the installer walks you through. A Claude Pro, Max, Team, Enterprise, or Console account all work; the free Claude.ai tier does not. Create or sign in to whichever you have at `claude.ai` or `console.anthropic.com`. You do not need to pre-generate an API key; the installer handles authentication on first run.

With those in place, proceed to the section for your platform.

macOS

Anthropic now publishes two install paths on macOS, and they are recommended in roughly this order:

1. **The native installer** — `curl -fsSL https://claude.ai/install.sh | bash` — which downloads a self-contained binary, places it on your `PATH`, and auto-updates itself. Does not require Node. This is the currently-recommended path for new installs and is the simplest to keep current over time.
2. **Homebrew (cask)** — `brew install --cask claude-code` — which installs the same native binary through the Homebrew package manager, so you can manage it alongside the rest of your Homebrew tooling.
3. **The npm path** — `npm install -g @anthropic-ai/claude-code` — which is the original distribution channel and still works fine. This is what the rest of this section walks through, because the npm semantics are what most of this book's prose assumes (global bin directories, `nvm` interplay, the version-pinning conventions in Chapter 4). If you take one of the first two paths, you can largely skim the npm-specific subsections below; the post-install verification, upgrade, and troubleshooting material at the end of this appendix applies regardless.

If you are coming back to this appendix to set up a brand-new machine, the native installer is the lowest-friction option and what I would do today. If you are working through Chapter 4's setup conversation alongside this appendix, the npm path keeps the commands matching the main text exactly, which is its only remaining advantage.

The npm walkthrough begins with Homebrew, which is the package manager the vast majority of Mac developers already have installed. If you do not have Homebrew yet, install it first; the npm install will work without Homebrew but Homebrew is the path with the fewest surprises.

Step 1: Install Homebrew (If You Do Not Have It)

Open Terminal or iTerm2. Run the official installer command, which you can find at the canonical URL `brew.sh`. The current recommended invocation is a single `curl | bash`-style line that downloads and executes Homebrew's install script. Follow the on-screen prompts. When the installer finishes, it will print two or three lines you need to run to add Homebrew to your shell's `PATH`. Run those exact lines. Then open a new terminal window and confirm Homebrew is available:

```
brew --version
```

You should see a version number. If the command is not found, the `PATH` additions did not take effect for your shell; the troubleshooting section at the end of this appendix covers this.

Step 2: Install Node.js

```
brew install node
```

Homebrew will pull in Node and npm together. Confirm:

```
node --version
npm --version
```

Both should print version numbers. Node should be 18 or higher.

Step 3: Install Claude Code

Claude Code is published as an npm package. Install it globally:

```
npm install -g @anthropic-ai/claude-code
```

The `-g` flag makes the `claude` binary available on your `PATH` globally. If npm complains about permissions on the global prefix, do **not** use `sudo`; that creates more problems than it solves. Instead, configure npm to use a user-owned global prefix (the troubleshooting section covers this), or use a Node version manager like `nvm` (see the Node Version Managers section below).

Confirm:

```
claude --version
```

You should see a version number.

Step 4: First-Run Authentication

Launch Claude Code from inside a working directory:

```
mkdir -p ~/workspace
cd ~/workspace
claude
```

On first run, Claude Code will detect the absence of credentials and walk you through a browser-based login flow. Follow the prompts. You will be asked to visit a URL, approve the login in your browser, and return to the terminal. The browser flow creates and stores an API key on your local machine for subsequent sessions; you do not need to copy or paste anything manually.

When the flow completes, Claude Code drops you into a session. Type `exit` to close it for now. We will return to it in Chapter 4, in context.

Windows

Claude Code runs on Windows, and the recommended path is **through WSL2** rather than directly in PowerShell. I know that is not the answer a lot of Windows users want, and I want to explain why before we go further.

Claude Code is a Node.js-based tool that invokes other command-line tools — `git`, `gh`, shell utilities, test runners, the lot — with the same POSIX-shaped assumptions that most open-source development tooling has made for decades. Those assumptions work cleanly under WSL2's Linux environment and less cleanly under Windows-native paths and shells. You can make Claude Code work in native PowerShell, and people do. But you will spend a noticeable amount of your early career on this method fighting small path-separator issues, line-ending issues, and permission-model mismatches that WSL2 users simply do not encounter. WSL2 is the path of fewest surprises. I recommend it.

The native-PowerShell path is documented further below, for readers who cannot use WSL2 for organizational reasons.

Recommended Path: WSL2

Step 1: Enable and install WSL2. Open PowerShell as Administrator. Run:

```
wsl --install
```

This command enables the WSL2 feature, installs the Linux kernel, and installs Ubuntu as the default distribution. Reboot when prompted. On first login to the Ubuntu environment, set a username and password — these are local to the WSL2 environment and have no relationship to your Windows account.

Step 2: Update the Ubuntu environment.

```
sudo apt update
sudo apt upgrade -y
```

Step 3: Install Node.js via NodeSource or nvm. The default Ubuntu repositories typically ship an older Node; use either NodeSource's current repository or install `nvm` (recommended) to get a current version. Using `nvm`:

```
curl -o- https://raw.githubusercontent.com/nvm-sh/nvm/master/install.sh |
bash
# restart shell, or source nvm as the installer instructs
nvm install --lts
nvm use --lts
```

Confirm `node --version` reports 18+.

Step 4: Install Claude Code.

```
npm install -g @anthropic-ai/claude-code
claude --version
```

Step 5: Open a working directory inside WSL2, not on the Windows filesystem. This one matters. Keep your workspace inside the WSL2 filesystem (for example, under `~/workspace` in your WSL2 home), not under `/mnt/c/Users/...`. Cross-filesystem access is dramatically slower, and git in particular behaves poorly across the boundary. Your VS Code or terminal can still see into WSL2 from Windows; just keep the files themselves on the Linux side.

Alternate Path: Native PowerShell

If WSL2 is not available to you, install directly on Windows. As on macOS, Anthropic offers a native installer as the lowest-friction path; if it works for your environment, prefer it:

```
winget install Anthropic.ClaudeCode
# or, equivalently, the install script:
irm https://claude.ai/install.ps1 | iex
```

If you'd rather install via npm (the distribution channel the rest of this book's prose assumes), the steps:

Step 1: Install Node.js. Use the official installer at `nodejs.org`, or if you have **winget**:

```
winget install OpenJS.NodeJS.LTS
```

Confirm in a new PowerShell window:

```
node --version
npm --version
```

Step 2: Install Claude Code.

```
npm install -g @anthropic-ai/claude-code
claude --version
```

Step 3: Use PowerShell 7, not Windows PowerShell 5.1. If `$PSVersionTable.PSVersion` reports a major version below 7, install PowerShell 7 via `winget install Microsoft.PowerShell` and use it as your default shell. Many small pieces of Claude Code's behavior assume a recent shell; the 5.1 compatibility story is not worth the debugging.

Step 4: First-run authentication. Same as macOS. From a new PowerShell 7 window:

```
New-Item -ItemType Directory -Path "$HOME\workspace" -Force
Set-Location "$HOME\workspace"
claude
```

Follow the browser login flow.

Linux (Generalized)

As on macOS, the easiest Linux install today is Anthropic's native installer:

```
curl -fsSL https://claude.ai/install.sh | bash
```

It downloads a self-contained binary, drops it on your `PATH`, and auto-updates. No Node needed. If that fits your environment, take it and skip ahead to *Post-install verification.*

The npm path below is the original distribution channel and is still what most of this book's prose assumes. It is conceptually identical to the macOS npm path but varies in detail depending on distribution. I will cover the three families most software engineers encounter: **Debian/Ubuntu**, **Arch-family** (including Manjaro), and **Fedora/RHEL-family**. If you are on a less-common distribution, the pattern is the same — install Node.js via your distribution's package manager or via `nvm`, then install Claude Code globally via npm.

Debian / Ubuntu

Prefer NodeSource for a current Node, since the Debian repositories typically lag. The NodeSource setup script is published at `deb.nodesource.com` and installs Node via apt in the way the Debian community expects:

```
curl -fsSL https://deb.nodesource.com/setup_lts.x | sudo -E bash -
sudo apt install -y nodejs
```

Confirm `node --version` reports 18+.

Install Claude Code:

```
npm install -g @anthropic-ai/claude-code
claude --version
```

Arch / Manjaro

```
sudo pacman -S nodejs npm
node --version
```

If your distribution ships a Node older than 18, consider using `nvm` instead (see below) rather than fighting the system package.

Install Claude Code:

```
npm install -g @anthropic-ai/claude-code
claude --version
```

Fedora / RHEL / Rocky

```
sudo dnf install -y nodejs npm
node --version
```

For a more current Node, NodeSource publishes RPM-family instructions parallel to the Debian ones. Install Claude Code the same way:

```
npm install -g @anthropic-ai/claude-code
claude --version
```

First-Run Authentication on Linux

Identical to macOS and Windows: create a working directory, `cd` into it, run `claude`, follow the browser login flow. If you are on a headless server without a browser, Claude Code offers a manual authentication path that prints a URL you can open from another machine; follow the on-screen instructions.

Node Version Managers

I have mentioned `nvm` a few times above. I want to make the recommendation explicitly here, because it is the single most-useful quality-of-life install tool for long-term Claude Code users, and it is almost always worth taking the extra fifteen minutes to set up.

A Node version manager — `nvm` on macOS and Linux, `nvm-windows` or `fnm` on Windows — lets you install multiple versions of Node side by side and switch between them. This matters for Claude Code for two reasons.

First, when Claude Code publishes a new major version that requires a newer Node, you can bump Node without having to reinstall your entire system's Node and all the globally-installed packages that depend on it. You simply `nvm install` the new version, `nvm use` it, and `npm install -g @anthropic-ai/claude-code` under the new Node. The old Node and everything installed under it remains untouched.

Second, a user-scoped Node version manager sidesteps the permission problems that come with `sudo npm install -g` on system Nodes. The `nvm`-installed Node lives under your home directory, owned by you, and `npm install -g` just works without elevated privileges.

To install `nvm` on macOS or Linux, run the official installer from `github.com/nvm-sh/nvm` (the README has the current one-line command). Restart your shell. Then:

```
nvm install --lts
nvm use --lts
nvm alias default lts/*
```

On Windows, `fnm` (github.com/Schniz/fnm) is a simpler alternative to `nvm-windows` and works cleanly in PowerShell 7.

After installing any Node version manager, install Claude Code the normal way:

```
npm install -g @anthropic-ai/claude-code
claude --version
```

Post-Install Verification

Regardless of how you got there, confirm the install is healthy before you move on.

1. The binary is on your PATH.

```
which claude
```

should print a path under your Node installation's global bin directory. If it prints nothing, `PATH` is misconfigured. See the troubleshooting section.

2. The version reports correctly.

```
claude --version
```

Should print a version number. Compare against the latest release at `docs.anthropic.com/claude-code` or `github.com/anthropics/claude-code`. If you are more than one minor version behind, upgrade now (see below).

3. A session starts cleanly.

```
mkdir -p ~/workspace
cd ~/workspace
claude
```

You should land in a Claude Code session. Type a trivial prompt, see a response, and exit.

4. The permission bypass flag, if you intend to use it, launches a session.

```
claude --dangerously-skip-permissions
```

A session should start with no permission-prompt prelude. Type `/exit` (or Ctrl+C twice) to leave. If Claude reports an unknown flag, your Claude Code is out of date — see *Upgrading Claude Code* below. Note that `claude --help` does not print every supported flag, so a flag's absence from `--help` is not a reliable signal that it doesn't exist.

Upgrading Claude Code

Claude Code updates frequently. A release every one to two weeks is normal. Upgrade with the same command you used to install, with the `@latest` tag:

```
npm install -g @anthropic-ai/claude-code@latest
```

If you are using `nvm`, make sure you are on the intended Node version before you run the upgrade; the global install is scoped to the active Node.

A useful habit: upgrade at the start of a session, never in the middle of one. Interrupted work in a slot that relied on a specific version's behavior is a small but real risk, and there is no benefit to upgrading mid-day rather than tomorrow morning before you start rotating.

Troubleshooting

These are the failure modes I have walked people through the most often. Each one has a diagnostic you can run and a specific fix.

`claude: command not found`

The binary installed but `PATH` does not include the global npm bin directory. Find where npm installs globally:

```
npm prefix -g
```

That directory should have a `bin/` subdirectory (or on Windows, the directory itself may contain the binaries). Add the bin directory to your shell's `PATH` by editing `~/.zshrc`, `~/.bashrc`, or your PowerShell `$PROFILE`, and open a new shell.

On macOS with Homebrew, the relevant path is typically `/opt/homebrew/bin` (Apple Silicon) or `/usr/local/bin` (Intel). On Linux, the path depends on whether Node was installed via your system package manager or `nvm`. On Windows, the path is typically `%APPDATA%\npm`.

`EACCES` Permission Errors During `npm install -g`

Do **not** reach for `sudo npm install -g`. That installs the package under root ownership and creates downstream permission problems that are a nightmare to unwind. The correct fix is to use a Node version manager, or to configure npm to install to a user-owned prefix. The official npm documentation has a page titled *Resolving EACCES permissions errors when installing packages globally* that walks through the user-prefix fix; follow it.

Browser Login Flow Does Not Complete

On first run, if the browser login flow opens but does not close out — the browser shows a success page but the terminal hangs — the common cause is a local firewall or proxy blocking the localhost redirect the flow relies on. Close the hung terminal session, check your firewall for localhost blocks, and try again. If you are on a corporate network that intercepts HTTPS, you may need to generate an API key manually in the Anthropic console and configure Claude Code to use it directly; the Anthropic docs cover this path under *authentication*.

Claude Code Runs but Cannot Find `git`, `gh`, or `jira`

Claude Code invokes your system's command-line tools by name. If a tool is not on your `PATH`, Claude Code cannot see it. Test each prerequisite directly:

```
git --version
gh --version
jira version
```

Any of these that does not print a version number is the one to fix. See Chapter 4's walkthrough, or the supporting-tools reference at the end of this appendix.

On WSL2: Operations on `/mnt/c/...` Paths Are Painfully Slow

This is not a Claude Code bug; it is the well-known cross-filesystem performance penalty in WSL2. Move your workspace onto the Linux filesystem (`~/workspace`, not `/mnt/c/Users/you/workspace`) and the slowness will vanish.

On Windows PowerShell: Scripts Refuse to Execute

Windows PowerShell's default execution policy blocks script execution. To allow npm global scripts to run:

```
Set-ExecutionPolicy -Scope CurrentUser -ExecutionPolicy RemoteSigned
```

This is safe for interactive development use. Do not set it to `Unrestricted`.

Node Version Mismatch After Upgrading

If you upgrade Node (via `nvm install` or a system package manager) and Claude Code suddenly says `command not found`, the issue is that global packages are scoped per-Node. Reinstall under the new Node:

```
npm install -g @anthropic-ai/claude-code
```

`nvm` has a helper command, `nvm reinstall-packages <old-version>`, that automates this for all globally-installed packages at once.

The `--dangerously-skip-permissions` Flag Behaves Differently Than Documented

This flag's exact spelling and behavior have evolved across Claude Code versions. If the documentation you are reading does not match what your `claude --help` shows, your Claude Code is out of date. Upgrade (see above) and re-check. The flag's *intent* — to suppress per-action approval prompts while you rely on your own workspace boundary, as described in Chapter 4 — is stable even if the spelling changes.

Corporate Laptop: Install Is Blocked by MDM or Endpoint Security

On managed work machines, `npm install -g` sometimes runs clean but the installed binary is quarantined by endpoint security the first time you invoke it. Symptoms include a long pause the first time `claude` is run, or a silent failure to launch. The fix, almost always, is to request an exception from your IT team for the Claude Code binary under your npm global directory. Be prepared to provide the exact path (`which claude`), the publisher (Anthropic), and a link to `docs.anthropic.com/claude-code`. Most IT teams grant this quickly once the ask is specific.

A Reference for the Supporting Tools

Chapter 4 walks through installing `git`, `gh`, and `jira` in context, using Claude Code itself to guide the install. This appendix's job is not to reteach that; it is to provide a condensed reference for readers who want the commands without the narrative.

git

Preinstalled on most systems. Confirm with `git --version`. If absent or outdated, install via your platform's package manager:

- **macOS:** `brew install git`
- **Windows:** `winget install Git.Git` (or install Git for Windows from `git-scm.com`)
- **Debian/Ubuntu:** `sudo apt install git`
- **Arch/Manjaro:** `sudo pacman -S git`
- **Fedora/RHEL:** `sudo dnf install git`

Configure identity:

```
git config --global user.name "Your Name"
git config --global user.email "you@example.com"
```

Use the email that is on your GitHub account.

gh (The GitHub CLI)

- **macOS:** `brew install gh`
- **Windows:** `winget install GitHub.cli`
- **Debian/Ubuntu:** follow `github.com/cli/cli/blob/trunk/docs/install_linux.md` for the official apt repository setup.
- **Arch/Manjaro:** `sudo pacman -S github-cli`
- **Fedora/RHEL:** `sudo dnf install gh`

Authenticate interactively:

```
gh auth login
```

Choose HTTPS unless you already use SSH keys against GitHub. Pick the browser flow when prompted for authentication method. Confirm with `gh auth status`.

jira (The Community CLI by Ankit Pokhrel)

There are several CLI tools named `jira`; the one this book uses is at `github.com/ankitpokhrel/jira-cli`. Install instructions are kept current in that repository's README. The typical paths:

- **macOS:** `brew install jira-cli` (the formula is in Homebrew core; the older `brew install ankitpokhrel/jira-cli/jira-cli` tap form still works as a fall-back)
- **Linux (via installer script):** the repository provides a one-line installer at its README.
- **Windows:** download the pre-built binary from the Releases page and place it on your `PATH`.

On first run, `jira init` walks you through configuration. You will need:

1. The base URL of your Jira instance.
2. Your Atlassian account email.
3. An API token generated at `id.atlassian.com` → Security → API tokens.

Treat the API token as a password. Store it in the CLI's config file, not in any git-tracked location.

A Closing Note

I hope by the time you are reading the later pages of this appendix, you have the binary installed and a clean `claude --version` in a terminal window, and you are ready to head back to Chapter 4 and begin the actual setup work. The install is, in some ways, the hardest part — not because the commands are complex, but because installation is the kind of task where small environment-specific surprises can eat half an hour and make you wonder whether the whole endeavor is worth it. If one of those surprises has cost you time today, I am sorry for it, and I want to promise you that the cost does not recur. You install Claude Code once per machine. After today, you will not think about any of this again until the day you get a new laptop, at which point you will work through this appendix in forty minutes instead of the couple of hours it took the first time, because you will know what you are building toward.

The method does not start until the install is done. But when the install is done, the method starts almost immediately, and the install becomes invisible within days. Push through whatever small friction this appendix has asked of you. The rest of the book is where you will spend the years.

Appendix C: Dark Mode and Terminal Setup

This appendix covers the pieces of the physical-and-visual setup that the main text glides past, because they would have slowed the narrative down in Chapter 4. They are not optional. You are going to spend most of your working day inside these settings, across multiple terminal panes at once, and the wrong choices here will quietly tax you for hours every week without your being able to name what's doing it.

The appendix is in four parts. First, a short honest section on what the research actually says about screen-related eye strain, so the recommendations that follow are grounded in something better than folklore. Second, dark-mode and font setup for each platform. Third, the recommended terminals and fonts themselves, with opinions. Fourth, a longer discussion of curved monitors, which deserves its own section because the comfort gains are larger and more sustained than most developers expect and the topic is rarely addressed in productivity writing aimed at engineers.

Part 1: What We Actually Know About Eye Strain

Before the settings, a ten-minute literacy briefing on the physiology you are trying to take care of. If you skip this, the rest of the appendix will read as cargo-culting, and it is not.

The Condition Has a Name

The phenomenon most developers call *eye strain* is more precisely called **Digital Eye Strain** (DES) or **Computer Vision Syndrome** (CVS). The American Optometric Association defines it as "the group of eye and vision-related problems that result from prolonged computer, tablet, e-reader and cell phone use," with symptoms including eyestrain, headaches, blurred vision, dry eyes, and neck and shoulder pain. The AOA's patient guide is the cleanest consumer-facing summary I know of and is worth bookmarking; see **aoa.org/healthy-eyes/eye-and-vision-conditions/computer-vision-syndrome**.

Prevalence is not marginal. A 2018 review published in *BMJ Open Ophthalmology* ("Digital eye strain: prevalence, measurement and amelioration," Sheppard & Wolffsohn) reported that DES prevalence estimates run **fifty percent or higher** among regular computer users, with later population studies pushing the upper bound considerably higher in some cohorts. If you are an engineer spending six or more hours a day in front of a screen, the baseline probability that you are experiencing some version of this is high enough that you should assume you are and calibrate your setup accordingly.

The Three Mechanisms That Actually Drive It

Most of what is sold as an eye-strain remedy is calibrated to one specific mechanism and silent on the other two, which is why none of them individually solves the problem. There are three.

Reduced blink rate. Normal blink rate is around fifteen to twenty times per minute. When people are concentrating on a screen, that rate drops — often by more than half. Reduced blinking evaporates the tear film unevenly across the cornea and produces the gritty-dry feeling most engineers associate with a long coding afternoon. This is well documented; the Mayo Clinic's patient education on eye strain (mayoclinic.org — search "eyestrain") identifies it as one of the top contributors, and the primary research trail goes back to work by Tsubota and Nakamori in the early 1990s (*New England Journal of Medicine*, 1993) showing measurably decreased blink rates during visual display use.

Prolonged near-focus. Your ciliary muscle — the one that bends the lens of your eye to focus on close objects — is not designed to hold one focal distance for eight hours. Holding a single near-focus for that long produces sustained muscular strain and, over years, is associated with the development of computer-related myopia and the symptom cluster clinicians call **accommodative fatigue**. The American Academy of Ophthalmology has a good lay summary at **aao.org/eye-health/tips-prevention/computer-usage**, and the canonical mitigation is the *20-20-20 rule*: every twenty minutes, look at something twenty feet away for twenty seconds. It is not magic; it is simply giving the ciliary muscle a break.

Glare and luminance mismatch. This is the part the dark-mode-versus-light-mode debates get wrong on both sides. The problem is not that light mode is intrinsically bad or that dark mode is intrinsically good. The problem is *mismatch* — between your screen's luminance and the ambient luminance of the room around it. Reading bright-white text on a nearly-black background in a bright office will strain your eyes from pupil oscillation as your gaze moves between screen and environment. Reading dark text on a bright background in a dim room will strain them just as badly from the screen itself being a small sun. The Sheppard & Wolffsohn review covers the ambient-versus-screen luminance ergonomics in some detail, and Ioanna Mylona's *Spotlight on Digital Eye Strain* (Clinical Optometry, 2023) is a good complementary entry point. The practical takeaway across this literature is the one this appendix leans on: **match your screen's luminance roughly to the ambient light of the room you are in.**

What This Means for the Rest of the Appendix

The three mechanisms give you three interventions: blink more (or set a timer), take focal breaks (or set a timer), and control luminance mismatch. The settings in the rest of this appendix address the third one directly and make the first two easier to maintain. The 20-20-20 rule and a glass of water on your desk address the first two without any tool help at all.

A last note before settings. If you are wearing an older prescription, are over forty, or have noticed genuinely new symptoms in the last few months, none of this substitutes for an actual eye exam. Many developers carry an outdated prescription for two years longer than they should, attribute the resulting strain to their monitor, change monitors, and continue to be miserable. The AOA recommends annual comprehensive exams for adults who spend significant time on screens. Take the appointment.

Part 2: Dark-Mode Setup by Platform

The goal of this section is to walk you through setting your system, your terminal, and the important UI surfaces you use to a dark color scheme, and to do it quickly. The whole procedure on any platform is about ten minutes the first time.

macOS

System appearance. Open **System Settings → Appearance** and set the appearance to *Dark*. This flips the menu bar, the Finder, and any native Cocoa app that respects the system setting (most of them do). Leave *Highlight color* and *Accent color* to taste.

Built-in Terminal.app. Open Terminal, then **Terminal → Settings → Profiles**. Pick a dark profile — *Pro* is the classic; *Homebrew* is green-on-black if you like the old-school look. Click *Default* to make that profile the one new windows open with. Under *Text*, set the font to whichever monospace face you chose from Part 3; 13 to 15 points is a good starting range.

iTerm2. If you use iTerm2 (I recommend it and we'll talk about why in Part 3), the equivalent path is **iTerm2 → Settings → Profiles → Default → Colors**. iTerm2 ships with a deep color-preset menu; *Smoothed* and *Dark Background* are both solid starting points. If you want to match a specific published theme (*Solarized Dark*, *Dracula*, *Nord*, *Tokyo Night*, *Gruvbox*), iTerm2 reads `.itermcolors` files — search "iterm2-color-schemes" on GitHub for the community library of them.

Browser. Chrome, Firefox, Safari, Edge, and Arc all have a system-matching theme as the default. With the system set to Dark, the browser chrome will follow. Site-level dark mode is separate and varies site by site. A reasonable middle ground is the **Dark Reader** extension (darkreader.org), which dark-modes the body of any site; it is not perfect on every page, but for the long-form reading you do in Zone 2 it is a significant comfort win.

Windows 11

System appearance. Settings → Personalization → Colors, set *Choose your mode* to *Dark*. This flips the taskbar and Start menu and any modern app that respects the system setting.

Windows Terminal. Windows Terminal is the right choice on modern Windows (install from the Microsoft Store if it's not already installed — it ships by default on Windows 11). Open it, then **Settings → Color schemes**. The built-in *One Half Dark* and *Tango Dark* are both good starting points; the *Campbell* theme is the default and is serviceable if slightly too blue for long sessions. Under **Profiles → Defaults → Appearance**, set your color scheme and font size.

PowerShell specifically. PowerShell's default colors can clash with dark backgrounds in a way that makes warning and error text unreadable. If you see low-contrast dark-blue-on-black anywhere in your shell output, run `Set-PSReadLineOption -Colors @{ Parameter = 'Cyan'; Operator = 'Cyan' }` in your PowerShell profile (`$PROFILE`) to override. A more comprehensive fix is the **PSReadLine** module's built-in `PredictionViewStyle` settings paired with a proper dark color palette; see **github.com/PowerShell/PSReadLine** for the current documentation.

Manjaro / Linux

Desktop environment. Manjaro ships in several flavors (KDE Plasma, GNOME, Xfce). The path to dark mode differs by flavor.

- **KDE Plasma: System Settings → Appearance → Global Theme**, pick *Breeze Dark* (or any of the community dark themes from the KDE Store).
- **GNOME: Settings → Appearance**, select *Dark*. GTK applications that respect the system setting will follow.
- **Xfce: Settings → Appearance**, pick a dark theme such as *Adwaita-dark* or *Arc-Dark*.

Terminal. Manjaro users are terminal-opinionated. The main options:

- **Konsole** (KDE default): **Settings → Configure Konsole → Profiles**. Edit the default profile, set the color scheme to *Breeze* or *Solarized Dark*, and set the font under *Appearance*.
- **GNOME Terminal: Preferences → Profile → Colors**, uncheck *Use colors from system theme*, and pick a dark palette.
- **Alacritty:** Edit `~/.config/alacritty/alacritty.toml`. Alacritty has no built-in themes; you paste a color block into the config. The community maintains a large catalog at **github.com/alacritty/alacritty-theme**.
- **Kitty:** Edit `~/.config/kitty/kitty.conf`. Kitty has a built-in theme chooser — run `kitty +kitten themes` in any Kitty window and pick one from the interactive menu.
- **WezTerm:** Edit `~/.config/wezterm/wezterm.lua`. WezTerm ships with a large built-in theme library — set `config.color_scheme = 'Tokyo Night'` (or any of hundreds of others) and restart.

Browser on Linux. Same as macOS — modern browsers follow the system setting, and Dark Reader handles the long-tail of sites.

What to Check After You Set Dark Mode

Three quick visual checks before you move on:

1. Your editor's color scheme. VS Code, JetBrains IDEs, Vim/Neovim, Emacs — all of them have dark themes; pick one that matches the terminal palette roughly so you are not oscillating between warm-dark and cool-dark every time you switch windows. Consistency reduces pupil-response churn.
2. Your communication apps. Slack, Teams, Gmail, and most calendars have a dark-theme setting buried one or two menus deep. Find it; flip it. Zone 2 is going to be in your peripheral vision for hours.
3. Your Jira / GitHub tabs. Both respect the browser's dark preference; GitHub has three levels (light, dark dimmed, dark) under **Settings → Appearance** and you might prefer *Dark dimmed* for long sessions — the off-black reduces the contrast-shock when you flip tabs.

Part 3: Recommended Terminals and Fonts

Terminals

These are the terminals worth considering for the workflow in this book. The core requirement is simple: tabs, splits, readable text, good Unicode support, and enough configurability to lock in your preferences.

iTerm2 (macOS). This is what I use. It has everything Terminal.app has plus a decade of thoughtful extensions — split panes, broadcast input across panes, session restoration, a hotkey window, and a preferences surface that is genuinely deep without being baroque. Free. **iterm2.com**. Install via Homebrew with `brew install --cask iterm2`.

Ghostty (macOS / Linux). Mitchell Hashimoto's GPU-accelerated terminal, open-source and genuinely fast. If you find iTerm2 or Terminal.app's rendering sluggish on a 4K or 5K display, Ghostty is the answer. **ghostty.org**.

Alacritty (cross-platform). The original GPU-accelerated terminal. Minimal by design — config is a single TOML file, no tabs or splits (use a terminal multiplexer like tmux if you need them). If you want a terminal that gets out of the way, this is it.

Kitty (cross-platform). Similar to Alacritty but includes native tabs, splits, and an extension system (*kittens*) for doing terminal-y things like inline image viewing. Excellent for the multi-pane layout this book recommends.

WezTerm (cross-platform). Wez Furlong's Lua-configured terminal. More featureful than Alacritty, more configurable than Kitty, and one of the few terminals with genuinely first-class Windows support. If you work across macOS / Linux / Windows and want the same terminal everywhere, WezTerm is the single best answer I know.

Windows Terminal (Windows). The default modern Windows terminal. Free, tabbed, supports PowerShell / cmd / WSL / SSH sessions side by side, and has caught up substantially with the Unix world in the last three years. If you're on Windows, start here.

Warp (macOS, Linux, recent Windows). Reinvents the terminal around blocks and AI assistance. Divisive in the developer community — some engineers love it, others find the reinvention gets in the way of established muscle memory. Worth trying; not worth forcing. **warp.dev**.

Fonts

Fonts matter more than most developers realize. A good monospace at 14 points on a dark background reduces accommodative load and prevents the character-ambiguity moments (0 vs. O, 1 vs. l, rn vs. m) that produce micro-distractions all day.

Pick any one of these and move on.

- **JetBrains Mono.** The one I recommend by default. JetBrains designed it explicitly for long reading sessions of code. Ligatures optional. Free. **jetbrains.com/lp/mono**.
- **Fira Code.** Mozilla's classic. Popularized programming ligatures (=>, !=, <=) and still one of the best-balanced faces for code. Free. **github.com/tonsky/FiraCode**.
- **Hack.** Clean, unornamented, designed specifically for source code. No ligatures — which some engineers prefer. **sourcefoundry.org/hack**.
- **IBM Plex Mono.** IBM's corporate family, but their monospace is quietly one of the best. Wider characters, very comfortable at larger sizes. Free. **ibm.com/plex**.
- **Cascadia Code.** Microsoft's answer to Fira Code. Ligatures, clean shapes. Ships with Windows Terminal by default. **github.com/microsoft/cascadia-code**.
- **Iosevka.** Narrow, which lets you fit more columns on screen — useful for side-by-side diffs. Highly customizable. **typeof.net/Iosevka**.

On sizing. 14 points is the right starting point on a standard-density external display; 13 on a high-density (Retina, 4K) display where each "point" is rendered at higher subpixel density. Your honest rule is: if you lean in to read, increase the size. If you pull back to see more, consider a larger monitor rather than a smaller font. The temptation to shrink to fit is one of the cheapest ways to acquire eye strain in a week.

On ligatures. Some engineers love them (=> rendered as a single arrow glyph). Some find them actively misleading, because the rendered glyph is not the same as the character sequence it represents, which can confuse what's actually in the file. This is personal taste. If you like ligatures, JetBrains Mono, Fira Code, and Cascadia Code all support them; turn them on in your terminal's font settings. If you dislike them, Hack and Iosevka are ligature-free, and the ligature-enabled fonts have no-ligature variants.

Part 4: Curved Monitors

This section is longer than the others because the topic is genuinely underdiscussed in developer productivity writing, the effect on daily comfort is larger than I expected when I first switched, and the purchase is one most engineers only make every four or five years — so it's worth getting right.

The Argument in One Paragraph

A curved ultrawide monitor matches the natural focal arc of your eyes better than a flat panel at the same physical width. Your cornea is itself curved; a flat monitor requires you to constantly re-focus as your gaze sweeps left and right, because the edges of the screen are physically farther from your eyes than the center. A curved monitor keeps every part of the screen at roughly the same focal distance, which reduces the continuous accommodation-reaccommodation cycle your ciliary muscle performs over an eight-hour working day. That reduction is not subtle, and it is compounded over months.

What the Research and the Vendors Actually Say

The cleanest independent study I know of is a Samsung-funded curved-versus-flat monitor study conducted at Harvard Medical School's Schepens Eye Research Institute / Mass Eye and Ear, presented at SID's Display Week. The study tracked participants performing sustained visual-search tasks on a 34-inch curved display versus a flat panel and measured symptom reports on a standard visual-discomfort scale. Participants on the curved display were significantly less likely to report worsening eyestrain, focus difficulty, blurred vision, or tired eyes across the working sessions. Neck and shoulder discomfort was also lower on the curved panel, likely because the geometry encourages a more neutral head posture for scanning wide content.

I will note honestly that this study was Samsung-funded, which means the usual caution applies: vendor funding does not invalidate findings but it warrants a grain of salt. The effect direction has since been reproduced in smaller independent studies in ergonomics journals, and the mechanism — matching screen curvature to natural visual geometry — is physiologically straightforward enough that the effect is not surprising. But it is fair to say the peer-reviewed literature on curved panels is thinner than the marketing volume around them would suggest, and if you are the sort of engineer who demands rigorous randomized controlled trials before changing anything, the honest answer is that the evidence is *directionally positive but not overwhelming*.

The practical question, though, is not *is the effect large in a controlled study* but *does the effect reliably help working developers in their actual work*. My own answer, and the answer of essentially every developer I know who has switched, is yes — strongly, sustainedly, and in a way that is noticeable within the first week.

Radius: What the Numbers Mean

Curved monitors are specified by a **radius of curvature**, written as *1800R*, *1500R*, *1000R*, and so on. The number is the radius in millimeters of the circle the monitor's curve is a segment of. **Smaller number equals more aggressive curve.** A 1000R monitor is more curved than an 1800R monitor.

The canonical recommendation from Samsung, LG, and most of the ergonomics literature is that the ideal curvature approximates the radius of the human visual field at normal seating distance — which lands somewhere around **1000R to 1200R** for a single-user workstation. At that radius, a 34- to 49-inch ultrawide panel wraps far enough to match the natural sweep of your gaze but not so far that text on the edges appears visibly tilted.

Practical guidance:

- **1800R.** The gentlest modern curve. If you come from a flat panel and find aggressive curves disorienting, this is the conservative starting point. Most 27-inch and 32-inch curved monitors are 1800R.
- **1500R.** A reasonable middle ground, common on 34-inch ultrawides.
- **1000R–1200R.** The ergonomic sweet spot for 34- to 49-inch ultrawides at a typical desk viewing distance. This is what I use and what I'd recommend to most developers.
- **800R and lower.** Aggressive. Reserved for very large (49"+) panels viewed from close range, and for gamers who want peripheral immersion. Not necessary for coding.

Size, Resolution, and DPI

The second choice after radius is physical size and resolution. For coding, three configurations are worth knowing about.

34-inch ultrawide, 3440×1440. The most popular coding ultrawide. Wide enough for two full-width browser/editor columns side-by-side with room for a third narrow column. Pixel density is comparable to a 27-inch 1440p monitor, which means text rendering is good-but-not-Retina. Widely available; LG, Dell, Samsung, and several others all make competitive 34" 1440p ultrawides. Works beautifully for the seven-slot layout if you configure it generously.

38-inch ultrawide, 3840×1600. A step up in both width and pixel density. The sweet spot for engineers who want an ultrawide with genuinely sharp text. LG's 38WN95C and its successors are the reference point. The extra column width is real and noticeable, but the price jump over 34" is nontrivial.

Dual 4K flats (27" or 32"). The alternative to ultrawide. Two 4K panels side-by-side give you more total pixels than any single monitor and beautiful text rendering, but you have a bezel down the middle of your visual field. If you care most about text sharpness and don't mind the seam, this is an excellent configuration. The seven-slot layout splits naturally across the two monitors — three zones on the primary, seven slots on the combined display.

Super-ultrawide, 49-inch, 5120×1440. The two-27"-monitors-fused-into-one panel. Samsung and Dell both make strong 49" options. Enormous horizontal space, no bezel seam, but it's a lot of monitor — overpowering in small home offices, and you may find you only use the middle third of it most of the time. For the seven-slot layout this is the most luxurious choice, with room to spare.

For the seven-slot layout specifically, **34" or 38" ultrawide at 1000R–1500R curvature** is the configuration most developers I know have converged on. It is wide enough to fit the layout comfortably, curved enough to reduce focal strain, and sized such that text at a sensible font size remains readable from a normal seating distance without leaning in.

Distance, Height, and Posture

A curved monitor that is positioned wrong will still strain you. The geometry assumes you are seated at roughly the monitor's intended viewing distance — which, for a 1000R 34-inch panel, is about 80 to 100 centimeters (roughly arm's length plus ten centimeters). Too close and the curve wraps past your peripheral vision awkwardly; too far and the ergonomic benefit collapses back toward a flat panel.

The same AOA guidance that applies to flat panels applies here: top of the screen at eye level or slightly below, screen tilted slightly upward (10–20 degrees), elbows at roughly 90 degrees when typing, feet flat on the floor, monitor arm adjusted to avoid hunching. Ergotron LX is the defensible default for monitor arms; there are cheaper ones that are fine and more expensive ones that are overkill.

Anti-Glare and Matte Coatings

Most curved monitors ship with either a matte or a glossy finish. For a working developer in an office with overhead fluorescent or daylight from a window, **matte is the right choice**, full stop. A glossy curved panel will reflect the entire room back at you, and because the reflection itself is curved, the distortion will make the reflections harder to ignore than on a flat glossy panel. Matte coatings reduce the contrast of reflections enough that they fade into background noise.

One caveat: some aggressive matte coatings (sometimes marketed as "anti-glare") impose a visible graininess on white backgrounds — "the sparkle effect." This is mostly gone on modern panels, but if you are buying used or budget, check reviews for mentions of it.

A Short List of Monitors Worth Considering

Not endorsements, just a starting list of panels that have been reliably well-reviewed by engineers who work at them all day. Models get refreshed annually; verify the current generation before buying.

- **LG 38WN95C** and successors — 38" curved ultrawide, nano-IPS, reference-quality for coding.

- **Dell U3423WE** and successors — 34" curved ultrawide, excellent matte coating, Dell's color accuracy.
- **Samsung Odyssey G9** — 49" super-ultrawide, 1000R, the gaming heavyweight that also works well for developers with the space.
- **Samsung ViewFinity S95B** — 34" curved ultrawide with Thunderbolt, good for MacBook-centric setups.
- **LG UltraGear 34GP83A** — 34" curved gaming ultrawide that also happens to be a fine coding panel.
- **Alienware AW3423DWF** — 34" curved OLED ultrawide. OLED text rendering is divisive for coding (subpixel layout differs from LCD, which some engineers find affects long-session text clarity); if you're curious, try one in person before buying.

When *Not* to Switch

Three honest cases where a curved ultrawide is the wrong answer:

1. **You work from a laptop 50% of the time.** The benefit is in the sustained sitting; if half your week is on the road, invest the money in a better laptop and a good portable monitor arm instead.
2. **Your work demands color-critical accuracy** (photo editing, color grading, print prepress). Curved panels have caught up significantly, but flat pro-grade panels (Eizo ColorEdge, NEC PA, high-end Dell UltraSharp) still lead at the top of color fidelity.
3. **You don't have the desk depth.** A 34" ultrawide needs at least 80 cm of viewing distance to sit right. If your desk is 60 cm deep, the monitor will be too close and the curvature will feel wrong rather than helpful.

Closing

A lot of this appendix is opinion about tools, and the opinions will age. Fonts will be superseded, terminals will consolidate, monitors will get better. The underlying physiology will not. Reduce luminance mismatch, take focal breaks, keep your blink rate up, and match your screen's geometry to your visual geometry — and the specific brand choices mostly don't matter.

If you do one thing from this appendix, set dark mode across your system, terminal, IDE, and communication apps. If you do two, buy a good curved ultrawide the next time you are due for a monitor refresh. If you do three, go get an eye exam. Your long career depends on the eyes you are using to read this sentence, and the tools in your toolbox are only as useful as the organs you are asking to look at them.

Appendix D: Parallel Instructions for Ubuntu, Bash, WSL2, Fish

The main text assumes you are on macOS or Manjaro running zsh, or on Windows running PowerShell 7. Those are not the only environments this method runs in. This appendix is for the other common cases — Ubuntu users, Bash holdouts, Windows developers working through WSL2, and Fish users — and its purpose is to give you a parallel set of instructions that match the main text example-for-example, plus the handful of syntax quirks and gotchas that will trip you up if you try to translate on the fly.

If your setup matches one of the main-text defaults, skip this appendix. If it doesn't, find your section below and keep it open in a second window while you work through Chapters 4, 5, and 6. Every command in the main text has an equivalent here, labeled by the relevant section and the approximate main-text page where it first appears.

I have organized this appendix by *environment* rather than by *task*, because in my experience readers in unusual shells flip to an appendix once and stay in it, not back and forth. Each of the four sections below is self-contained. You should not have to reassemble the method from fragments.

One meta note before we start. The commands in this book are boring on purpose. They are the same shell builtins and standard utilities that have worked for decades, and the reason they show up in this book in their most conservative form is so that they keep working when the rest of your toolchain does not. If you are the kind of engineer who has a ninety-line `.zshrc` with a plugin manager, seventeen aliases, and a custom prompt — and many of you are, and I was for years — you already know how to translate. This appendix is written for the case where you are not that engineer, or where you are setting up a new machine and have not rebuilt your dotfiles yet.

Section 1: Ubuntu With zsh

Ubuntu is one of the most common Linux distributions in developer use, and zsh is the shell most main-text examples are written for. The good news is that the main-text commands work almost unchanged on Ubuntu; the differences are in package management, in a few default-path assumptions, and in a handful of permission and systemd specifics that do not come up on macOS.

Setting zsh as Your Login Shell

Ubuntu ships with Bash as the default login shell. If you want zsh — and for the main-text examples to work verbatim you do — install it and make it your login shell:

```
sudo apt update
sudo apt install -y zsh
chsh -s "$(which zsh)"
```

Log out and back in. The next terminal you open should greet you with zsh's first-run configuration wizard. Pick option 2 (the reasonable defaults) unless you already know exactly what you want. You can refine `~/.zshrc` later.

A note: on some cloud-hosted Ubuntu images, `chsh` is disabled or the password requirement fails quietly. If `chsh` appears to succeed but your shell does not change on next login, check `/etc/passwd` — your user line should end in `/usr/bin/zsh` rather than `/bin/bash`. If it still ends in bash, edit the file as root with `sudo vipw` and correct the trailing shell path. This is the only time in the whole appendix I will tell you to edit a system file directly.

Installing Core Command-Line Tools

The main text assumes you have `git`, `gh` (GitHub CLI), `jq`, and `ripgrep` (`rg`) available. On Ubuntu:

```
sudo apt update
sudo apt install -y git jq ripgrep curl build-essential
```

The GitHub CLI is reachable from Ubuntu's universe repositories via `sudo apt install gh`, but the packaged version typically lags GitHub's official releases by months. For a current `gh`, install from the official source:

```
curl -fsSL https://cli.github.com/packages/githubcli-archive-keyring.gpg |
sudo dd of=/usr/share/keyrings/githubcli-archive-keyring.gpg
sudo chmod go+r /usr/share/keyrings/githubcli-archive-keyring.gpg
echo "deb [arch=$(dpkg --print-architecture) signed-by=/usr/share/keyrings/
githubcli-archive-keyring.gpg] https://cli.github.com/packages stable main" |
sudo tee /etc/apt/sources.list.d/github-cli.list > /dev/null
sudo apt update
sudo apt install -y gh
```

After install, authenticate with `gh auth login` and pick the browser flow. The main text assumes `gh` is authenticated; if you skip this step, every gh command later in the book will fail silently or with unhelpful errors.

Creating the Workspace

The main text has you create `~/workspace` with `mkdir -p`. The command is identical on Ubuntu:

```
mkdir -p ~/workspace
cd ~/workspace
```

The home-folder shorthand ~ resolves to `/home/yourname` on Ubuntu rather than `/Users/yourname` on macOS. The main text uses ~ throughout specifically so the path works on both. If you are reading an example that hard-codes `/Users/...`, substitute `/home/...`. There should be very few such examples in the book, and they are mistakes if they appear.

Font and Terminal Setup

The main text recommends a 14-point terminal font. Ubuntu ships with GNOME Terminal by default, and you can set the font from Preferences → Profile → Text → Custom font. If you are running a window manager that uses a different terminal — Tilix, Alacritty, Kitty, WezTerm, Terminator — each has its own configuration mechanism, and I will not walk you through all of them. The rule is simple: large enough to read without leaning in. Every terminal emulator supports this; find the setting and set it.

Claude Code on Ubuntu

Appendix B covers Claude Code installation on the main-text platforms. For Ubuntu, the most reliable path at the time of writing is the official installer, which handles the dependency chain correctly regardless of your distribution specifics. Follow the instructions Anthropic publishes at the time you are reading this; I will not transcribe version-specific installer URLs here because they age poorly.

After install, the launch command is identical to the main text:

```
claude --dangerously-skip-permissions
```

The flag behaves identically on Ubuntu. The boundary it creates — "Claude can do whatever it wants inside `~/workspace` and nothing outside it" — relies on the same filesystem permission primitives on Ubuntu as on macOS. The main-text warnings about not navigating the Claude session outside the workspace directory before the boundary is set up apply without modification.

Git Global Config

The main text has you set your global git identity early. On Ubuntu the commands are identical:

```
git config --global user.name "Your Name"
git config --global user.email "your.email@example.com"
git config --global init.defaultBranch main
```

One Ubuntu-specific gotcha: if you have `git` installed from apt and you are also using a Python virtualenv that brought in its own `git` wrapper (some tools do this), the `git` that Claude invokes may not be the one you just configured. Check with `which -a git` and make sure `/usr/bin/git` is first in your `PATH`.

The Seven Environment Directories

The main text walks you through creating the seven Norse environments. On Ubuntu the commands are identical:

```
cd ~/workspace
for env in thor freya tyr loki odin heimdall baldr; do
  mkdir -p "$env"
done
```

The `for` loop syntax above works in both bash and zsh. If you have already switched to Fish, see Section 4 of this appendix.

`ENVIRONMENT_LEASES.md` and PID Liveness

The main text's lease file uses the PID liveness check described in Chapter 4. On Ubuntu, the check that the main text writes as `kill -0 $PID 2>/dev/null` works identically, because `kill -0` is the POSIX signal-sending idiom for "test whether the process exists without actually signaling it," and Linux honors it the same way macOS does. No change needed.

One small difference: on Ubuntu systems with aggressive PID recycling (short PID space, many short-lived processes), there is a theoretical risk that a PID in `ENVIRONMENT_LEASES.md` gets recycled between the moment your agent wrote it and the moment a later check reads it. In practice I have never seen this happen on a developer laptop, but if you are running this method on a shared multi-tenant server with thousands of process cycles a day, adding a start-time field alongside the PID would defeat the recycling risk; on a single-developer laptop we don't bother.

The `.claude/` Directory

No difference. The main text has you create `~/workspace/.claude/` and populate it with the configuration files. On Ubuntu, the directory behaves identically, the file permissions (`chmod 755` for directories, `chmod 644` for markdown files) are identical, and the `.claude/settings.json` format is identical across platforms.

systemd and Process Persistence

This is an Ubuntu-specific topic that doesn't appear in the main text. If you want your seven agents to survive a terminal closing — say, you SSH into a workstation from home and want to leave the agents running after you disconnect — the main text's answer is `tmux` or `screen`, and that answer works identically on Ubuntu. I recommend `tmux` for new users; the learning curve is not steep and the payoff is real.

What the main text does *not* cover, and what you might be tempted to try on Ubuntu because Ubuntu makes it accessible, is running your agents as systemd user services. Don't. For this method, you specifically do not want the agents to survive a reboot or a crash as if they were daemons. They are not daemons. They are interactive developer tools that should die cleanly when you close their terminal, so that the lease-file PID liveness check has reliable signal. A systemd-managed agent that respawns automatically will leave stale leases that the protocol cannot detect. Use tmux for attach/detach, not systemd.

Clipboard

One small quality-of-life difference on Ubuntu: the clipboard CLI is `xclip` (X11) or `wl-copy` (Wayland), not `pbcopy`. If any of your personal aliases in the main text use `pbcopy`, substitute:

```
# X11
sudo apt install -y xclip
alias pbcopy="xclip -selection clipboard"
alias pbpaste="xclip -selection clipboard -o"

# Wayland
sudo apt install -y wl-clipboard
alias pbcopy="wl-copy"
alias pbpaste="wl-paste"
```

If you do not know whether you are on X11 or Wayland, run `echo $XDG_SESSION_TYPE`. If you are on Ubuntu 22.04 or later on a standard desktop install, you are almost certainly on Wayland.

Section 2: Bash on Any Platform

The main text is written for zsh. Most of the commands work identically in bash, because both shells are POSIX-compliant in the ways that matter for developer scripting. A handful of things differ, and the differences will bite you if you do not know them.

When to Stay in Bash and When to Switch

If you are already comfortable in bash and do not want to switch to zsh, you can run the entire method in bash. The main text's commands will work. You will need to account for the handful of zsh-specific syntax choices the main text uses, which I will enumerate below.

If you are on a fresh machine and have no preference yet, I would gently suggest zsh, because the main text's examples work without translation and the macOS default is now zsh. But there is no engineering reason to abandon bash. The method does not care which shell you use.

Array Syntax

The single most common source of bash-vs-zsh confusion is array indexing. The main text contains a handful of examples that iterate arrays. In zsh, arrays are 1-indexed by default. In bash, arrays are 0-indexed. If you see a main-text example like this:

```
envs=(thor freya tyr loki odin heimdall baldr)
echo "The first environment is ${envs[1]}."
```

In zsh, that prints `thor`. In bash, it prints a blank, because `envs[1]` in bash is the *second* element. The correct bash version is:

```
envs=(thor freya tyr loki odin heimdall baldr)
echo "The first environment is ${envs[0]}."
```

The main text tries to avoid array-index examples for exactly this reason, preferring `for` loops that do not rely on indexing. But if you see one, know that 1-indexing is the zsh default and translate accordingly.

Globbing and Null Matches

When a glob pattern matches nothing, the two shells behave differently. In bash, an unmatched glob is passed through literally as a string:

```
for f in /no/such/path/*.md; do
  echo "$f"
done
# prints: /no/such/path/*.md
```

In zsh, the default behavior is to abort with a "no matches found" error. This matters in a handful of places in the main text where the script is iterating over files that may not exist yet. The safe cross-shell idiom is:

```
for f in /some/path/*.md; do
  [ -f "$f" ] || continue
  # ... do something with $f
done
```

If you are running the main-text examples in bash and see odd behavior on first runs of a script (before the files it expects exist), this is usually why. Use the guarded idiom above.

`read` and Terminals

Both shells have `read`, but the flag surface differs. The main text's `read -p "prompt: " varname` works identically in both. The flag `read -e` (enable readline) is bash-specific; zsh's equivalent is `vared`. This matters in Appendix A's prompt-template script, if you are editing it to run in bash; replace `vared` with `read -e` in a bash environment.

Parameter Expansion

A zsh feature the main text leans on occasionally is the `${(j: :)array}` expansion for joining arrays. Bash has no direct equivalent. The portable alternative is `IFS` manipulation:

```
arr=(a b c d)
joined=$(IFS=' '; echo "${arr[*]}")
```

If you see a main-text example using the `(j: :)` flag, the equivalent in bash is the IFS approach above.

`.bashrc` vs. `.zshrc`

The main text occasionally asks you to add an alias or an environment variable to your shell's startup file. For zsh, that file is `~/.zshrc`. For bash, it is `~/.bashrc` on most Linux distributions and `~/.bash_profile` on macOS (with some caveats about login vs. interactive shells that have filled bookshelves and that I will not reproduce here). A safe default in bash is to put your customizations in `~/.bashrc` and ensure `~/.bash_profile` sources it:

```
# ~/.bash_profile
[ -r ~/.bashrc ] && . ~/.bashrc
```

After adding an alias or export, reload the file with `source ~/.bashrc` or close and reopen the terminal. The zsh `source ~/.zshrc` in the main text translates to exactly this.

`source` vs `.`

Both work in both shells. The main text uses `source` because it is more readable; `.` is the POSIX-portable form. If you are scripting for strict POSIX (say, for `/bin/sh` portability), prefer `.`. For the main-text method, use whichever you prefer; it makes no operational difference.

Prompt String and Terminal Title

The main text does not configure the shell prompt beyond the defaults. If your bash prompt is plain and you want to know which environment directory you are in at a glance — which is genuinely useful when you have seven terminals open — add this to your `~/.bashrc`:

```
PS1='\[\e[1;34m\]\w\[\e[0m\]\n\$ '
```

That prints the full working directory in bold blue on its own line, with the prompt beneath. The directory name alone tells you which Norse environment you are in, which satisfies the main text's goal of always knowing which slot you are looking at.

Editor Integration

No difference. `$EDITOR` works identically in bash and zsh. Set it in `~/.bashrc` as `export EDITOR=vim` (or whatever you use), and every main-text example that invokes `$EDITOR` works.

Section 3: Windows via WSL2 With Ubuntu and zsh

This is the setup I recommend to Windows-based readers of this book. Native Windows tooling for this method exists — PowerShell 7 is capable, and the main text covers it — but the parallel-agent workflow specifically is smoother inside a Linux environment, and WSL2 is the path of least resistance to that.

If you are a Windows developer who has never touched WSL2, this section will walk you through the full path from zero to a working seven-environment workspace inside an Ubuntu-in-WSL2 shell. If you already have WSL2 running, skip to the Ubuntu Section 1 above; the parallel instructions there apply once you are inside your WSL2 shell.

Installing WSL2

WSL2 is built into Windows 11 and recent Windows 10 updates. Open PowerShell 7 as Administrator and run:

```
wsl --install -d Ubuntu
```

This downloads and installs the Windows Subsystem for Linux (WSL2 variant) along with an Ubuntu distribution. The install takes several minutes. When it completes, Windows will reboot; after the reboot, a setup dialog will appear asking you to create a Linux username and password.

Pick a username. Pick a password you will remember. These credentials are separate from your Windows credentials. Ubuntu inside WSL2 is, for most practical purposes, a separate machine; the username you pick here becomes the owner of your Linux home directory, and the password is what `sudo` will ask for.

If the install fails — which happens on older Windows builds — update Windows fully (Settings → Windows Update → Check for updates) and try again. If it still fails, the manual install path in Microsoft's documentation is the next step; I will not reproduce it here because the exact commands change with Windows updates.

Confirming WSL2 Is the Default

After install, in PowerShell 7:

```
wsl --list --verbose
```

You should see Ubuntu listed with `VERSION 2`. If it says `VERSION 1`, set it to 2:

```
wsl --set-version Ubuntu 2
```

WSL1 will technically run the method, but performance on filesystem operations is noticeably worse — and the method is filesystem-heavy — so stay on WSL2.

Entering Ubuntu

From PowerShell 7, PowerShell ISE, Windows Terminal, or the Start menu's Ubuntu entry:

```
wsl
```

Or just click the Ubuntu icon. Either way, you land in a bash prompt inside Ubuntu. From here, the Ubuntu Section 1 of this appendix applies: install zsh, switch shells, install the tools, create the workspace, and proceed.

Where Your Files Actually Live

This is the single most important WSL2-specific thing to understand, and getting it wrong is the source of 90% of the "it's slow" complaints WSL2 receives.

WSL2 has *two* filesystems that are accessible from the Linux shell. One is the Linux filesystem itself, which lives inside a virtual disk image Windows manages for you and which is accessed by the usual Linux paths (`~`, `/home/yourname`, `/etc`). The other is your Windows filesystem, which is mounted inside Linux at `/mnt/c/` (for the C drive), `/mnt/d/` (for the D drive), and so on.

These two filesystems behave very differently.

Operations inside the Linux filesystem are fast. They are as fast as they would be on native Linux. `git clone`, `npm install`, `make`, `rg` — all of the common developer operations — run at native Linux speed.

Operations that cross the boundary — reading Windows files from the Linux shell — are *much slower*. This is not a configuration mistake. It is a consequence of the 9P protocol WSL uses to bridge the two filesystems, and no amount of tuning will make it fast.

Put your workspace in the Linux filesystem. Specifically, put `~/workspace` at `/home/yourname/workspace`, which is `~/workspace` in your WSL2 shell. Do not put it at `/mnt/c/Users/yourname/workspace`. You will regret it within a week. The seven-environment method involves a lot of parallel filesystem access, and crossing the WSL boundary seven times per second will pin one of your CPU cores and grind the whole workflow.

This is worth saying twice: **do not put your workspace anywhere under `/mnt/c/`**. Put it in your Linux home directory.

Editing From Windows Tools

You can, of course, use Windows-side editors — VS Code, Cursor, JetBrains IDEs — to edit files that live inside your WSL2 Linux filesystem. The standard way to do this with VS Code is the `code` command from inside WSL2:

```
cd ~/workspace
code .
```

VS Code opens on the Windows side but connects to a server running inside WSL2, so all file operations happen on the Linux side at Linux speed. Your terminal inside VS Code is a Linux shell. Your git operations are Linux git operations. This is the correct way to use Windows-side editors with WSL2.

Do not open `\\wsl$\Ubuntu\home\yourname\workspace` from a Windows-side editor that has no WSL integration. That path works, technically, but it routes every file read through the 9P protocol, which is the slow path. Use the WSL-integrated mode of whatever editor you prefer.

Clipboard Integration

WSL2 has functional, though not perfect, clipboard integration with Windows. The commands are:

```
alias pbcopy='clip.exe'
alias pbpaste='powershell.exe -NoLogo -NoProfile -Command "Get-Clipboard"'
```

`clip.exe` is a Windows executable that WSL2 can invoke directly. Piping text into it puts the text on the Windows clipboard. The `pbpaste` analog shells out to PowerShell because there is no direct WSL equivalent of the Windows `Get-Clipboard` command. Both aliases work in practice.

One quirk: `clip.exe` adds a trailing newline. If you are piping text that must not end in a newline (rare, but it comes up), the fix is `printf "%s" "$text" | clip.exe`.

Networking, Ports, and Localhost

The seven-agent method does not require any specific networking configuration, but your agents will frequently start dev servers, and the question of how to reach those dev servers from a Windows browser confuses a lot of WSL2 newcomers.

The short answer: if your dev server is listening on `localhost:3000` inside WSL2, you can reach it at `http://localhost:3000` in a Windows browser. WSL2 and Windows share localhost transparently for TCP traffic. You do not need to expose ports, set up port forwarding, or configure the Windows firewall for the standard case.

The exception is when the dev server binds specifically to `127.0.0.1` rather than `0.0.0.0`. Most dev servers are fine either way; a few stubborn ones are not. If your Windows browser cannot reach a server you know is running in WSL2, check the bind address of the server and set it to `0.0.0.0` if it was set to `127.0.0.1`. That usually resolves it.

File Permissions

WSL2's handling of file permissions on Linux-native files is the same as Linux. However, if you ever copy files from `/mnt/c/` into your Linux home directory, the permissions will often come through as `777`, which makes git angry on subsequent operations. The fix is `chmod` the files back to `644` for regular files and `755` for directories. Better: avoid copying across the filesystem boundary in the first place. Clone your repos directly into the Linux filesystem using git.

Windows Terminal and the Seven-Slot Layout

Windows Terminal (the one from the Microsoft Store, not the legacy console host) supports multiple tabs and split panes, which is the Windows-side primitive for arranging the seven-slot layout. Each pane can run a separate WSL2 session. The keyboard shortcuts for splitting (`Alt+Shift+D` for an automatic split in whichever direction has more space, `Alt+Shift++` for an explicit vertical split, `Alt+Shift+-` for horizontal) and navigating (`Alt+arrow-keys`) match the Chapter 5 mental model closely.

For the full screen layout from Chapter 5 — two rows, seven slots, three zones — you will combine Windows Terminal panes with a tiling window manager, or you will arrange Windows Terminal windows manually. The main text covers the macOS window-manager path (Moom, Rectangle, Amethyst). For Windows, the equivalents are FancyZones (from Microsoft PowerToys), Komorebi, or glazewm. FancyZones is the least opinionated and the easiest to learn; I recommend starting there.

WSL2 Gotchas

A few things that bite WSL2 users and that the main text cannot anticipate:

- **Memory**. WSL2 can consume a lot of RAM over time, especially if you run seven long-lived agents. Create `%UserProfile%\.wslconfig` with a memory cap (e.g. `memory=8GB`) if you notice Windows becoming unresponsive.
- **Time drift**. After the Windows host sleeps, the WSL2 clock can drift. If you see TLS errors on git operations after waking from sleep, run `sudo hwclock -s` inside WSL2 to resync.
- **Systemd**. WSL2 now supports systemd, but it is opt-in via `/etc/wsl.conf`. You almost certainly do not need it for this method. If you enable it for other reasons, see the note in Section 1 about not using it to manage the agents.

- **Automatic shutdown**. WSL2 shuts its VM down after a period of inactivity. If you are relying on long-running agents with tmux, a short period where all shells are closed can result in the VM exiting. Counter-intuitive but true: at least one shell must remain open to keep the VM alive.

Section 4: Fish Shell Notes and Gotchas

Fish is a thoughtful, user-friendly shell with syntax choices that differ from both bash and zsh in ways that will break main-text examples if you paste them in uncritically. I want to be honest: for this book specifically, fish is the hardest shell to translate into, because fish's scripting syntax was intentionally designed to be different from POSIX, and "intentionally different" is exactly the condition under which a parallel-instructions appendix earns its keep.

If you are a fish user, you can absolutely run the method. You will simply need to translate a modest number of examples as you go. This section enumerates the translations.

Setting Fish as Your Login Shell

```
sudo apt install -y fish                # Ubuntu / Debian
brew install fish                       # macOS
sudo pacman -S fish                     # Manjaro
chsh -s "$(which fish)"
```

Log out, log back in. The next shell is fish.

Variable Assignment

The largest single syntax difference. In bash and zsh:

```
NAME="Blake"
echo "Hello, $NAME"
```

In fish:

```
set NAME "Blake"
echo "Hello, $NAME"
```

The `=` sign in fish is not used for variable assignment. It is only used for numeric comparisons. A main-text line like `FOO=bar` will fail in fish with a syntax error. Substitute `set FOO bar`.

For exported environment variables, the fish equivalent of `export FOO=bar` is:

```
set -x FOO bar
```

To make it persist across sessions:

```
set -Ux FOO bar
```

`for` Loops

Fish loops are close to bash/zsh but the opening is slightly different:

```
for env in thor freya tyr loki odin heimdall baldr
  mkdir -p $env
end
```

Note: `end` instead of `done`, and no `do` keyword. Also no trailing semicolons are required on any of the lines.

Command Substitution

In bash/zsh:

```
latest_tag=$(git describe --tags --abbrev=0)
```

In fish:

```
set latest_tag (git describe --tags --abbrev=0)
```

Fish uses parentheses for command substitution instead of `$(...)`. The main-text examples that assign command output to variables all need this translation.

`&&` and `||`

Fish recently added `&&` and `||` support; older versions required `; and` and `; or`. If you are on fish 3.0 or later (which you almost certainly are, in 2026), `&&` and `||` work the same as in bash. You do not need to translate them. This was a major pain point in fish 2.x and is resolved.

Conditionals

The `if` statement differs:

```
if test -f ENVIRONMENT_LEASES.md
  echo "Lease file exists"
else
  echo "Lease file missing"
end
```

Note `test` is used explicitly instead of square-bracket syntax, and `end` closes the block. Single brackets (`[ ... ]`) also work but `test` is the idiomatic form in fish.

Startup Configuration

Fish's equivalent of `~/.zshrc` or `~/.bashrc` is `~/.config/fish/config.fish`. Any aliases or environment variables from the main text go there:

```
set -x EDITOR vim
alias pbcopy="xclip -selection clipboard"
```

Note the quoted alias target; fish's alias syntax is slightly less forgiving than bash's about bare command strings with flags.

Globbing

Fish's default globbing is closer to zsh's than bash's — an unmatched glob errors by default rather than passing through the pattern literally. The guarded-iteration idiom from Section 2 still applies:

```
for f in /some/path/*.md
  test -f $f; or continue
  # ... do something with $f
end
```

The Lease File PID Check

The main text's PID-liveness idiom `kill -0 $PID 2>/dev/null` works in fish with a small syntax adjustment:

```
if kill -0 $PID 2>/dev/null
  echo "Process $PID is alive"
else
  echo "Process $PID is dead"
end
```

The `2>/dev/null` is fish-compatible; the redirection syntax is shared with bash.

Arrays

Fish arrays (which fish calls "lists") are 1-indexed by default, like zsh, not 0-indexed like bash. Most main-text array examples transfer cleanly to fish without changing the index values. Array slicing syntax differs (`$arr[1..3]` in fish instead of `${arr[@]:0:3}` in bash), but the main text uses this sparingly.

`.` and `source`

Modern fish supports both `source` and the POSIX `.` alias, but `source` is the documented form and the one fish's own scripts use. Any main-text example using `. ~/.zshrc` should be translated to `source ~/.config/fish/config.fish` for fish — the form is clearer and ages better across fish versions.

Process Substitution

Fish does not have the `<(command)` process-substitution syntax from bash/zsh. The main text uses it rarely, but if you encounter an example like:

```
diff <(cmd1) <(cmd2)
```

The fish translation is to use `psub`:

```
diff (cmd1 | psub) (cmd2 | psub)
```

This is one of the few cases where fish is more awkward than bash/zsh, and I would not try to generalize beyond this example.

Prompt Configuration

Fish ships with a sensible default prompt. If you want the chapter-5-style prompt that shows the current directory in color on its own line, run:

```
fish_config
```

This opens a browser-based prompt configuration tool. Pick or customize a prompt that shows the full path (`pwd`) rather than just the basename.

Auto-Suggestions and History Search

Fish has excellent built-in auto-suggestions and history search. These are not specific to the method but worth calling out because they are the reason some developers prefer fish, and they work beautifully alongside the seven-agent workflow. When you are rotating between terminals and typing nearly the same command into environment after environment, fish's up-arrow history search is a genuine quality-of-life improvement.

When Fish Gets Hard

There are a small number of situations in the book where fish genuinely does not fit and where I would recommend temporarily dropping into bash for the specific task:

- Running shell scripts that live on the internet (install scripts, GitHub Actions local-run tools, etc.) — most are written for bash/zsh and will not work in fish. Run them with `bash scriptname.sh` rather than translating them.

- Sourcing environment files from third-party tools (nvm, pyenv, conda) — these are almost always bash/zsh scripts. Fish-compatible versions exist for many of them; search for `{toolname}.fish` before assuming you need to drop out to bash.
- Writing automation for your workspace that other people on your team will also run — unless your whole team uses fish, write these in bash for portability and run them with `bash script.sh` from your fish shell.

Fish is, for interactive work, excellent. It is, for portable automation, a poor choice. The main text's scripts should be written in bash if you intend to share them with teammates; your interactive usage can remain in fish.

A General Note on Translation

Across all four of these parallel tracks, the meta-principle is the same: the *method* is shell-agnostic, but the *examples* in the main text are not. When in doubt, the method is:

- Seven environment directories, under `~/workspace`.
- A `CLAUDE.md` file with permanent memory.
- An `ENVIRONMENT_LEASES.md` file with PID-checked leases.
- A `NOTES.md` per environment for wind-down.
- A terminal-driven agent, launched with its skip-permissions flag, scoped to a single environment.

If you can reproduce those five things in your shell of choice, the main text's concepts work unchanged. The things this appendix is for are the small plumbing details — how you set an environment variable, how you iterate a list, how you check whether a process is alive, how you arrange windows on your screen — that differ between shells and platforms and that would clutter the main text if listed there.

Keep this appendix open while you work through Chapters 4 through 6. Close it afterward; by Chapter 7, you will have internalized the translations, and the method itself will carry you from there.

Appendix E: Workspace File Reference

This appendix is a reference, not a chapter. It's meant to be consulted, skimmed, and jumped around in. Each section describes a specific file or directory in the workspace — what it holds, who writes to it, what you can safely edit by hand, and what you should leave to the agents. Most sections include an annotated example, so you can compare what you see on your own machine to a known-good version.

Use this appendix when:

- Something in your workspace looks wrong and you want to check it against a reference.
- You want to hand-edit a file and aren't sure what's safe to touch.
- You're about to migrate the workspace to a new machine and you want to know which files matter.
- You're teaching a teammate the method and you need a compact description of the filesystem layout.

The conventions here mirror the setup from Chapter 4. If your workspace was built following that chapter, the paths and filenames will match one-for-one.

The Workspace at a Glance

The whole method lives inside a single directory tree rooted at `~/workspace/`. Nothing Claude does escapes this tree, with the two narrow exceptions the boundary rule names — files you explicitly link in a prompt, and anything under `/tmp`. Inside the tree, the agent is free to move; outside, it refuses. Here is the canonical layout, with every file and directory you should expect to see:

```
~/workspace/
├── CLAUDE.md                   ← permanent memory (human-readable, version-
controlled)
├── ENVIRONMENT_LEASES.md       ← live coordination file (agent-written)
├── .claude/                    ← Claude Code configuration
│   ├── settings.json           ← allow/deny rules, permission config
│   └── (other tool-generated files)
├── repos/                      ← canonical clean clones of each repo
│   ├── service-a/
│   ├── service-b/
│   └── ...
├── thor/                       ← ephemeral working environment 1
├── freya/                      ← ephemeral working environment 2
├── tyr/                        ← ephemeral working environment 3
├── loki/                       ← ephemeral working environment 4
├── odin/                       ← ephemeral working environment 5
├── heimdall/                   ← ephemeral working environment 6
└── baldr/                      ← ephemeral working environment 7
```

Every file and directory in this tree has one of three roles, and knowing the role is the key to knowing what's safe to edit:

1. **Human-authored, agent-read.** You write it; the agent reads it and obeys. `CLAUDE.md` is the flagship example. Edit these directly whenever you want, or ask the agent to edit them for you.
2. **Agent-authored, agent-read.** The agent maintains it as part of the running system. `ENVIRONMENT_LEASES.md` is the flagship example. You can read it anytime; hand-editing it is discouraged because the agents expect a specific format.
3. **Ephemeral working state.** Anything inside the Norse environment directories falls here. Treated as disposable; the environment gets reset between tickets.

The three roles correspond, loosely, to the book's color conventions — yellow for permanent memory, green for persistent coordination, purple for ephemeral work. Keep that mapping in your head as you move through this appendix.

`CLAUDE.md`

`CLAUDE.md` is the file Claude Code reads automatically at the start of every session launched from inside `~/workspace/`. It is your permanent memory for the whole workspace — the letter you leave for every future instance of the agent that walks into the room.

What It Contains

A well-maintained workspace `CLAUDE.md` has roughly these sections, in this order. The exact headings vary; the content categories are what matters.

1. **Identity and scope.** One or two sentences naming the workspace and the kind of work done in it.
2. **The boundary rule.** The explicit statement that the agent may not read or write outside `~/workspace/`, with its two exceptions (files the user links explicitly, and `/tmp`).
3. **The deny list pointer.** A line telling the agent where to find the deny rules (normally `.claude/settings.json`). Keeping the actual rules in the settings file rather than in prose means the agent enforces them structurally, not by remembering to check the prose.
4. **The lease protocol.** The full seven-step procedure for acquiring and releasing an environment, as laid down in Chapter 4, Step 7.
5. **The work-on-a-ticket protocol.** The full eight-step procedure for starting work on a ticket, as laid down in Chapter 4, Step 8, with your company's branch naming convention substituted in.
6. **House rules.** Project-specific rules you've accumulated over time — coding conventions, testing preferences, commit-message style, things that have bitten you twice and are now written down.
7. **Per-repo pointers, if applicable.** Occasional "for repo X, see repos/X/CLAUDE.md" breadcrumbs so the agent knows to descend into repo-specific memory when it's working in that repo.

Annotated Example

Here is a minimal but complete `CLAUDE.md`, as it might look after Step 1 of Chapter 4 is finished and a few weeks of use have refined it:

```
# Workspace CLAUDE.md

This is the permanent memory for my software-engineering workspace at
~/workspace/. Any Claude session launched from inside this directory reads
this file at startup and is bound by the rules in it.

## Filesystem boundary

You may not read, modify, or create files outside this workspace, with two
exceptions:
  1. Files the user explicitly links in a prompt.
  2. Anything inside /tmp or its subdirectories.

## Deny list

Destructive commands are blocked in .claude/settings.json. Consult that
file for the canonical list. Do not attempt to bypass it.

## Lease protocol

(Full protocol from Chapter 4, Step 7. Reproduced verbatim.)

## Work-on-a-ticket protocol

When I say "Work on ABC-1234":
  1. Acquire a lease per the lease protocol.
  2. jira issue view ABC-1234; summarize the ticket back in 2-3 sentences
     and wait for my confirmation.
  3. Identify the correct repo in the leased environment.
  4. Create a branch named feature/ABC-1234-short-kebab-description.
  5. Make an initial empty commit: "ABC-1234: begin work". Push.
  6. gh pr create --draft --title "ABC-1234: <ticket title>" with body
     linking the Jira ticket and describing the work you understand.
  7. Do the work. Commit logical chunks. Push after each commit.
  8. When you believe the work is complete, tell me. Do not promote the
     PR out of draft yourself.

## House rules

- Always run the test suite before declaring work complete.
- Never use --no-verify on commits. If a hook fails, fix the root cause.
- Commit messages follow Conventional Commits (feat:, fix:, chore:, ...).
- Prefer editing existing files over creating new ones.
- Don't write code comments that just restate what the code does.
- When in doubt about a naming convention, grep existing code first.
```

What You Can Safely Edit

Almost everything. `CLAUDE.md` is yours. Edit it directly in a text editor, or ask the agent to edit it for you. The only rules:

- Keep the boundary rule and the deny list pointer intact. Weakening either is how people end up with agents touching files they shouldn't.
- Don't let the file grow past about 400 lines. If it's getting longer than that, split repo-specific rules into per-repo `CLAUDE.md` files inside each repo under `repos/`, and have the workspace file point at them.
- Re-read the whole file occasionally (see the refinement cadence in Chapter 8). Prune rules that aren't earning their keep.

Who Writes to It

You do, primarily. The agent will append to it when you ask it to ("please add this rule to permanent memory"), and it's appropriate to trust the agent to make that edit — `CLAUDE.md` is in version control if the workspace is a git repo, and edits are reviewable.

The `.claude/` Directory

The `.claude/` directory sits next to `CLAUDE.md` at the workspace root. It holds Claude Code's own configuration — the machine-readable counterpart to the human-readable `CLAUDE.md`. Hidden by default because of the leading dot; use `ls -la` on macOS/Linux or `Get-ChildItem -Force` in PowerShell to see it.

What's in It

The shape of this directory evolves with Claude Code's version, but the file you care about is reliably present:

```
~/workspace/.claude/
└── settings.json        ← permission and tool configuration
```

Newer versions of Claude Code may add subdirectories for caches, session history, or hooks. Those are agent-managed; leave them alone unless you have a specific reason.

`settings.json`

This is where the deny list actually lives — not in `CLAUDE.md` prose, but in a structured permission config that Claude Code enforces before a command runs. A typical workspace `settings.json` looks like this:

```json
{
  "permissions": {
    "deny": [
      "Bash(git push --force*)",
      "Bash(git reset --hard*)",
      "Bash(cdk deploy*)",
      "Bash(aws dynamodb *)",
      "Bash(aws s3 rm*)",
      "Bash(aws s3 mv*)",
      "Bash(aws s3 cp*)",
      "Bash(aws lambda delete*)",
      "Bash(aws lambda update-function-configuration*)",
      "Bash(aws lambda create*)",
      "Bash(aws cloudformation delete*)",
      "Bash(aws iam *)",
      "Bash(aws cognito*)",
      "Bash(aws apigateway*)"
    ]
  }
}
```

The `Bash(...)` wrapper tells Claude Code the rule applies to shell commands. The glob patterns inside each string match families of dangerous commands, not single spellings. The reasoning for each pattern is covered in Chapter 4's deny-list detour.

What You Can Safely Edit

- **`permissions.deny`**: add, remove, or adjust entries as your stack changes. New Kubernetes cluster? Add `kubectl delete*`. Stopped using AWS IAM directly? Remove the IAM line. Edits take effect at the start of the next Claude Code session from this workspace.
- **`permissions.allow`** (if present): a list of commands explicitly allowed without per-run approval. Useful for safe, frequently-used tools (see Appendix A for recommended entries).

What to Leave Alone

- Any subdirectory named with a leading underscore, a dot, or a version-ish name (`_cache`, `.history`, `v2`). Those are agent-internal.

- Hook configuration files, if you haven't deliberately added them. Hooks are a powerful feature but out of scope for the base method.

`ENVIRONMENT_LEASES.md`

This is the coordination file that lets multiple Claude instances share the seven Norse environments without stepping on each other. It is a live document, constantly being edited by the agents themselves as they acquire and release environments.

Format

The file is a simple markdown table. A freshly created one (end of Step 6 in Chapter 4) looks like this:

```
# Environment Leases

| Environment | PID  | Leased At            | Ticket     |
|-------------|------|----------------------|------------|
| thor        | -    |                      |            |
| freya       | -    |                      |            |
| tyr         | -    |                      |            |
| loki        | -    |                      |            |
| odin        | -    |                      |            |
| heimdall    | -    |                      |            |
| baldr       | -    |                      |            |
```

A live one, mid-workday, with four environments in use and three free, looks like this:

```
# Environment Leases

| Environment | PID   | Leased At            | Ticket     |
|-------------|-------|----------------------|------------|
| thor        | 48213 | 2026-04-24 09:04:12 | ABC-1847   |
| freya       | 48741 | 2026-04-24 09:11:08 | ABC-1852   |
| tyr         | -     |                      |            |
| loki        | 48812 | 2026-04-24 09:18:44 | ABC-1853   |
| odin        | -     |                      |            |
| heimdall    | 49032 | 2026-04-24 09:26:01 | ABC-1861   |
| baldr       | -     |                      |            |
```

The Three States an Entry Can Be In

- **Free.** PID column shows `-`. No agent holds this environment. Available for lease.
- **Held.** PID column shows a number that corresponds to a live process. The environment is in active use; another agent must pick a different one.
- **Stale.** PID column shows a number, but that process is no longer running (the session crashed, the machine rebooted, someone closed a terminal without cleanly releasing). The lease protocol treats this as an environment that is *reclaimable* — the next agent that tries to lease will detect the dead PID, evaluate any uncommitted work left behind, and ask you whether to continue it or discard it.

What You Can Safely Edit

Almost nothing, in normal operation. The agents own this file. But two hand-edits are legitimate and occasionally necessary:

- **Forcibly free a stuck entry.** If you are certain an environment is free — you killed every Claude process, restarted your machine, and the PID definitely isn't live — you can replace the PID with `-` and clear the other columns. The next agent will take it fresh.
- **Add the `Ticket` column annotation.** Some workspaces include a `Ticket` column (as shown above); some don't. If yours doesn't and you want it, add the column and the agents will start populating it once you ask the work-on-a-ticket protocol to include a ticket-recording step.

What to Leave Alone

- Do not delete rows. The seven environments must always be present as rows, even when free. The agents look up environments by name.
- Do not reorder rows. The canonical Norse order (thor, freya, tyr, loki, odin, heimdall, baldr) is the order used throughout the book and the muscle-memory you've built in Chapter 5. Reordering it here will not break the mechanism, but it will make a future you briefly confused, which is the kind of small friction the method is specifically designed to eliminate.

The `repos/` Folder

`repos/` is the persistent home of clean clones of every repository you work in. It is not where work happens. It is where work *starts from.*

Structure

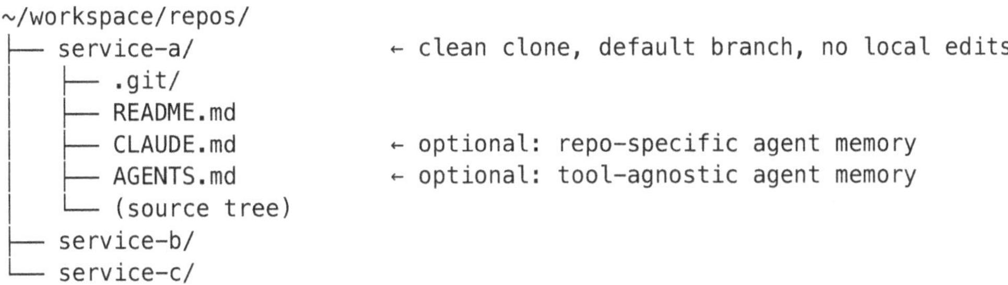

```
~/workspace/repos/
├── service-a/              ← clean clone, default branch, no local edits
│   ├── .git/
│   ├── README.md
│   ├── CLAUDE.md           ← optional: repo-specific agent memory
│   ├── AGENTS.md           ← optional: tool-agnostic agent memory
│   └── (source tree)
├── service-b/
└── service-c/
```

The Rule

Repos in `repos/` stay on their default branch (normally `main` or `master`), with no local edits, no stashed changes, no half-finished work. When you start a ticket, the lease protocol copies the relevant repo from here into the leased environment (for example, `~/workspace/thor/service-a/`), and the messy work happens there. If a working environment gets hopelessly tangled, you can reset by wiping the environment and copying a fresh clean tree from `repos/`.

What You Can Safely Edit

- **Pull updates** (`git pull` on the default branch) whenever you like. Doing so keeps the source-of-truth current.
- **Clone new repos** into `repos/` as your work requires. No ceremony needed.
- **Remove repos** you no longer work in. They're just clones; there's no state loss.

What to Leave Alone

- Don't commit work-in-progress inside `repos/`. That's what the Norse environments are for. If you ever find yourself about to `git commit` inside `~/workspace/repos/whatever/`, stop and move the work into a leased environment first.

The Norse Environment Directories

The seven directories named thor, freya, tyr, loki, odin, heimdall, and baldr are the ephemeral working environments. Each one hosts the active work of a single ticket at a time.

Expected Contents During a Ticket

While a ticket is in flight in, say, `thor`, the directory looks roughly like this:

```
~/workspace/thor/
├── NOTES.md                ← wind-down notes for the next session
└── service-a/              ← copy of the repo, checked out to the ticket's
branch
    ├── .git/
    ├── (source tree with in-progress edits)
    └── (any scratch files the ticket required)
```

Expected Contents When Idle

When the environment is free (unleased), it should be essentially empty:

```
~/workspace/thor/
└── NOTES.md                ← may contain notes from the last completed tick-
et
```

Some teams keep `NOTES.md` around between tickets as a lightweight log; others wipe it on release. Either is fine. The *code* always leaves on release — checkouts, build artifacts, scratch files. An idle environment should not be holding the state of a completed ticket.

`NOTES.md`

This file is the heart of the wind-down ritual from Chapter 6. Thirty seconds at the end of each session, per active environment, written by the agent at your direction. A representative one looks like this:

```
# thor — NOTES

## 2026-04-24, end of afternoon session
Ticket: ABC-1847 (refactor ingestion pipeline batch size)

### What got done today
- Read existing BatchProcessor. Identified the hardcoded 500 limit
  at ingestion/processor.py:142.
- Added a config flag `INGESTION_BATCH_SIZE` with a sensible default.
- Updated two of three call sites. Third call site (stream_handler.py)
  is trickier because it batches inside a retry loop; started the
  refactor but didn't finish.

### What's in progress
- stream_handler.py changes are half-done. The retry loop needs to
  read the new config flag but I haven't threaded it through yet.
  See the uncommitted diff.

### Pick up first tomorrow
- Finish stream_handler.py.
- Run the integration tests (`pytest tests/integration -k batch`).
- If they pass, push and summarize for Blake.
```

What You Can Safely Edit

- `NOTES.md` : freely. It's a note to your future self. Write whatever helps.
- **Nothing else.** Everything else under a Norse environment is either a git checkout (use git commands) or a build artifact (let the build tool manage it).

What to Leave Alone

- Do not manually hand-edit files inside the leased repo checkout under an environment. The agent is working there. Coordinate through prompts, not through silent file edits that will confuse the agent on its next commit.
- Do not try to share state between environments. Each one is isolated on purpose. If two tickets genuinely need to see the same intermediate artifact, that artifact belongs in `/tmp` — one of the two carve-outs in the workspace boundary's exception list, and the correct scratch area for cross-environment state.

Per-Repo Memory: `repos/<repo>/CLAUDE.md` and `AGENTS.md`

Many teams keep AI-readable instructions inside their repositories. If your company or open-source project does, those files end up in `repos/<repo>/` alongside the source tree, and the lease protocol instructs the agent to read them on the way in.

- **`CLAUDE.md` inside a repo**: Claude Code reads this automatically, just as it reads the workspace-level one. Contains repo-specific conventions: testing patterns, service architecture, the location of the one cursed file everyone forgets to update.
- **`AGENTS.md`**: an emerging tool-agnostic convention. Same content, different filename, read by multiple agent tools (Claude Code, Codex, others). Use `AGENTS.md` if your team is multi-tool, `CLAUDE.md` if Claude-only, or both if you can't decide.

These files are owned by the repo, not the workspace. Treat them like any other source file: edit through PRs, reviewed by the team.

A Summary Table

Path	**Role**	**Who writes**	**Safe to hand-edit?**
~/workspace/CLAUDE.md	Permanent memory	You + agent	Yes, freely
~/workspace/.claude/settings.json	Permission config	You	Yes, carefully
~/workspace/.claude/ (other files)	Agent internal	Agent	No
~/workspace/ENVIRONMENT_LEASES.md	Live coordination	Agents	Rarely
~/workspace/repos/<repo>/	Clean source clone	You (git)	Via git only
~/workspace/<norse>/NOTES.md	Session wind-down	Agent	Yes, freely
~/workspace/<norse>/<repo>/	Active ticket working	Agent	Coordinate via prompt
/tmp/	Scratch (outside workspace)	Either	Yes, freely

Migrating the Workspace to a New Machine

If you ever set up a new laptop, the files that matter — the ones worth carrying across — are a small and specific set. Bring these:

1. `~/workspace/CLAUDE.md`
2. `~/workspace/.claude/settings.json`
3. Optionally, `~/workspace/repos/` if re-cloning from remote is slow or bandwidth-constrained.

Do *not* bring:

- `~/workspace/ENVIRONMENT_LEASES.md` — regenerate it empty on the new machine. The PIDs from the old machine mean nothing on the new one.
- The Norse environment directories — they're ephemeral by definition. Let them be empty.
- Anything under `.claude/` other than `settings.json` — those are machine-local caches that the tool will rebuild.

This is also a natural moment to prune `CLAUDE.md` (see Chapter 7's cold-start notes). Rules you wouldn't bet money still matter: leave them behind.

Appendix F: Troubleshooting Common Failure Modes

How to Use This Appendix

This is the appendix you will open on a Thursday afternoon when something has gone quietly wrong and you need someone to walk through it with you without making you feel stupid for hitting it. Most of what's in here has happened to me, often more than once, and none of it is a sign that you are doing the method wrong. Systems are physical things. Physical things occasionally break. The point of a troubleshooting appendix is to shorten the distance between *something is off* and *it's fixed and I'm back to work.*

Each failure mode below follows the same shape. First, what you'll actually see when it happens — the symptoms in plain language, not the error code. Then, what's usually causing it. Then, a response ladder that starts with the cheapest possible fix and escalates only as needed. Climb from the bottom. Most problems are solved on rung one or two. The nuclear option exists, for the rare case, but you will almost never need it.

A general note before we start. When something goes wrong with the method, the first instinct for many engineers is to blame the method itself — to rip out the seven-slot structure, revert to a single terminal, and start over. Resist that urge for at least an hour. Almost every failure mode in this appendix is a small, local, boring fix that doesn't require touching the structure at all. Tighten the one loose screw. Don't take the table apart.

F.1 Claude (or Codex) Ignores the Boundary Rule

What you'll see. You ask the agent to read or edit a file outside `~/workspace`, expecting a refusal, and instead the agent cheerfully complies. Or you notice, in session output, a path beginning with `~/Documents` or `/etc` where there shouldn't be one. Or a colleague looks over your shoulder and asks, mildly, *should it be reading that?*

This is the most important failure mode in the whole book, because the boundary is the rule every other rule depends on. If it's leaking, everything else is running without a net. Fix it first. Fix it completely. Do not continue working until you have.

What's probably happening.

- The boundary rule never made it into `CLAUDE.md` / `AGENTS.md` during setup.
- The rule made it in, but the current session was launched before the rule was written, so the agent is running on an in-memory context that predates the rule.
- The rule is in the memory file but phrased ambiguously enough that the agent is interpreting it as a soft preference rather than a hard constraint.
- You edited a *repo-level* `CLAUDE.md` inside one of the Norse environments instead of the workspace-level one, so the rule is scoped to a single repo rather than the whole workspace.
- A later edit to the memory file accidentally removed the rule.

Response ladder.

1. **Read the memory file with your own eyes.** Open `~/workspace/CLAUDE.md` (and `AGENTS.md` if you use Codex) in a text editor. Search for the word *workspace* and the word *boundary*. The rule should be there, stated unambiguously, under a clear heading. If it isn't, that's your answer — re-run the Step 3 prompt from Chapter 4 and watch the file get written this time.
2. **Restart the session.** Exit the agent (`Ctrl+C` twice in Claude Code; the equivalent in Codex), relaunch it from inside `~/workspace`, and run the same probe again. A fresh session loads the memory file cleanly and usually binds the rule.
3. **Ask the agent to self-diagnose.** Paste this prompt: *I just asked you to read a file outside the workspace and you did. You were supposed to refuse. Read the*

memory file, tell me what it currently says about the boundary, and tell me exactly what to change to make the rule bind. The agent will inspect its own configuration and report back. This trick — having the agent audit the rules the agent is supposed to be following — is one of the most useful patterns in the book.

4. **Check you edited the right file.** Run `find ~/workspace -name CLAUDE.md -maxdepth 2` and confirm the boundary rule lives in `~/workspace/CLAUDE.md`, not in a repo-level `CLAUDE.md` nested two directories deep.
5. **Nuclear option: re-run the setup conversation.** If the rule keeps failing to bind after the above, walk through Chapter 4's Step 1 conversation again from the top. It takes forty-five minutes. That is annoying but not catastrophic, and a clean rebuild is much better than six months of work on a foundation that doesn't hold.

Prevention. After any non-trivial edit to `CLAUDE.md` or `AGENTS.md`, run the probe once — ask the agent to read a file outside the workspace — just to confirm the rule still binds. Twenty seconds of verification, and you never find out three weeks later that the rule was lost in a wording cleanup.

F.2 Lease File Out of Sync With Running Processes

What you'll see. You go to lease an environment and find every slot marked as taken — but half of the PIDs in `ENVIRONMENT_LEASES.md` belong to sessions that closed yesterday. Or two agents both claim they hold `thor`. Or you come back from lunch and find a lease that you don't remember creating, in an environment you don't remember touching.

What's probably happening.

- A session crashed or was force-closed without running the release step.
- The machine rebooted overnight and everything in the lease file is now stale.
- You closed a terminal window (⌘W, or the little red dot) without telling the agent to release first.
- Two agents raced to claim the same environment in a narrow window.
- A PID got reused by the operating system — rare, but it happens on long-running machines.

Response ladder.

1. **Ask any running agent to reconcile.** Launch a fresh Claude (or Codex) session in any environment, and give it this prompt: *Read `ENVIRONMENT_LEASES.md`. For each lease, check whether the recorded PID is still running on this machine. Produce a report of which leases are live and which are stale. Do not modify the file.* Read the report. If all the stale ones are obviously stale — sessions from yesterday, or from before the last reboot — have the agent clear them in a second step.
2. **Verify manually.** If the automatic reconciliation feels off, run `ps -p <PID>` for each PID in the lease file. A PID that isn't running will give you a blank result. Clear those entries by hand — `ENVIRONMENT_LEASES.md` is just a markdown file, and editing it in a text editor is always safe.
3. **Resolve double-leases.** If two live sessions both claim the same environment, pick one to yield. In the yielding session, tell the agent to release the lease (check out the default branch, `git stash`, clear the entry). The other session keeps working.

4. **Handle uncommitted work in a stale-leased env.** If a stale lease points at an environment that has uncommitted work, do not blow it away. Run `git log`, `git status`, and `git stash list` in that directory, summarize what's there, and decide whether to resume the work or abandon it. Abandoned work should be stashed with a dated label (`git stash push -m "abandoned-2026-04-24-thor"`) rather than discarded outright — disk is cheap and occasionally the abandoned work turns out to have been the good version.
5. **Nuclear option: clear the lease file entirely.** If everything is stale — for example, after a reboot with no sessions running — open `ENVIRONMENT_LEASES.md` and empty the lease section. The structure stays; the entries go. Any new session can now lease cleanly.

Prevention. The NOTES.md wind-down ritual from Chapter 6 is the main defense. If every session writes its notes and releases its lease at the end of every working session, the lease file is self-correcting. Crashes and reboots will still desync it occasionally; that's fine. The ritual prevents the slow accumulation of stale entries that makes a Thursday morning harder than it needed to be.

F.3 Draft PR Fails to Open

What you'll see. The agent runs `gh pr create --draft` and gets back an error. Common messages include *HTTP 401*, *HEAD is not yet pushed*, *no commits between main and the current branch*, *this repository does not support draft pull requests*, or the particularly confusing *could not determine base repository*.

What's probably happening.

- `gh` authentication expired or was invalidated (covered in more detail in F.4).
- The branch was created locally but never pushed, so GitHub has nothing to open a PR against.
- The initial empty commit step was skipped, so there are no commits distinguishing the feature branch from the base.
- The agent is sitting on the default branch rather than on a feature branch.
- The repository is on a private plan that doesn't support draft PRs (older Team plans, some legacy Enterprise configurations).
- The local `origin` remote is pointing at a fork rather than the canonical repo, and `gh` can't figure out which repo to target.

Response ladder.

1. **Read the exact error.** `gh` error messages are, unusually for CLI tooling, mostly accurate. Read what it said before guessing.
2. **Confirm the branch is pushed.** Run `git branch --show-current` to see which branch you're on, then `git log origin/<branch>` to see whether the remote knows about it. If the remote doesn't: `git push -u origin HEAD`.
3. **Confirm you're not on the default branch.** If the agent ended up on `main` or `master` by accident — this happens when a step of the work-on-a-ticket protocol gets skipped — create the feature branch properly: `git checkout -b <branch-name>` following your company's naming convention, then push.
4. **Confirm the empty commit exists.** `git log --oneline` should show at least one commit on the branch that isn't on the base. If the branch is empty, run the

empty commit step: `git commit --allow-empty -m "TICKET-NUMBER: begin work"` and push.

5. **If the repo doesn't support drafts, fall back to a normal PR with `[WIP]` in the title.** `gh pr create --title "[WIP] TICKET-NUMBER: title"` — this gives you the same continuous-publishing target as a draft, just without the native draft state. Ask a teammate with admin access whether the organization can enable draft PRs; it's almost always worth turning on.
6. **Fix the `origin` confusion.** If `gh` is confused about which repo is canonical, run `gh repo set-default` inside the repo and pick the right one. This writes a config entry that sticks for the whole checkout.
7. **Ask the agent to retry with verbose output.** `GH_DEBUG=api gh pr create --draft ...` will show you the actual API call, which sometimes surfaces a cause that the summary error hid.

Prevention. Keep the empty-commit-then-push-then-draft-PR sequence in permanent memory as a single unit, so it never gets partially executed. Most PR-creation failures are really step-skipped failures.

F.4 `gh` Authentication Breaks

What you'll see. Any `gh` command — `gh pr create`, `gh pr view`, `gh auth status` — returns an authentication error. *401 Unauthorized, bad credentials, token expired*, or the more subtle *you are not authenticated to the correct account.*

What's probably happening.

- Your `gh` token expired. Tokens have lifetimes, and SSO-backed tokens in particular expire on the organization's SSO cadence.
- Your account's 2FA or SSO session ended — for example, because you have been away for a week.
- You are logged into `gh` with a personal account but the repo requires organization SSO.
- A `GITHUB_TOKEN` environment variable is set somewhere in your shell (a lingering leftover from a scripted task, perhaps) and is silently overriding the stored credentials.
- The `~/.config/gh/hosts.yml` file got corrupted — rare, but possible after a crash during a previous `auth login`.

Response ladder.

1. **Run `gh auth status`.** It will tell you exactly which accounts are logged in, which token is being used, and what's wrong. Read the whole output; it is unusually useful.
2. **Refresh the token.** `gh auth refresh -h github.com -s repo,workflow,read:org` will attempt a token refresh with the scopes this workflow needs. For most expired-token cases, this is all you need.
3. **Check for a `GITHUB_TOKEN` env var.** Run `echo $GITHUB_TOKEN`. If it's set and you didn't mean it to be, unset it (`unset GITHUB_TOKEN`) and try `gh` again. A stale environment variable will silently win over a correct `gh auth login`, and you can chase this one for an hour before noticing.
4. **Confirm SSO authorization.** If your organization uses SAML SSO, even a valid token won't work against org repos until it's been SSO-authorized. Run `gh auth status` and look for the SSO line; if it's missing or marked *not authorized*,

visit the token's settings page in GitHub and click *Authorize* for your organization.

5. **Full logout and back in.** `gh auth logout` followed by `gh auth login`. Choose *HTTPS* for the protocol (unless your org requires SSH), authenticate through the browser, and grant the scopes the prompt asks for. Confirm with `gh auth status` and then retry whatever failed.
6. **Nuclear option.** If `gh auth login` itself fails in strange ways, delete `~/.config/gh/hosts.yml` (after saving a copy somewhere safe), then re-run `gh auth login` for a truly fresh slate.

Prevention. At the start of each working week — first thing Monday morning — run `gh auth status` once before your first ticket. It takes two seconds and catches the expired-over-the-weekend case before it bites mid-rotation.

F.5 `jira` CLI Loses Credentials

What you'll see. `jira issue view ABC-1234` returns an authentication error, or the CLI claims it doesn't know your Jira domain, or it silently returns an empty result where you know a ticket exists.

What's probably happening.

- Your Jira API token expired or was revoked in Atlassian account settings.
- The CLI config file (`~/.config/.jira/.config.yml` on most setups — location varies by CLI version) is corrupted or missing.
- Your organization switched SSO providers and the API token format changed.
- The Jira domain in the config points at an old subdomain that was renamed.
- You are logged in as the wrong account (this happens most often on shared machines).

Response ladder.

1. **Confirm the CLI is finding a config.** Run `jira me` (or `jira issue list --limit 1`) to trigger a live call. Whatever error comes back will usually point you at the cause.
2. **Inspect the config file.** Open the CLI's config (location varies — check `jira --help` or the relevant `man` page for your CLI distribution). Confirm the domain, the user email, and the token path look right.
3. **Regenerate the API token.** In Atlassian → Account Settings → Security → API tokens, revoke the old one (cleanup) and issue a new one. Paste it into the CLI's config, then test.
4. **Reconfirm the domain.** If your organization moved from `yourcompany.atlassian.net` to something else, every Jira CLI on the team will break on the same day. Update the domain, re-auth, move on.
5. **Try `jira init` fresh.** Most Jira CLIs have an interactive `init` command that rewrites the config from scratch. If the config is subtly wrong in a way you can't diagnose, a fresh init is faster than forensic work.

Prevention. When Atlassian sends the *your token will expire* warning email — and it will — handle it the same day. Deferred token renewals are a reliable source of Monday-morning frustration.

F.6 Branch Gets Hopelessly Tangled

What you'll see. The feature branch has a merge commit you didn't expect. Conflicts piled up during a rebase that went sideways. Claude made commits you don't recognize. `git log --oneline --graph` shows a shape that looks like a family tree. You are staring at the terminal and feeling the specific flavor of panic that git alone can produce.

What's probably happening.

- A rebase was attempted, failed, and was partially abandoned, leaving the branch in a mid-rebase state.
- Main (or master) was accidentally merged *into* the feature branch (instead of the feature branch being rebased onto main), creating a merge commit that obscures the history.
- An interaction between a manual git command and the agent's git commands produced a state neither expected.
- A force-push from a partially-recovered state overwrote good commits.
- The branch was created from the wrong base and has accumulated drift.

Response ladder.

Before anything else: **stop. Do not push. Do not run any destructive git command.** Panic git is how small problems become large problems. A tangled branch is almost always recoverable. A tangled branch *after three force-pushes* is sometimes not.

1. **Assess what you actually have.** Run `git status`, `git log --oneline --graph --all -20`, and `git stash list`. Read all three outputs carefully before forming a plan. Nine times out of ten, the state is weirder-looking than it is bad — you can see the good commits sitting there, and the recovery is just a matter of isolating them.
2. **Stash any uncommitted work.** `git stash push -m "tangled-<date>-<env>"` — give it a descriptive message so you can find it later. This gets your working tree clean so you can maneuver.

3. **Abort any in-progress rebase or merge.** If `git status` mentions *rebase in progress* or *merge in progress*, run `git rebase --abort` or `git merge --abort`. This drops you back to the state before the operation started, which is usually the right place to reconsider.
4. **Try the minimum recovery.** If the tangle is recent and small, `git reset --hard <good-commit-sha>` back to the last commit you trust, then re-apply the stashed work. Do this only on a branch that isn't already pushed, or on a feature branch where you are the only author — never on `main` or a shared branch.
5. **Extract the good commits as patches.** For worse tangles, the reliable move is: `git format-patch <base>..HEAD` to save every commit on the branch as a patch file, *then* delete the branch, create a fresh one from the correct base, and `git am <patch-files>` in the order you want. This is the escape hatch that almost always works because it separates *what you wrote* from *the graph it's sitting in.*
6. **Ask the agent to describe the situation before acting.** Before running any recovery command, give the agent this prompt: *Describe the exact state of this branch, what you believe went wrong, and the steps you propose to recover. Do not run any commands yet.* Read the plan. Approve or amend it. Then let it run. This one habit prevents more cascading git disasters than any other.

Prevention. Two rules. First, resolve conflicts in small bites — if a rebase throws you into three conflicts in a row, stop, abort, and ask whether a merge would have been simpler. Second, never run destructive git from panic. If your heart is racing, the worst possible time to type `git reset --hard` is right now. Stash, breathe, look at the graph, make a plan.

F.7 All Seven Agents Stuck at Once

What you'll see. You return from getting coffee, glance at the screen, and every slot is *doing something* but nothing is finishing. Spinners in four. A prompt in two. One that's been "thinking" for longer than feels right. The screen looks busy and is producing nothing.

This one sounds dire and usually isn't.

What's probably happening.

- All seven agents are blocked on the same external resource — a shared CI system, a rate-limited API, a flaky staging environment, a test database that went down.
- Several agents are waiting on confirmations from you that you didn't notice, because the ask was a single polite sentence scrolled halfway up in a slot you weren't watching.
- One agent hit a rate limit on an API, and the others depending on the same token are queued behind it.
- The laptop itself is under load — swap pressure, background index rebuild, or a runaway process outside the method — and every session is crawling because the machine is.
- You are the bottleneck. Six of them finished and are waiting; one is doing real work; and the shape you're reading as *stuck* is actually *waiting for you*.

Response ladder.

1. **Run the glance cycle properly.** Click into each slot, one at a time, and identify the actual state: typing, finished, asking, stuck. Chapter 6 covers this. Most *all seven are stuck* moments resolve the moment you realize four of them are actually in the *asking* state and you've been ignoring the questions.
2. **Find the common blocker.** If the stuck ones really are stuck, look for what they share. Same repo? Same external service? Same test suite? A single external dependency being down will look like seven independent failures until you notice the pattern.

3. **Check machine load.** `top` or Activity Monitor or Task Manager. If CPU is pegged or memory is swapping hard, the agents aren't stuck — the machine is. Kill whatever's hogging resources, or drop from seven slots to four until whatever's loading the machine finishes.
4. **Answer questions sequentially.** If multiple agents are genuinely asking you questions, answer them one at a time, not in parallel. Trying to multi-thread your own attention across seven simultaneous questions is the specific over-parallelization trap Chapter 11 warned about. Pause. One at a time. Move on.
5. **Step down for the rest of the session.** If the pile-up happens often in a single week, that's a signal — not that the method is failing, but that today's conditions don't support seven slots. Drop to four or five for the day and pick back up tomorrow.

Prevention. Note rate-limited or flaky shared resources in permanent memory (*the staging database is unreliable between 2 and 3 PM UTC; do not depend on it*). Keep an eye on machine load during rotation; if the fans spin up, something is wrong that is not about the method.

F.8 Environment Performance Degradation

What you'll see. Terminals lag when you type. Branch switches that used to be instant now take noticeable seconds. Test suites run slower than they used to. The laptop fans are on more than they were a month ago. Nothing is broken; everything is just a little worse.

What's probably happening.

- `node_modules` — or the equivalent for your stack — has accumulated separately inside each of the seven environments, and the cumulative size is hundreds of gigabytes.
- Test artifacts, coverage reports, and build outputs are piling up in environments that never got cleaned.
- Git repositories have grown large without the occasional `git gc`.
- Docker containers from past tickets are still running in the background, or images are stacked up in the Docker cache.
- macOS Spotlight (or the equivalent) is trying to index a growing workspace and losing.
- Swap is being hit hard because seven concurrent agent sessions plus seven IDEs plus seven browsers is genuinely a lot for 16 GB of RAM.

Response ladder.

1. **Check disk space.** `df -h` on macOS/Linux; `Get-PSDrive` on Windows. If you're under 10% free on the volume that holds `workspace`, that alone will cause the whole system to feel gummy.
2. **Find the heavy environments.** `du -sh ~/workspace/*` will tell you which of the seven is bloated. Usually one or two are significantly larger than the others — those are the candidates for a wipe-and-reset.
3. **Reset the worst environment.** Pick a moment when it isn't leased, and do a clean reset: back up any notes, delete the working directories inside it, and let the next lease re-clone fresh from `repos`. This is exactly what the environments were designed to allow; use the affordance.

4. **Prune Docker.** `docker container prune && docker image prune && docker volume prune` (answering *yes* to each). If you use Docker at all, this is the single highest-impact cleanup you can run.
5. **`git gc` the big repositories.** Inside each large checkout under `repos` and inside the Norse environments, `git gc --aggressive` occasionally. It will take a few minutes. It will make everything faster afterward.
6. **Restart the machine.** This one sounds trivial. It is trivial. It is also genuinely the right answer more often than engineers like to admit. A reboot clears swap, closes stuck processes, and resets whatever indexer was chewing on something in the background. If the system has been up for more than two weeks and is feeling slow, a reboot is the cheapest experiment.
7. **Nuclear option.** Wipe `~/workspace/thor` through `~/workspace/baldr` entirely (not `repos`, not `CLAUDE.md`, not the lease file), and let the seven environments rebuild organically from the next seven ticket leases. This loses nothing — the environments are, by design, disposable — and it guarantees a fresh state.

Prevention. Release-reset each environment at the end of each ticket, as the lease release protocol prescribes. `git stash`, check out the default branch, clear the working tree. An environment that gets reset per-ticket doesn't accumulate bloat in the first place. It is one of those small disciplines that costs nothing per use and pays back steadily over the quarters.

A General Philosophy of Troubleshooting

Three things I've learned after a lot of these.

Read the error before forming a theory. Software error messages have gotten much better over the past decade. The first guess of a tired engineer is almost always worse than the plain text of the error. Read what it said.

Fix the smallest thing first. A restart fixes more problems than any amount of clever debugging, and costs almost nothing. A fresh session loads memory cleanly. A reboot clears stuck state. Try the boring, cheap thing before the interesting, expensive one.

Ask the agent to describe before acting. When you're about to run something you're not sure about — a git reset, a file deletion, a config rewrite — ask the agent to describe in plain English what it's about to do and why, *before* it does it. This one habit has saved me more hours than I can count, and it costs one extra prompt. The agent is, in this mode, a very patient rubber duck who also happens to know your repo.

And — this is the closing note of the appendix, and of this book's practical reference material — if something is genuinely, deeply broken, and none of the ladders got you home, it is always okay to stop for the day. The tickets will be there tomorrow. The workspace will be there tomorrow. A tired engineer making panicked fixes at 6 PM on a Thursday is, statistically, where most of the worst failures in this appendix were born in the first place. Close the laptop. Have dinner. The problem will be easier in the morning, and you will be better at solving it.

Appendix G: Glossary

A quick reference for the terms used throughout the book. Some of these are standard industry vocabulary that I've defined here because I know not every reader has the same background. Others are terms this book coined, or uses in a specific methodological sense that might not match how you've heard them elsewhere. Where a term is introduced in the main text, the chapter is noted in parentheses.

Agent / Agentic. A program that takes actions in your name — writing files, running commands, calling APIs — based on instructions you give it in natural language. "Agentic" is the adjective describing software built around that idea, as opposed to software that merely answers questions. Claude Code and Codex are both agentic; a chatbot that can only talk to you is not.

AGENTS.md. The tool-agnostic permanent-memory file that an emerging cross-vendor convention asks coding agents to read at the start of every session. Codex reads it. Other tools are starting to. If a team has written careful rules about how their codebase works, those rules belong either in `AGENTS.md` (if the tools support it) or in a tool-specific file like `CLAUDE.md` (if they don't), with the two kept in sync. *(Chapter 9.)*

AI burnout. A specific strain of burnout — distinct from classical overwork burnout — caused by the decision-load of collaborating with an AI agent for hours at a time. Its defining feature is that it is not fixed by rest, because the exhausted faculty is the *judgment* faculty, not the *effort* faculty. Chapter 11 treats it in detail.

Allow list. The opposite of a deny list: a list of commands or paths that the agent is explicitly permitted to use, with everything else forbidden by default. This book generally uses the deny-list approach, because an allow list in a real developer's environment ends up being so long that the cost of maintaining it exceeds the cost of simply naming the small number of things that are dangerous.

Attention residue. The finding, from Sophie Leroy's 2009 research, that a portion of your attention remains on the previous task for a measurable period after you switch to a new one. Every AI interaction is, in effect, a small task switch — which is why sustained AI-assisted work imposes a different kind of fatigue than sustained solo coding. *(Chapter 11.)*

Blast radius. How far the consequences of a command reach. `rm -rf` inside a disposable environment has a small blast radius; a force-push to a shared branch has a large one. The deny list exists to catch commands whose blast radius extends outside the sandbox even when the command itself originates inside it. *(Chapter 4.)*

Boundary. The filesystem sandbox that scopes where the agent is allowed to read and write. In this book, the boundary is the `workspace` directory. Inside it, the agent moves freely. Outside it, the agent refuses to act. Replacing per-action approval prompts with a structural boundary is the move that makes the whole method possible. *(Chapter 4.)*

Branch naming convention. Your company's standard for how feature branches are named — typically something like `feature/ABC-1234-short-description`. Taught to Claude once in the work-on-a-ticket protocol so you never have to specify it again.

Buzzy. The old carpenter in Chapter 1's opening story — the one who drove a nail in a single blow because he knew how to blunt the tip first. Buzzy is the book's recurring archetype for quiet, unshowy mastery: good at his craft, not preoccupied with proving it, drinking beer and smoking between jobs. The aim of this method is, eventually, to be Buzzy.

Checking. What your brain does when it is verifying a finished or near-finished proposal against criteria you already hold — *is this right? does this match the codebase? does this break anything?* Cheap, externalizable, sustainable across many slots in parallel. Contrast *Holding*. *(Chapter 11.)*

CLAUDE.md. The permanent-memory file that Claude Code reads at the start of every session in a directory. The workspace-level `CLAUDE.md` holds your boundary rule, your deny list, your lease protocol, your work-on-a-ticket protocol, and any codebase-specific conventions you've taught Claude over time. It is the single most load-bearing file in the system. *(Chapter 4.)*

Claude Code. Anthropic's terminal-based coding agent. The specific tool this book uses in Chapters 4 through 8. The method is not bound to it; Chapter 9 reimplements the same method with Codex to prove the point.

Codex. OpenAI's terminal-based coding agent, and the alternative tool this book ports the method to in Chapter 9. Similar enough to Claude Code that the port is straightforward; different enough that the exercise genuinely tests the method's portability.

Cold start. Coming back to the workspace when nothing has survived — new laptop, fresh install, empty disk. The procedure is the full Chapter 4 + Chapter 5 rebuild: workspace first, desktop layout on top of it. Happens once or twice a year for most people. *(Chapter 7.)*

Context (AI sense). The working memory an agent has access to during a single session — the conversation history, the files it has read, the system prompts and memory files it was given at startup. When "context" appears near words like *window*, *size*, or *rot*, this is the sense in play.

Context (methodology sense). One of the seven working streams the method manages in parallel — a ticket, its draft PR, its environment, the Claude session working on it, and the slot that holds all of them. "Seven contexts" is a shorthand for this; the book's title is a direct reference to it.

Context rot. The slow deterioration of an agent's usefulness over a long-running session, as its working memory fills with stale information, accumulated misapprehensions, and now-irrelevant tangents. The fix is to release the session and start fresh, letting the NOTES.md and the branch state carry the important context forward. *(Chapter 7.)*

Context stack. The browser-and-terminal pair that together fill a single slot. The browser sits on top of the terminal (or vice versa) with each one's edge peeking out past the other's, so both are always clickable. "Stack" here refers to z-order, not to geometry — clicking an edge flips which window is in front, but nothing moves. *(Chapter 5.)*

Decision load. The cumulative cost of making evaluative micro-decisions across a day — accept, reject, modify, retry, clarify. Distinct from workload. The operating claim of Chapter 11 is that AI burnout is a decision-load problem, not a workload problem, which is why rest-based interventions often fail to fix it.

Deny list. A list of commands the agent is forbidden from running, regardless of context. Maintained in the workspace's `.claude/` directory. The three guiding principles for what belongs on a deny list — anything destructive and irreversible, anything that touches production, anything that costs significant money to undo — are given in Chapter 4.

Dish by the door. The book's shorthand for the principle that every kind of work should have a fixed permanent home — on your screen, in a permanent-memory file, in a wind-down routine — so your hand or your eye or your habit knows where to go without your brain having to decide. Named after the habit of putting your keys in a specific dish when you walk in your front door, so you never lose them. The layout in Chapter 5 is a dish-by-the-door for the whole working day. *(Chapter 3.)*

Draft PR. A pull request marked as *not ready for review* — visible on GitHub, pushed to a branch, but not yet requesting anyone's attention and not mergeable until promoted. Claude opens a draft PR *before* writing any code on a ticket, so that every commit thereafter flows into a living document of the work. You promote it out of draft only after validation. *(Chapter 4, Chapter 6.)*

Empty commit. A git commit that changes no files (`git commit --allow-empty`). Used on a fresh branch so there is something to push, which is what lets the draft PR exist from minute one of a ticket's work instead of having to wait for Claude to produce code first. *(Chapter 4.)*

Environment. One of the seven Norse-named sub-directories inside `workspace` — **thor, freya, tyr, loki, odin, heimdall, baldr** — where actual ticket work happens. An environment holds a copy of the relevant repo and is leased to one Claude session at a time. Disposable by design: when a ticket is done, the environment is reset and reused. Distinct from a *slot*, which is a screen position rather than a directory. *(Chapter 3, Chapter 4.)*

ENVIRONMENT_LEASES.md. The single markdown file at the workspace root that tracks which Claude session currently holds which environment. Four columns of text — environment name, PID, timestamp, state — and that is the whole of the coordination primitive. *(Chapter 4.)*

Ephemeral environment. An environment whose state is not meant to last. The seven Norse directories are ephemeral — filled, used, and reset constantly. Compare to the `repos/` folder, which is persistent. *(Chapter 4.)*

Execution. The second of the three phases: the day-to-day running of the method. Triage, load seven tickets, rotate across slots, validate, finalize, lunch, repeat. Chapter 5 starts the engine; Chapter 6 is the daily workflow; Chapter 7 is coming back.

Fourteen tickets a day. The sustainable throughput target the method is designed for: seven tickets in a morning session, seven in an afternoon session, separated by a real lunch. "Sustainably" is the operative word. Higher numbers are possible for short stretches and always come due as exhaustion in later weeks. *(Chapter 6.)*

Ghost. A Claude terminal that looks normal on screen but whose underlying process has died. The cursor blinks, the window is there, but nothing is home. Caught by the liveness-check probe in the warm-start ritual. *(Chapter 7.)*

Glance cycle. The two-to-three-minute visual sweep across all seven slots that keeps you in the air-traffic-controller's seat. Each glance classifies a slot into one of four states: *typing* (leave alone), *finished* (validate), *asking* (answer), or *stuck* (intervene). *(Chapter 6.)*

Glob pattern. A simple wildcard syntax used in deny-list entries. The asterisk in `git push --force*` matches anything that starts with that string — `--force`, `--force-with-lease`, `--force origin main`, and so on. Globs let a short rule cover a whole family of related commands. *(Chapter 4.)*

Holding. What your brain does when it is actively maintaining a problem in working memory while a partial solution is being co-constructed. Expensive, non-externalizable, the load-bearing mechanism behind decision-load AI burnout. Contrast *Checking*. *(Chapter 11.)*

Jira Master. Zone 3, bottom-right of the screen. A browser window holding the Jira board, one Pull Requests tab per repo, and one Actions tab per repo. The command center for "what am I doing today" and "where is everything standing." *(Chapter 5.)*

Keep-or-release decision. The judgment call, made at the start of a warm-start morning, about whether a surviving Claude session should be continued or released in favor of a fresh one. The rule of thumb is to keep two or three and release the rest. *(Chapter 7.)*

Lease / leasing. The protocol by which a Claude session claims an environment for the duration of a working session. A lease is recorded in `ENVIRONMENT_LEASES.md` with the claimer's process ID. The PID serves as a built-in liveness check: if the recorded PID is no longer running, the lease is stale and any new Claude can reclaim the environment. *(Chapter 3, Chapter 4.)*

Liveness check. A small probe prompt — *"are you still there?"* — used after any idle period to confirm a Claude session is alive rather than a ghost. Takes two seconds; saves you thirty seconds of waiting for a response that was never going to come. *(Chapter 7.)*

Maslach burnout model. The three-dimension framework of occupational burnout — emotional exhaustion, cynicism and depersonalization, reduced sense of accomplishment — developed by Christina Maslach (with Susan Jackson) from the early 1980s onward, operationalized as the Maslach Burnout Inventory, and reflected in the World Health Organization's 2019 ICD-11 entry classifying burnout as an occupational phenomenon. Referenced in Chapter 11 as the classical baseline against which AI burnout is a specific variant.

Method / methodology. The system described in this book, considered independently of any particular AI tool. The method is what stays the same across Chapters 4, 9, and any future port to a different agent.

Norse environments. The canonical names of the seven working directories: **thor, freya, tyr, loki, odin, heimdall, baldr**. The specific mythology is arbitrary; the *distinctness* of the names is not. Human memory is better at vivid names than at numbered slots, so pick any scheme whose names your brain will lock onto. *(Chapter 4.)*

NOTES.md. The short per-environment note — three to ten lines — that each Claude writes at the end of every session, summarizing what got done, what is in progress, and what to pick up first next time. Read in ninety seconds at the start of the next working day. The single cheapest-to-produce, most-valuable-in-practice file in the whole system. *(Chapter 6, Chapter 7.)*

Permanent memory. The set of persistent files — `CLAUDE.md` or `AGENTS.md`, plus the `.claude/` or `.codex/` configuration directory — that an agent reads at the start of every session. Rules written here survive across sessions, across reboots, and (mostly verbatim) across tool swaps. The proposition behind Principle 2 in Chapter

10: a rule you teach an agent once and it remembers forever is worth vastly more than the cleverest prompt you type fifty times.

PID. Short for *Process ID.* Every running program on your computer has one — a unique integer assigned by the operating system, gone the moment the process exits. Recorded with each entry in the lease file so other Claudes can tell at a glance whether a lease is still active or stale. *(Chapter 4.)*

Prompt template. A short, reusable prompt for a situation you hit often — *"summarize where you are," "review your own work as a skeptical senior engineer," "diagnose what went wrong, don't fix anything yet."* Templates are not elaborate configuration; they are sentences you say a lot, written down once. *(Chapter 8.)*

Refinement. The third of the three phases: the small, specific, observed adjustments you make to permanent memory, prompt templates, and the environment/layout, based on what using the system has actually taught you. Refinement is *not* rebuilding, rearchitecting, or constant tinkering. Rule of thumb: more than once a day is tinkering; less than once a week is letting problems accumulate. *(Chapter 8.)*

Release (of a lease). The cleanup that happens when a Claude session is done with an environment. Checks out the default branch, clears any local state (typically via `git stash`), removes the lease entry, and leaves the environment free for the next session. Can be triggered deliberately or reclaimed automatically by the PID liveness check. *(Chapter 4, Chapter 7.)*

Repos folder. The `workspace/repos/` directory, a library of clean, untouched checkouts of the repositories you work in. You never edit inside `repos/` directly; when a ticket needs a repo, Claude copies the relevant repo into one of the seven environments and works there. Keeps the library pristine and the work disposable. *(Chapter 4.)*

Reset ritual. The five-step protocol for bringing a kept-from-yesterday Claude back into a usable state after a warm start: self-summary, branch-state check, pull and rebase main, check CI, resume. Takes three or four minutes per kept session. *(Chapter 7.)*

Rotation rhythm. The air-traffic-controller's pattern of attention across seven slots — glance, classify, move on, answer the slot that needs attention, return to the sweep. You control; the Claudes fly. Any morning where you catch yourself watching one Claude for more than five minutes is a morning where you stopped rotating and started piloting. *(Chapter 6.)*

Sandbox. In software, a controlled area where a program can do whatever it wants without affecting anything outside that area. Named after the physical sandbox at a playground: children can build, knock down, and make a mess, but the mess stays in the box and the grass around it stays clean. The `workspace` directory is the agent's sandbox. *(Chapter 4.)*

Second-Claude pattern. The practice of asking a *different* Claude session, in a separate slot, to review the first Claude's PR as a skeptical senior engineer — useful on production-critical, security-sensitive, or otherwise pager-adjacent tickets. The reviewing Claude has no emotional attachment to the code because it did not write it, and it often catches real issues as a result. *(Chapter 6.)*

Setup. The first of the three phases: the one-time work of building the workspace, teaching the agent to live in it, and installing your command-line tools. Chapter 4 is Setup with Claude Code; Chapter 9 is Setup with Codex, as a portability demonstration.

Slot. A permanent position on your screen where a browser and a terminal live together, side by side, in the same geometry every day. Seven slots total. Slots do not move. They are the *dishes by the door*, always waiting for keys — where the keys in this case are whichever environment gets leased into the slot on any given morning. *(Chapter 5.)*

Stack. See *Context stack.*

Stagger. The deliberate alternation pattern across slots on the bottom row of the layout — each terminal peeking out on the opposite side from its neighbor — that lets seven slots pack horizontally into one monitor's width. Borrowed, in spirit, from dovetail joinery. *(Chapter 5.)*

Stuck. One of the four glance-cycle states: a Claude running the same command repeatedly, or oscillating between two approaches without converging, or generating output that contradicts output from three prompts ago. Diagnosed by the absence of *convergence* — thinking narrows the problem; stuck circles it. The standard intervention is a single prompt asking Claude to summarize where it is and what is blocking it. *(Chapter 6.)*

Team of one. The operating scale of this method: an individual using the system locally, inside an organization that's doing its own thing. You don't need permission, buy-in, or a pilot program to start. The only externally visible artifacts are the PRs you open and the tickets you close. *(Chapter 2, Chapter 12.)*

Three phases. The shape of the method over time: **Setup** (one-time), **Execution** (daily), **Refinement** (continuous). Setup is Chapters 4 and 9. Execution is Chapters 5, 6, and 7. Refinement is Chapter 8. *(Chapter 3.)*

Three zones. The three fixed regions of the screen reserved for human work: **Zone 1** (Validation, upper-left), **Zone 2** (Communications, upper-right), **Zone 3** (Jira Master, bottom-right). Distinct from the seven *slots*, which are where the Claudes work. *(Chapter 5.)*

Tidy start. The simplest of the three shapes of starting up: everything still running, all seven Claudes alive, the lease file matches reality. Takes about two minutes. The most common case. *(Chapter 7.)*

Triage. The ten-minute morning sort through your Jira queue — titles and first paragraphs only — that separates tickets into *system fit, needs my thinking,* and *needs a conversation.* Only the first bucket gets loaded into the seven slots. The other two are handled the old-fashioned way, with the actual inside of your actual skull. *(Chapter 6.)*

Two-session day. The canonical rhythm: a morning session (roughly 9:00 to noon), a real lunch, an afternoon session (roughly 1:00 to 4:00), and a hard stop. Seven tickets per session times two sessions equals fourteen tickets a day, sustainably. *(Chapter 6, Chapter 11.)*

Validation. The work of reviewing what Claude produced — not line-by-line, but shape-level — and either promoting the draft PR or feeding Claude specific feedback and letting it iterate. The method's productivity depends on validation being fast and mostly delegated; if you had to read every diff line by line, the whole parallel structure would collapse under its own weight. *(Chapter 6.)*

Validation ritual. The six-step validation protocol: (1) ask Claude to self-critique; (2) ask Claude to run the tests; (3) ask Claude to check the diff against acceptance criteria; (4) review the PR yourself at the shape level; (5) if satisfied, promote out of draft; (6) if not, give specific feedback and iterate. About five minutes per ticket. *(Chapter 6.)*

Warm start. The interesting case of starting up: some Claudes alive, some ghosts, browsers mostly in place, the lease file partly fictional, overnight CI failures waiting. The bulk of Chapter 7 is about walking through it. Usually fifteen minutes, start to finish.

Wind-down ritual. The end-of-session habit of asking each active Claude to write a short note to its environment's `NOTES.md` before you walk away. Thirty seconds per environment. Pays itself back every time you come back to the workspace. *(Chapter 6.)*

Work-on-a-ticket protocol. The procedure triggered by the phrase *"Work on ABC-1234."* Claude leases an environment (if needed), reads the Jira ticket, summarizes it back, waits for confirmation, creates a branch following your company's naming convention, makes an empty commit, pushes, opens a draft PR, and only then begins actual work. Eight numbered steps, taught once, recorded in permanent memory, followed forever. *(Chapter 4.)*

Workspace. The top-level sandboxed directory — conventionally `~/workspace` — that contains everything the agent is allowed to touch. Holds `CLAUDE.md`, `ENVIRONMENT_LEASES.md`, the `.claude/` directory, the `repos/` folder, and the seven Norse environments. Outside the workspace, the agent cannot see or modify anything. *(Chapter 4.)*

Zone. One of the three fixed screen regions (Zone 1: Validation, Zone 2: Communications, Zone 3: Jira Master) reserved for the human's own work. Zones never move. They are the dishes by the door for your non-Claude tools. Distinct from slots, which hold Claude sessions. *(Chapter 5.)*

Appendix H: Layout Variants

When to Consider a Variant

The canonical layout in Chapter 5 is recommended for most people. It's the layout I use, and it's the one I've validated through years of my own work and with every team I've helped adopt this method. Before you consider any variant, use the canonical layout for at least three weeks. Most people who try variants prematurely come back to the canonical layout anyway; the friction of re-learning isn't worth the marginal preference.

That said, the layout is yours, and some people genuinely work differently. Here are the variants I've seen people adopt successfully.

Variant 1: Swapped Zone 2 and Zone 3

Figure H.1 — Layout Variant: Swapped Zone 2 and Zone 3

Same skeleton as Figure 5.1 — Validation still spans two stacks below — but Jira sits upper-right and Communications drops to bottom-right.

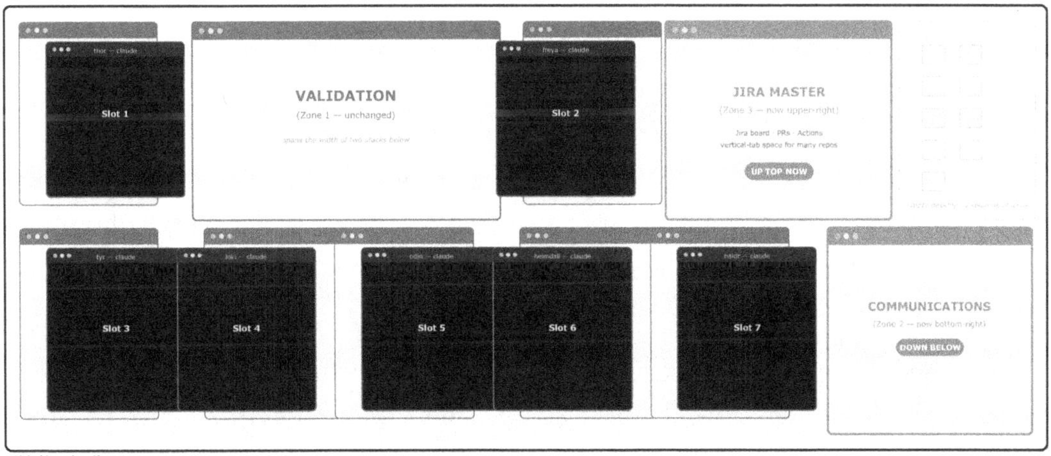

Pick this variant if:

• you spend more time looking at Jira/PRs than at Slack/email · • you work across many repos (need vertical-tab space) · • your team does async comms (short messages OK)

The canonical layout from Chapter 5 is recommended for most people. Only switch to this variant after you've used the canonical layout for several weeks and know you want it.

This variant flips the positions of Zone 2 (Communications) and Zone 3 (Jira Master). Jira takes the larger upper-right space and communications moves to the smaller bottom-right corner. See Figure H.1.

Pick this variant if:

- You spend significantly more of your day looking at Jira boards and PRs than at Slack or email. This is common in teams where async communication is the norm — email and chat messages are short and infrequent, but Jira is always full of context.
- You work across many repositories (five or more) and want the extra horizontal space for a longer list of PR and Actions tabs, especially if you use left-side vertical tabs.
- You find yourself repeatedly resizing the Jira window in the canonical layout because it feels too cramped.

Don't pick this variant if:

- Your team communicates primarily in Slack or Teams and you need to see chat messages at a glance. Making Communications smaller will make you miss things.
- You work in a single repository most of the time. The smaller Jira space in the canonical layout is plenty for one repo.

The trade-off is simple: bigger Zone 3 means smaller Zone 2. If you can live with less Slack and more Jira, swap them. If not, stick with canonical.

Variant 2: Multiple Desktops for Multiple Projects

Some developers work on two or three different projects in a given week — maybe a main project at their day job plus a side project or an open-source contribution, or two different products within the same company. The canonical layout supports this with virtually no change.

Every modern operating system supports multiple virtual desktops (macOS calls them Spaces, Windows calls them Desktops, most Linux environments call them Workspaces). Set up the canonical seven-slot layout on Desktop 1 for Project A, then switch to Desktop 2 and set up another seven-slot layout for Project B. Each project has its own workspace directory on disk (`~/workspace-a`, `~/workspace-b`) with its own seven environments and its own lease file. You flip between projects with a keyboard shortcut — one button press and your whole screen rearranges itself.

This works because slots are positional, not tied to any specific project. The slot in the top-left of Desktop 1 is completely independent of the slot in the top-left of Desktop 2, even though they're in the same screen location. Your muscle memory still works — every kind of work on every project has the same physical home.

A few tips if you adopt this variant:

- Keep your three fixed zones (Validation, Communications, Jira Master) the same across both desktops if the apps overlap. If you use the same email and Slack for both projects, don't duplicate them; just have them visible from both desktops. Some operating systems let you "pin" specific windows to all desktops — use that for your communications window.
- Each project needs its own Jira Master, because each project has its own board and its own repos. That one varies by desktop.
- Don't try three projects at once. Two is already demanding. Three is where people burn out.

Variant 3: Fewer Than Seven Slots

Some developers genuinely do better with five or six slots instead of seven. Reasons vary:

- You work primarily with enormous codebases where Claude takes much longer per ticket, so you can't actually feed seven of them at once without getting ahead of yourself.
- You have heavy meeting load on certain days and can't realistically supervise seven parallel streams.
- You're new to the methodology and learning seven contexts at once is too much. Start with three, get comfortable, scale up.

If you use fewer slots, leave the unused positions as empty space rather than resizing everything to fill in. The mental map of where each slot lives is valuable even when some slots are empty, and when you scale back up to seven, the slots are still where they belong.

I do *not* recommend more than seven slots. Past seven, the cognitive cost of tracking additional contexts rises faster than the throughput benefit. If you feel like you could handle more, you're probably either (a) working on tickets that are too small, in which case combine them, or (b) underestimating the overhead you're carrying, in which case you'll burn out in a few weeks.

Index

A

B

C

D

S

T

U

V

W

Y

Z

About the Author

Blake Tullysmith has been writing software for thirty of the last thirty-two years. He lives in Santa Rosa, California, with his family.

He has worked at every scale a software career can take you to — five-person startups where he wrote every line himself, and sprawling enterprise systems where he was the senior technical authority across architectures owned by multiple engineering vice-presidents. Along the way he has shipped cybersecurity platforms processing tens of millions of events per second, led international engineering teams, designed message-bus systems and developer platforms that quietly made the working lives of hundreds of his colleagues measurably easier, helped organizations cross from monolithic to microservice architectures without the usual wreckage, and put his name on a few patents in networking and security. Twenty-some programming languages, the usual long shelf of cloud and distributed-systems acronyms — the kind of breadth you only accumulate by actually using the things, in production, under deadline, for years.

Before any of that, Blake served in the United States Air Force Security Forces, where he held a Secret-level Department of Defense clearance with Personnel Reliability Program (PRP) qualifications. A medical discharge ended his service and made him a disabled veteran. It also drew him back to technology, which has had his attention ever since. The discipline, the situational awareness, the habit of caring for the people on either side of you — none of that left when the uniform did. It shaped how he later led engineering teams, and it shapes how he writes about software now.

Over the last several years he has watched developers — from people on their first job to staff peers a decade into their careers — pick up some version of the method in this book and quietly change the shape of their working week with it. *The Book of Seven Contexts* is the first written record of the method, and the book he wishes he could have handed his younger self the first time an AI tool landed on his desk and nobody had any idea what to do with it. He is at work on the second.

When he isn't engineering or writing, Blake enjoys time with his family and the slow accumulation of books on topics that have nothing to do with software.

www.ingramcontent.com/pod-product-compliance
Lightning Source LLC
Chambersburg PA
CBHW081207130726
47997CB00009B/2588
9798904176945